Recent Contributions to Fluid Mechanics

Editor
W. Haase

With 199 Figures

Springer-Verlag
Berlin Heidelberg New York 1982

Dr. Werner Haase
c/o DORNIER GmbH
D-7990 Friedrichshafen 1

ISBN-13: 978-3-642-81934-6 e-ISBN-13: 978-3-642-81932-2
DOI: 10.1007/978-3-642-81932-2

Softcover reprint of the hardcover 1st edition 1982

2061/3020 - 5 4 3 2 1 0

Scientific Committee

W. HAASE	DORNIER GmbH, Friedrichshafen
H.-G. KNOCHE	MBB, München
J. ROTTA	DFVLR, Göttingen
W. SCHMIDT	DORNIER GmbH, Friedrichshafen
P. THIEDE	MBB - UT, Bremen
F. THIELE	Technische Universität Berlin

Sponsors

DFVLR	Deutsche Forschungs- und Versuchsanstalt für Luft- und Raumfahrt
DORNIER	
MBB	Messerschmitt-Bölkow-Blohm
MBB - UT	Messerschmitt-Bölkow-Blohm, vormals VFW

Professor Dr.-Ing. Alfred Walz

Preface

The present volume entitled "Recent Contributions to Fluid Mechanics" is dedicated to Professor Dr.-Ing. Alfred Walz in honour of his 75th birthday.

Alfred Walz, born on 11 May 1907, began his outstanding career as an electrical engineer. A few years after obtaining his university-degree he became extremely engaged in fluid dynamics. Walking in the footsteps of Prandtl he was able to direct the development of theoretical activities in an inimitable way. He had the great opportunity to work both as an engaged fluid dynamicist - always trying to get to the bottom of things - and as a popular and patient teacher. To all of these things - in his own words - he gave his heart.

Consequently, it is a great pleasure to publish the following 34 contributions summarizing the efforts of 56 authors. These articles in total cover the wide range of experimental as well as theoretical fluid dynamics and reflect the present state of the art.

Moreover, all colleagues and friends of Alfred Walz wish that he may be able to continue his work and his influence on the work of all of us via his enlightening ideas.

Friedrichshafen, August 1982

Werner Haase
Chairman of the Scientific Committee

Table of Contents

SURVEY PAPER

Shear Layer Studies - Past, Present, Future
P. Bradshaw 1

INDIVIDUAL PAPERS

On Heat Transfer by Superimposition of a Creeping Flow of a Fluid with its Stationary Free Convection
G. Appel, V. Lippig, D. Straub, R. Waibel 10

Half Model Testing Applied to Wings above and below Stall
H. Bippes, M. Turk 22

On the Turbulence-Modeling Requirements of Three-Dimensional Boundary-Layer Flows
T. Cebeci, K. C. Chang 31

Topological Analysis of Computed Three-Dimensional Viscous Flow Fields
G. S. Deiwert 40

Instability and Coherent Structures in Jet Flames
H. Eickhoff 50

Finite-Difference Solutions for Laminar Boundary Layer Flows with Separation
E. Elsholz, U. Schade, F. Thiele 58

Turbulent Flow Simulation: A Large Eddy Simulator's Viewpoint
J. H. Ferzinger 69

Reduced Turbulence Production by Increased Elongational Viscosity
B. Gambert, P. Wagner 78

Some Aspects to the Structure of Separated Flows
D. Geropp 89

Isentropic Magnetogasdynamic Flow of a Perfect Plasma
K. Gersten, D. Glück, K. Oberländer 98

Investigation of the Three-Dimensional Transonic Flow around an Air Intake by a Finite-Volume Method for the Euler Equations
J. Grashof 109

Calculation of a Laminar-Turbulent Two-Dimensional Turbine Blade Boundary Layer Including Surface Curvature Effects
R. Grundmann 121

Considerations of the Vorticity Fields on Wings
E. H. Hirschel 129

Slender Wings with Leading Edge Vortex-Separation
St. M. Hitzel 138

Turbulent Heat Transfer
M. Jischka, H. B. Rieke 151

Optical Flow-Velocity Measurement in Irregularly Shaped Cavities
J. Larsen, K. G. Roesner 161

Measured Velocity Fluctuations Inside the Mixing Layer of a Supersonic Jet
H. Oertel sen. 170

Mathematical Model of the Single-File Flow of Red Blood Cells in Capillaries
H.-D. Papenfuss, J. F. Gross 180

Experimental Method for Pressure-Velocity Correlation Measurements in Three-Dimensional Turbulent Boundary Layers
L. Paulsen, K. O. Felsch 196

Random Walk and Diffusion in Two-Dimensional Lagrangian Systems
N. Peters, H. J. Thies 205

Mesh Influence on Vortex Shedding in Inviscid Flow Computations
A. Rizzi 213

A Numerical Method to Solve the Steady-State Navier-Stokes Equations for Natural Convection in Enclosures
Y. Schkalle, F. Thiele, H. Wagner 222

Computational Study of the Magnus Effect on Boattailed Shell at Supersonic Speeds
W.B. Sturek 235

Viscid-Inviscid Interaction Analysis on Airfoils with an Inverse Boundary Layer Approach
P. Thiede, G. Dargel, E. Elsholz 244

A Turbulent Boundary Layer Approaching Separation
B.E. Thompson, J.H. Whitelaw 253

Considerations on the Transport-Theorem for Extensive Flow Properties
E. Truckenbrodt 265

The Structure of Flows with Unsteady Boundary Conditions
H. Viets, R.J. Bethke 274

Boundary Layer Calculations for Cryogenic Wind Tunnel Flows
B. Wagner 283

An Inverse Integral Computational Method for Compressible Turbulent Boundary Layers
D.L. Whitfield, T.W. Swafford, T.L. Donegan 294

Transition of Rotational Flow into Outer Irrotational Flow
K. Wieghardt 303

The Design of Optimum Diffusers for Incompressible Flow
K.G. Winter, L.F. East 312

Relaxing Evolution Systems for the Solution of an Ill-Posed Design Problem
H.J. Wirz 321

What can be Achieved by Similarity Laws?
J. Zierep 329

Shear Layer Studies - Past, Present, Future

Peter BRADSHAW

Imperial College, London, SW7 2BY, England

Summary

Assuming that the Navier Stokes equations can be taken as accurate, it is strictly correct to say that fluid dynamics is just a computing problem! However, the first laminar boundary layer calculations were carried out by Blasius only 75 years ago, full turbulent-flow simulations have only recently been achieved in highly restricted conditions, and some people doubt whether transition from laminar to turbulent flow can ever be treated as a well posed mathematical problem. Our inability to solve the time dependent Navier Stokes equations in cases of engineering interest has led to (1) the flourishing branch of science called "experimental fluid dynamics" (2) equally flourishing theoretical work on simplifications of the Navier Stokes equations, of which the foremost is based on Prandtl's concept of the thin shear layer. This paper is a look back down the road along which shear layer calculations have marched, and a forward look to see where we go from here.

Introduction

Alfred Walz was born on 11 May 1907, at about the same time that Blasius was starting the laminar boundary layer calculations presented in his 1911 doctoral thesis. Therefore, Professor Walz's life began at the same time as the practical applications of the boundary layer concept, and it is an honour and a pleasure to be invited to celebrate both in one paper. Some people foresee that fluid dynamics in general, and shear layer studies in particular, will eventually descend into routine, uninteresting electronic computing, and it is amusing to note that Alfred Walz began his career as an electrical engineer, although when he began to study the subject in 1925 at the TH Karlsruhe there was little hint that the obscure electrical properties of silicon and germanium would transform our world. After some years in the electrical engineering industry, Alfred Walz joined the AVA in Gottingen in 1937, beginning the part of his career which is most familiar to us by a study of potential flow, which is indeed a common ground of electricity and of fluid dynamics. In 1941 Alfred Walz promovierte (untranslatable term!) at the TH Braunschweig, with his paper "Graphical methods for the calculation of laminar and turbulent boundary layers". Although Professor Walz's main activities, at Braunschweig, Gottingen, Karlsruhe and Berlin, have been concerned with fluid dynamics and in particular with boundary layer theory, he has maintained his interest in electrical engineering and its practical applications, even to the extent of contributing a paper on electrokinetics in capillary flow to ZfW [1]

Professor Walz's contributions to fluid dynamics have included work on potential flow, aerofoil theory, and boundary layer control, but the only area that I am competent to discuss is that of shear layer calculations. Here, his contributions have included a series of approximate calculation methods, each carefully tailored to the computational facilities available at the time, a well known text book on boundary layers with heat transfer - to remind us that for many engineers outside the aerospace industry it is heat or mass transfer, rather than momentum transfer, which is the justification for studying boundary layers at all - and, starting as early as 1950, a number of papers on the more complex shear layers that occur when the boundary layer approximation breaks down because of significant pressure gradients normal to the wall. The present paper is a review of methods of calculating boundary layers and other shear layers, with special emphasis on the "complex" flows in which the boundary layer approximation, though qualitatively useful, is quantitatively inaccurate - and, of course, concentrating upon turbulent flows because the calculation of laminar flows <u>is</u> now purely a computing problem, although still in some cases a formidable one.

The Boundary Layer Approximation

This author prefers to call it the "Thin Shear Layer" approximation, because (1) it applies to jets and wakes as well as to layers on a boundary, (2) the approximations used rely on the shear layer being <u>thin</u> - that is, its thickness δ increases only slowly in the general direction of flow, x. If $d\delta/dx$ is small, streamwise gradients of viscous or turbulence stresses are small, and so is the pressure difference across the thickness of the shear layer. The effect of neglecting the corresponding terms in the equations of motion is to remove the possibility of "upstream influence" - the effect which the solution at $x=x^*$ has on the solution for $x<x^*$ - so that the "type" of the partial differential equations changes from elliptic to parabolic. It is sometimes thought paradoxical that neglect of the y-wise gradient of pressure should lead to the disappearance of x-wise (upstream) influence, but the reason is that if dp/dy is neglected everywhere the pressure is forced to be equal to its value at the outer edge of the shear layer - specified, in effect, as part of the boundary conditions - so that p and dp/dx cannot be altered by changes within the boundary layer. Upstream influence can, of course, be transmitted via the external flow, so that a change in displacement thickness at x^* does lead to a change in the pressure distribution upstream of x^*. Upstream influence via pressure gradients and viscous stresses is, of course, less strong than upstream influence due to recirculation of the flow, and it is all too tempting to assume that the thin shear layer approximation is valid up to the point of flow reversal (separation) or even beyond. It is well known [2] that the thin shear layer equations have a singularity at a separation point, because the failure to conserve v-component momentum implies that a sudden change in <u>surface</u> streamline direction at a separation point can be propagated through the shear layer. Of course, if the streamline at the edge of the boundary layer is prevented from suddenly changing direction, by specifying the displacement thickness rather than the external flow velocity for instance, the kink in the streamline direction at the surface is suppressed and the solution becomes regular. However, this merely suppresses the worst effects of neglecting upstream influence via the pressure; and downstream of separation, the even stronger upstream influence via the convective terms leads to further difficulties in the solution. Methods of avoiding this include (1) the FLARE approximation [3], in which the term UdU/dx is neglected altogether when U is negative (2) the two layer scheme of Phillips and Ackerberg [4], in which the downstream going part of the flow is marched downstream and matched iteratively to an upstream marching solution of the upstream going part of the flow near the surface, and (3) the DUIT scheme of Williams [5] which uses downstream and upstream marching solutions for the whole shear layer alternately. Methods (2) and (3) numerically simulate the required upstream influence by the velocity field whilst still, by using the thin shear layer equations, neglecting upstream influence via the pressure or viscous stresses.

The simple rule governing attempts to extend the thin shear layer equations to flows with upstream influence without going through a full Navier Stokes solution is that, after an initial downstream marching "sweep" of the calculation, the quantities that contribute to upstream influence must be <u>stored</u> so that, on the next iteration, the "old" stored values at $x=x^*$ may be used to influence the solution at $x=x^*-\Delta x$. That is, if the flow is not reversed we may be able to get away with storing only the pressure field - upstream influence by stress gradients, laminar or turbulent, being almost always negligible if the shear layer is <u>fairly</u> thin - while if the flow <u>is</u> reversed we must store the velocity as well (almost always, it will be better to work out V from the continuity equation than to store it). We can classify the practical examples of upstream influence and thus the steps necessary to reproduce them in numerical calculations, as follows

(1) flows without separation but with significant upstream influence via the pressure (for instance flows with rapid changes in direction, as near the trailing edges of aerofoils or over flap "knees")

(2) flows with significant upstream influence via reversed flow, <u>large</u> compared with upstream influence via pressure variations within the layer so that the latter can be neglected

(3) reversed flows with significant pressure gradients

normal to the flow direction.

Class (2) is probably restricted to laminar separated flows, because a turbulent flow's greater ability to resist separation implies that, when it does separate, the normal pressure gradient, and the change in flow direction, may be considerable. The amount of work that has been done on this problem may therefore be regarded as disproportionately large compared to its importance in engineering practice. There are now several examples of solutions for type (1) in which only the pressure field need be stored from one iteration to another. Pratap and Spalding [6] discuss the case of internal flows, where the boundary conditions are simply the consequences of the constancy of mass flow down the duct. Mahgoub and Bradshaw [7] showed that, in internal or external flows using the viscous inviscid interaction technique, the repeated sweeps of the shear layer calculation required to propagate upstream influence via the pressure could be identified with the repeated sweeps needed for the viscous inviscid matching procedure itself, and therefore achieved at virtually no extra cost (the solution of the normal component momentum equation for p, given the velocity field and turbulent stresses, being quite trivial). Problems of type (3) require what are generally known as "full Navier-Stokes" solutions, with storage of the velocity field as well as the pressure. Almost invariably, the "Navier-Stokes" - i.e. separated - region is confined to only part of the flow, and it is unnecessarily expensive in computer time and storage to use a Navier-Stokes solution for the whole flow: therefore, it may, again, be economic to use a viscous inviscid matching technique, with one Navier-Stokes iteration taking place in each viscous inviscid matching cycle. However, the only "imbedded Navier Stokes" solution method known to the author is that of Seginer and Rose [8], and this does not appear to optimise the iterations as described.

In summary, there is, or can be, a whole range of calculation methods intermediate between the conventional thin shear layer "marching" solution and the conventional and expensive Navier Stokes solution with storage of all variables. The choice depends upon which types of upstream influence are thought to be important in the problem considered.

Turbulence

It remains to call attention to the chief outstanding difficulty of our subject. The last sentence is taken from the 1916 edition of Lamb's "Hydrodynamics" and alas remains true. It is probable that, by the time the present author has reached Professor Walz's age, complete time dependent calculations of the larger eddies of turbulent motion will be routinely used by engineers to calculate the mean properties of turbulent flow. It is unlikely to be possible, and should certainly not be necessary, for the calculations to be extended to the smallest eddies which effect the turbulent dissipation but do not contribute significantly to the Reynolds stresses. At the present time, however, "large eddy simulations" have merely reached the stage of being useful research tools for idealized flows, in that one can deduce, from the time dependent calculations of u, v, w, and p, any statistics which cannot be conveniently measured in the laboratory, particularly those involving the fluctuating pressure. Therefore, most engineering calculation methods for turbulent flows make use of "turbulence models" for the Reynolds stresses, turbulent heat flux rate, and so on. Now it seems to be one of the rules of the Universe that any simplification of an exact equation is either less accurate, or less widely valid, than the original equation. A special case of this unfortunate general truth is that a turbulence model is likely to be less accurate, or less widely valid, than the Navier Stokes equations. That is, we are unlikely to find a turbulence model that can compute a whole range of flows to acceptable engineering accuracy without making the empirical data used in the turbulence model dependent on the type, or geometry, of the flow. As Kline[9] has recently pointed out, there is nothing shameful in adjusting one's constants to suit the flow, providing that the adjustment is included in the computer program and not left to the discretion of the user of the program, who is likely to be an engineer in industry with many things on his mind other than the niceties of turbulence modelling.

One of the main contributions of the recent "Stanford" conference on Complex Turbulent Flows was, indeed, to demonstrate that none of the existing turbulence models was adequate over a wide range of flows, a demonstration which prompted Kline's remarks quoted above. The detailed lessons of the conference are still being absorbed, but it is to be hoped, that, over the next few years, logical schemes will be developed for "fine tuning" the empirical data in well founded turbulence models so as to improve their range of accuracy. It seems to have been generally accepted by the attendees at the Stanford conference that "well founded" probably means "founded on the exact Reynolds stress transport equations": although calculation methods like the "k,e" method, which solve transport equations for eddy viscosity but then relate the Reynolds stresses to the local mean strain rates, have been useful over the last few years, they are manifestly inaccurate in rapidly distorted flows in which the mean velocity gradients change quickly but the turbulence stresses do not. It has to be admitted, however, that the overall performance of present day methods based on the Reynolds stress transport equations is not significantly better than that of the eddy viscosity transport methods.

Complex Turbulent Flows

Some years ago this author [10] presented a conference paper suggesting that a wide range of complex flows can be classified as either

(1) shear layers interacting with another turbulent flow
(2) perturbed shear layers with small extra rates of strain (3) perturbed shear layers with large extra rates of strain or other strong perturbations.

The classification of complex flows is not just an idle exercise, but seeks to reduce the rather large range of flows found in engineering hardware to a comparatively small range of fundamental flow phenomena. It seems essential for the success of Kline's zonal modelling proposals that the dependence on flow geometry should be minimised, while allowing the coefficients in the turbulence model equations to depend on flow phenomena. To take a trivial example, it is obviously better that any correction for the effects of streamline curvature should be related to the local value of dV/dx rather than to the curvature of any solid surface which is guiding the flow. However, one of the main problems of turbulence modelling at present is the effect of a solid surface on a nearby turbulent flow, via the <u>normal</u> component constraint at the surface: many turbulence models contain a "wall effect" correction which is significant in boundary layers but negligible in free shear layers, entirely in the spirit of Kline's proposal - <u>some</u> dependence on flow geometry as such is obviously necessary. It is in the nature of zonal modelling that no very spectacular break-through is likely: the grand synthesis has been sought but not found. Therefore, the next few years are not likely to be very exciting ones in turbulence modelling, but they can be years in which a number of small contributions to zonal modelling can aggregate into a real improvement in calculation methods for complex turbulent flows.

Conclusions

For another generation or longer, engineers will have to avoid full time-dependent Navier Stokes solutions for turbulent flow by "modelling" equations for the Reynolds stresses. The approximations needed are necessarily arbitrary and empirical and must be based firmly on experimental data <u>or</u> on large-eddy simulations for a few representative flows. For the next generation, too, Prandtl's boundary layer approximation, the first example of matched asymptotic expansions, will remain in use in viscous-inviscid matching methods, again to avoid full solution of the time-averaged Navier-Stokes equations: higher-order approximations, intermediate between Prandtl and Navier-Stokes, will come into wider use. Thus the subjects to which Professor Walz has made such notable contributions will continue to flourish.

References

1. Walz, A.: Elektrokinetik - ein Prinzip der Wechselwirkung zwischen mechanischer und elektrischer Energie. ZFW <u>14</u>

(1966) 25-30.

2. Bradshaw, P.: Singularities in unsteady boundary layers. AIAA.J.17 (1979) 790-793.

3. Reyhner, I.: Flugge-Lotz,I: The interaction of a shock wave with a laminar boundary layer. Int.J.Non-Lin.Mech.3 (1968) 173-199.

4. Phillips, J.H.: Ackerberg, R.C.: A numerical method for integrating the unsteady boundary layer equations when there are regions of backflow. J.Fluid Mech. 58 (1973) 561-579.

5. Williams,P.G.: A reverse flow computation in the theory of self- induced separation. Lecture Notes in Physics 35, (1975) 445-451

6. Pratap, V.S.; Spalding, D.B.: Numerical computation of the flow in curved ducts. Aero.Quart 26 (1975) 219-228

7. Mahgoub, H.E.H.; Bradshaw,P.: Calculation of turbulent inviscid flow interactions with large normal pressure gradients. AIAA.J. 17 (1979) 1025-1029.

8. Seginer, A.; Rose, W.C.: A numerical solution of the flow field over a transonic airfoil including strong-shock-induced separation. AIAA paper 76-330 (1976)

9. Kline, S.J.: Universal or zonal modeling - the road ahead. Proceedings 1980-81 AFOSR-HTTM-Stanford Conference on Complex Turbulent Flows. Stanford University Thermosciences Div. 1982

10. Bradshaw, P.: Variations on a theme of Prandtl. AGARD Conf. Proc. 93, 1971.

On Heat Transfer by Superimposition of a Creeping Flow of a Fluid with its Stationary Free Convection

G. Appel, V. Lippig, D. Straub and R. Waibel

Hochschule der Bundeswehr München
Institut für Thermodynamik im
Fachbereich Luft- und Raumfahrttechnik
D-8014 Neubiberg

Summary:

Under special circumstances heat energy has to be transferred from a bundle of parallelly arranged finned cylindrical pipes to a fluid coolant, which can be moved only in a creeping flow between the heated pipes in the direction of the cylinder axes.

The experimental and theoretical results presented here demonstrate that, by stationary heat transfer and an increasing mass flow of coolant, the effective mean temperature difference at first weakly increases behind a characteristic minimum, then reaches a pronounced maximum and finally decreases asymptotically towards zero in a high coolant mass flow. This behaviour offers the possibility of transferring large rates of heat with small amounts of coolant, an application which has, for example, increasingly interesting ecological aspects.

Introduction and formulation of problem

The stimulus for this work originated from a contract to design an underground mining heat transfer unit with a relatively high capacity. Restricted spatial conditions as well as a very small mass rate of cooling water in comparison to the recooling airflow at a pressure of 40 bars were among the project's difficult specifications. The complicated distribution geometry for the accurate coolant guide, resulting from the very low obtainable flow velocities, excluded a conventional heat transfer unit from the beginning due to the high costs. Our examined alternative to make the heat transfer possible by means of so-called gravitational heat pipes, however, offered an acceptable solution only if the required total rate of energy transfer could be realized by free convection on the side of the cooling water.

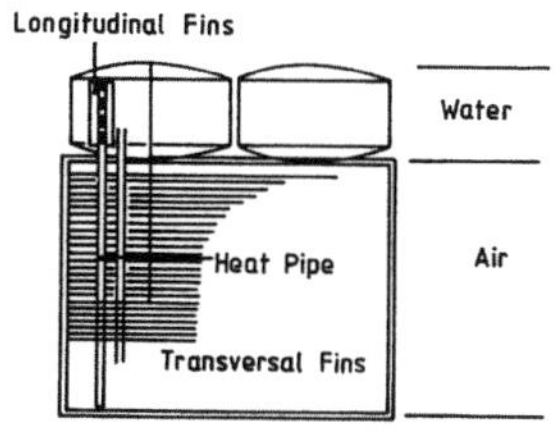

Fig. 1

In fig. 1 the described heat transfer unit is schematically outlined; important construction features are noted.

In order to check the theoretical estimations, a preliminary experimental programm was carried out. The experimental method allowed us on the one hand to determine the number of the quantities influencing the design of the heat transfer unit. On the other hand the relatively simple geometric configuration of the testing stand enabled us to compare the experimental results with the evaluated data from the selected physical model used for a theoretical description of the superimposition of the creeping bulk flow of the coolant over the natural convection along the heating rods. The results not only fulfil their basic purpose, but also indicate new possibilities of transferring large heat fluxes by small amounts of cooling water. In our opinion, the process can be applied not only in special-purpose engineering, but also in such cases where the circulating coolant flow has to be kept low for safety reasons. Increasingly, ecological considerations are likely to question the use of large supplies of cooling water required for conventional high-performance heat exchangers. We could not find any literature on the subject. Problems in mixed convection heat transfer are offen treated in nuclear engineering [1], but pertinent solutions are not relevant to this work.

This paper covers the presentation of the physical model, a short description of the simple testing equipment and the relevant measuring results, a brief summary of the arithmetical methods, as well as a summarized valuation of all results.

Physical Model

The heat transfer unit consists of a box-shaped container with a double bottom and a number z_p of vertical tubes extending into the container. The double bottom has inlet slots to ensure an uniform flow of the coolant $\dot{m}$ along each tube. As a rule, the coolant flows in at temperature T_i and flows out at the top of the container at temperature T_o. The coolant is charged with the total heat flux through the areas of the more or less regularly arranged tubes. Thus the temperature difference ϑ_p between the average temperature T_p of the single tube surface and the bulk temperature T_∞ of the coolant flow can be considerably influenced by attaching a number z_f of suitable fins to each tube.

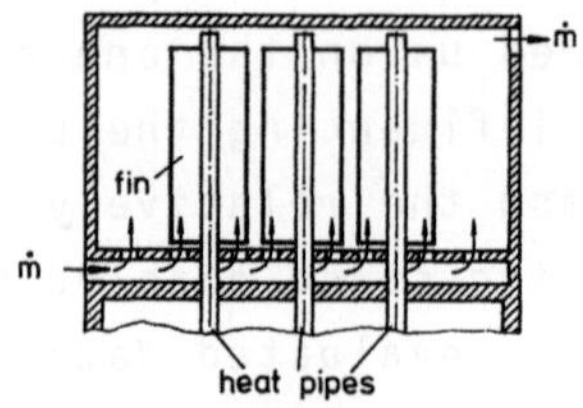

Fig.2: Principle of heat exchanger unit

In fig. 2 the model of the device is illustrated. As the simplest form, axially arranged rectangular fins (longitudinal fins), with the figured dimensions plotted in fig. 3, have been chosen in order to ensure the desired strong influence of the free convection flow on the immediate surroundings of the heating tubes.

Fig. 3

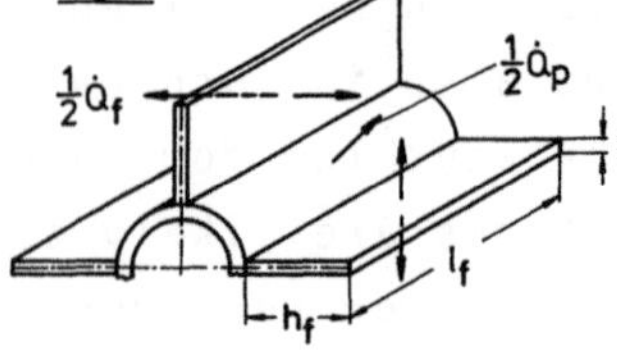

The choice of a sufficiently large free cross section A for the container results in a very low value of averaged flow velocity v of the coolant (with the average mass density ρ), according to the continuity equation

$$\dot{m} := \rho\, A\, v. \qquad (1)$$

In principle this also holds at a relatively high heat flux $\dot{Q}$ and at a relatively slight permissible temperature increase $\Delta T_c := T_o - T_i$ of the coolant. The desired total rate of the heat exchanger $\dot{Q}$ is linked by the energy balance

$$\dot{Q} = \dot{m}\; c\; \Delta T_c \qquad (2)$$

to the required coolant flow $\dot{m}$ (with the average specific heat c). In contrast to this creeping flow a significant flow velocity v_{fc}^{max} will be observed in the immediate neighbourhood of the wall at the upper end of each heated tube and of the soldered longitudinal fins, respectively. This results as an integral mean value for the local velocity profile of the boundary layer flow parallel to the heated surfaces. For instance, along the two sides of one longitudinal fin (width h_f), the mass flow $\dot{m}_f$ is thus transported due to free convection. This mass flow, defined as

$$\dot{m}_f := 2\, h_f\, \delta_s^{max}\, v_{fc}^{max}\, \rho \tag{3}$$

is related to both the properties v_{fc}^{max} and δ_s^{max}, representing the assigned thickness of the hydrodynamic boundary layer. Including all finned pipes, the total mass $\dot{m}_{fc}$ is transported per time unit by the effect of free convection. Here we examined the special case that the total mass flow $\dot{m}_{fc}$ is considerably larger than the overall coolant flow $\dot{m}$. In a stationary state, a back flow occurs between the heated pipes, as partly and schematically outlined in fig.4, of two opposing longitudinal fins. This flow reversal produces a considerable change in the local heat transfer conditions: in simple words, the natural circulation loops, known from other examples, decrease for the heat flux

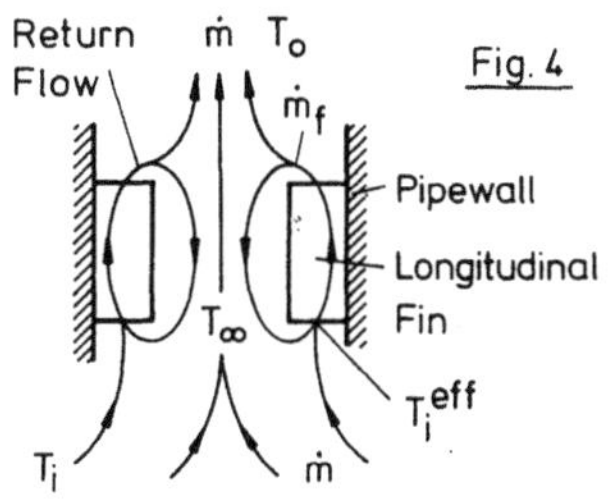

Fig. 4

$$\dot{Q}_f := \alpha_f\, A_f\, \vartheta_f^{eff} \quad \text{resp.} \quad \dot{Q}_p^{red} := \alpha_p\, A_p^{red}\, \vartheta_p^{eff} \tag{4}$$

the actually effective temperature difference ϑ_f^{eff}, respectively ϑ_p^{eff} between the coolant and the fin surface A_f, respectively the (free) pipe surface A_p^{red} between the fins. The corresponding average heat transfer coefficients α_f and α_p result as a consequence of the definition $\alpha := \lambda/\delta_T$ from a boundary layer calculation for the average thickness $\bar{\delta}_T$ of the thermal boundary layer along a longitudinal fin, respectively of the (free) pipe surface; λ is the mean coefficient of thermal conduction of the coolant.

In the case of direct proportionality between ϑ_f^{eff} and ϑ_p^{eff}, the reference temperature T_*, is of considerable practical importance: for instance, in the difference $\vartheta_p^{eff} := T_p - T_*$, if one determines experimentally the average heat transfer coefficient α_{pf} according to

$$\dot{Q}_p := \alpha_{pf} A_{pf} \vartheta_p^{eff} \qquad (5)$$

for one finned pipe with the area A_{pf} of the total surface. That portion of the heat flux due to one pipe is denoted $\dot{Q}_p$, with the relation $\dot{Q} := z_p \dot{Q}_p$ to the total rate of the heat transfer unit. A simple energy balance leads to the equation

$$T_i^{eff} = T_o - \frac{\dot{m}}{\dot{m} + \dot{m}_{fc}} \Delta T_c \qquad (6)$$

which can also be expounded as a definition for the corrected inlet temperature T_i^{eff} of the coolant as a result of the uplift movement. The two limiting cases

- disappearing influence of free convection:

$$\dot{m}_{fc} \rightarrow 0 : T_i^{eff} \rightarrow T_i \quad ,$$

- dominating influence of free convection:

$$\dot{m}_{fc} >> \dot{m} : T_i^{eff} \rightarrow T_o$$

elucidate the considerable influence of the flow reversal on the effective inlet temperature T_i^{eff}, which apparently can assume values between T_i and T_o as well as the usually arithmetic-mean bulk temperature of the coolant flow $T_\infty := (T_i+T_o)/2$. The subject of our investigation was the heat transfer with the dominating influence of free convection as an elementary effect. Thus the simple rule for determining the reference temperature of the coolant

$$T_* := T_o \qquad (7)$$

is favourable, since the outlet temperature T_o of the coolant is easily accessible arithmetically (by equation (2)) as well as experimentally.

Testing equipment and experimental results

Fig. 5 shows the scheme of the testing equipment used by us; further details are available in the test report [2]. The energy influx to the cooling water is supplied by cartridge heaters,

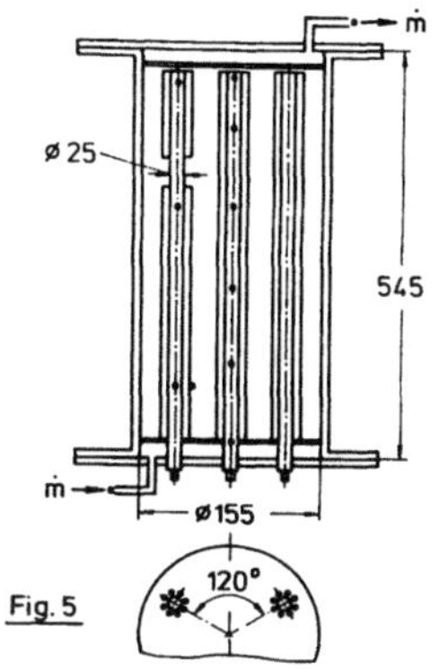

Fig. 5

• Temperature measuring points

which were formed as cylindrical rods (diameter d_p), each one permitting a relatively simple mounting of rectangular longitudinal fins. The container was insulated on the outer walls. Trial tests have shown that the electrical power $\dot{W}_{el}$ fed to the container by the cartridge heaters at a given temperature difference ΔT_c is sufficiently proportional to mass flow $\dot{m}$ of the coolant. According to the First Law of Thermodynamics

$$\dot{W}_{el} = \dot{m}\ c\ \Delta T_c - \Delta\dot{Q} \qquad (8)$$

the rate of heat transfer $\Delta\dot{Q}$ from the system to the surroundings was always negligible.

The power $\dot{W}_{el}$ was experimentally determined by measuring the current voltage. The temperatures T_i and T_o were measured with resistance thermometers. The determination of the average surface temperature T_p of the cartridge heaters was done by employing thermocouples which had been installed azimuthally on several points and on some places of the surface along a generating line. The mass flow $\dot{m}$ was measured by a rotameter and - more accurately - by weighing the water flow in uniform time intervals. Finally, the number of the longitudinal fins per cartridge heater was systematically increased from 2 to 8. The measurements were easily reproduced, their total error amounted to 4 % at the most (aside from some very small $\dot{m}$-values). The nonlinear profile of the measured temperature along a generating line of the cartridge heaters excluded a simple mean-value-procedure. To simplify matters, therefore, the heat transfer according to equation (5) between the finned rod surface and the coolant was referred to the excess temperature $\vartheta_p^{eff} := T_p^{max} - T_o$, with T_p^{max} as the highest wall temperature.
The most important experimental results are summarized in fig.6. The parameter for each curve is the constant electrical power $P := \dot{W}_{el}/z_p$ per rod equipped with fins.

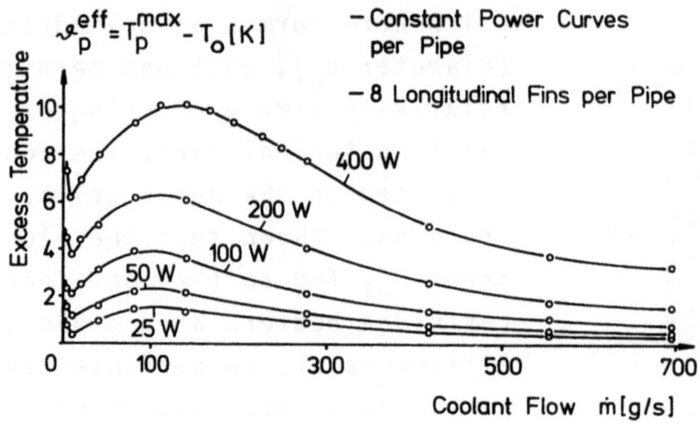

Fig.6 : Experimental Results

Accordingly, the sketched representation equals a plotting of the reciprocal product of the heat transfer coefficient α_{pf} times the heating surface A_{pf} (respectively the reciprocal Nusselt number) over the mass flow $\dot{m}$ (respectively a suitably defined Reynolds number). The typical shape of each curve shows (for A_{pf} = constant) in no way a monotonous trend from small α-numbers at free convection to large α-numbers at turbulent forced convection. The connection between ϑ_p^{eff} and $\dot{m}$ is, at first, characterized for each parameter P by a nearly linear shape behind a sharp minimum at small $\dot{m}$-values, then by a maximum, which appears to be more distinctive the larger the P-values. Finally, all curves tend asymptotically toward zero, independent of P for very large $\dot{m}$-values.

Apparently, the positions of the extreme values clearly demarcate the region of predominant influence of two autonomous physical effects. The phenomenology of the complete ϑ_p^{eff}-$\dot{m}$-function, however, suggests a relatively complicated interaction between both effects, but, so far, their undivided theoretical description is not known. It is not really necessary, since for the exergetic efficiency of the heat transfer, the small temperature differences ϑ_p^{eff} in both these branches of dominating free or forced convection are more favourable than the comparatively high ϑ_p^{eff}-values in the extremal area of the interaction of both effects. The fact, however, that for a fixed given power P of a heating rod at the same value of the temperature difference ϑ_p^{eff} in both the functional branches, only a mass flow $\dot{m}$ up to 2 orders of magnitude lower in the free convection branch is necessary, opens up new perspectives for numerous applications.

Arithmetical Methods

Construction of an adequate practical heat exchanger based on the physical model would mean, in effect, that even in a stationary operation the geometric conditions exclude a detailed theoretical description of the local energy-momentum transfer processes. However, as long as the investigations are concentrated on the branch of the dominating free convection, even very simplified mathematical methods yield results which adequately confirm the experimental results to a sufficient degree, and explicate all important quantities, e.g. for design work; so far, they are entirely sufficient, in most cases, for engineering purposes. The objective is to figure out a arithmetical procedure for determining the effective temperature difference ϑ_p^{eff} and the heat transfer coefficient α_p which depend on the most important parameters, especially on the required total rate $\dot{Q}$ of the heat exchanger and the mass flow $\dot{m}$. The extensive calculations, however, can only be roughly outlined here.

Starting from the relevant partial differential equation, one obtains

$$\vartheta_f\ (z) = \vartheta_{p*}^{eff}\ \frac{\cosh\ (\mu\ \xi)}{\cosh\ (\mu\ h_f)} \tag{9}$$

for the local distribution of the excess temperature $\vartheta := T - T_*$. In contrast to the measured excess temperature ϑ_p^{eff} the ambient reference value is the bulk temperature T_∞. In equation (9) the local fin temperature T_f is averaged over the thickness b_f: an optimization of the heat flux density with respect to the width h_f of the fin has also been taken into consideration [3]. The fin parameter is denoted by $\mu := (2\alpha/\lambda b)_f^{1/2}$ and the transformed longitudinal coordinate is denoted by $\xi := h_f - z$. All quantities, such as the thermal conductivity λ, refer to a single fin.

Fig. 7 Longitudinal Fin

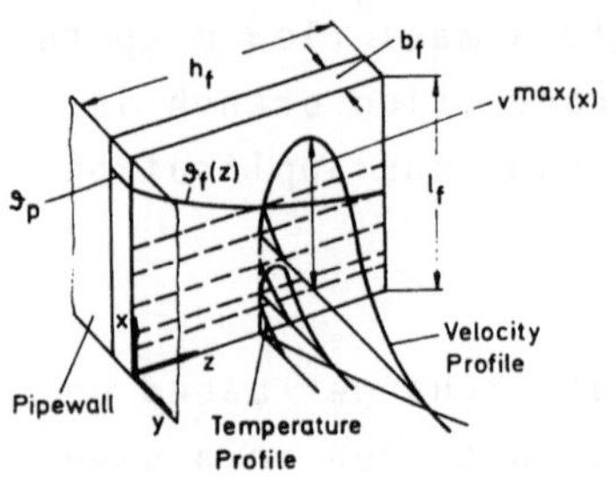

The profile ϑ_f (z) for a longitudinal fin is sketched in fig. 7. Integrating equation (9) along ξ yields

$$\vartheta_{f*}^{eff} = \vartheta_{p*}^{eff} \frac{\tanh(\mu h_f)}{\mu h_f} \qquad (10)$$

as average surface temperature of the fin. This solution expresses explicitly the above-mentioned proportionality between ϑ_f^{eff} and ϑ_p^{eff}.

In fig. 7, the local velocity- and temperature profiles are also illustrated as they are set up at free convection and constant wall temperature ϑ_{f*}^{eff} along a vertical plate. These profiles are only of interest so far as their approximate knowledge is important for the local thickness δ_T of the thermal boundary layer. Selecting a cubic ansatz for the velocity profile and a parabolic ansatz for the temperature profile with respect to the y-coordinate (compare fig. 7), the integration of the momentum and heat flux equations yields for the local thickness δ of the hydrodynamic boundary layer the relation with the length coordinate x - equation (11) -

$$\delta = 3{,}94\ Pr^{-1/2} \frac{[(\frac{20}{21} + Pr) + 6\ (\frac{5}{3} + Pr)\ \gamma + 12\ (\frac{20}{3} + Pr)\ \gamma^2\ ]^{1/4}}{(1 + 4\gamma)^{1/2}}\ Gr_x^{-1/4}\ x$$

on the right there appear the local Grashof number

$$Gr_x := g\ \beta\ \vartheta_{f*}^{eff}\ \nu^{-2}\ x^3 \qquad (12)$$

and the Prandtl number $Pr := \nu/a$, aside from the earth acceleration g, the mean values β (coefficient of thermal expansion); ν (kinematic viscosity), and a (thermal diffusivity) occur. The strong dependence of β and ν on temperature must be taken into consideration.

In equ. (11) arises the superimposition-parameter γ combining the creeping velocity v - equ. (1) - with v_{fc}^{max} - equ. (3) - in such a way that γ represents the theory only in the region $-1/2 < \gamma \leq 1$ and disappears, if $\dot{m}$ decreases towards zero. For this limiting case, equ. (11) is identical with ECKERT's well-known solution [4]. By applying the integral method this parameter γ can be conveniently evaluated, using the quadratic equation

$$\frac{\dot{m}}{\rho\ A} = \frac{5{,}16\ \nu\ \gamma\ Gr_{x=l_f}^{1/2}\ l_f^{-1}}{\{(\frac{20}{21} + Pr) + 6(\frac{5}{3} + Pr)\gamma + 12(\frac{20}{3} + Pr)\gamma^2\}^{1/2}} \tag{13}$$

together with the requested coolant mass flow $\dot{m}$ according to equ. (1). As a consequence of the connection between the two parameters δ and δ_T, defined as $\delta := \delta_T\ Pr^{1/3}$, the one adjustable parameter is the exponent n in the formula

$$\Delta\vartheta := T_o(1 - \frac{1 + \tau^{-1}}{2})\ e^{n(1-\tau)}\quad ;\quad \tau := \frac{T_o}{T_i}\ . \tag{14}$$

Equation (14) serves to correct the calculated value ϑ_{p*}^{eff} in comparison with the measured value ϑ_p^{eff}:

$$\vartheta_p^{eff} := \vartheta_{p*}^{eff} - \Delta\vartheta\quad ;$$

for n, values in the range n = 1 to n = 10 are recommended.

By our calculations we find the relevant values on the order of magnitude of $\gamma = 10^{-3}$ on the left side of the minimum and of $\gamma = 10^{-1}$ on the right. Therefore, for understanding the decreasing functional part of the free convection regime, equations (11) and (13) may be linearized for γ.

The local heat transfer coefficient α_f follows from the reciprocal thickness δ_T according to equ. (11) and equ. (14) with the computed value γ. Furthermore, the velocity v_{fc}^{max} for the mass flow $\dot{m}_f$ due to free convection is obtained from the simple relation

$$v_{fc}^{max} = \frac{\dot{m}}{\rho\ A}\ (\frac{1}{12\gamma} + \frac{2}{3})\ ,\qquad (\dot{m} \neq 0)\ ; \tag{15}$$

for $\dot{m} = 0$ this velocity follows from ECKERT's equation [4]. Finally, a helpful relation for the comparison with experimental data is obtained from the First Law of Thermodynamics, equation (8). For the electrical power $\dot{W}_{el}$ the well-known equation $\dot{W}_{el} = R_{el}\ I_{el}^2$ (electrical resistance R_{el}, electrical current I_{el}) is used and for the complete enthalpy difference of the coolant within the heat exchanger the equivalent definition $\dot{Q} := z_p\ (z_f\ \dot{Q}_f + \dot{Q}_p^{red})$ is introduced.

With the permissible assumption that both the mean heat transfer coefficients α_p and α_f are nearly equal, the two equations (4) yield with $\alpha_p := \lambda/\bar{\delta}_T$ the implicit relation

$$z_p\ l_f\ \alpha_f\ (2z_f\ \frac{\tanh(\mu\ h_f)}{\mu} + \pi\ d_p)\ \vartheta_{p*}^{eff} + \Delta\dot{Q} - R_{el}\ I_{el}^2 = 0 \quad (16)$$

for the wall temperature ϑ_{p*}^{eff}, which can iteratively evaluated, if you take account of the functional dependence of the values α_f and μ on ϑ_{p*}^{eff} and γ, especially. To get the required mean value $\bar{\delta}_T$ equ. (11) must be averaged over the length coordinate x with the result $\bar{\delta}_T = (3/4)\ \delta_T$ for the argument $x = l_f$ in equ. (11).

The performed calculations (with $\Delta\dot{Q} = 0$) for the mass flow $\dot{m}$ (g/s) of $2,5 \leq \dot{m} \leq 100$ have shown some deviations but the most significant deviation in respect to the relevant monotonous course of the function between the two extreme values will probably be not greater than 5 %.

Conclusions

The results of this work may open some possibilities of transferring big heat rates together with small amounts of coolant in stationary states. The superimposition of a creeping fluid flowing in the free areas of the heat exchanger with the free convection along the heat transferring finned areas results in simple design and construction. Well suited fins will also allow a high flexibility in operating conditions. This is true especially in cases where the exit temperature of the coolant can be increased more than in usual heat exchangers.
Another advantage is the simple handling of both the experimental and theoretical methods, previously described : as both elementary effects will not be influenced by each other in practice, small and simple experiments can be performed for scaling

up even higher efficiency heat exchanger units. The introduction of the so-called superimposition parameter γ allows a simplified but adequate theoretical description of the relevant functional behaviour in both the interesting regimes of free convection heat transfer. Further details will be published.

Acknowledgment: we are indebted to Mrs. S. Zankl, Mr. W. Aschenbrenner and Lt. P. Enzi for their engaged support.

References

[1] G.F. Jackson and T.A. Williams:
The effect of buoyancy on heat transfer in a bundle of roughened rods at low forced flow.
Nato Advanced Study Institute, Ankara, 1981
in: S. Kakač, R.K. Shah and A.E. Bergles (eds):
Low Reynolds Number Flow Heat Exchangers
Hemisphere Publ. Corp., New York, 1982 (to be published)

[2] G. Appel, E. Oswald and D. Straub:
Theoretische und experimentelle Untersuchungen zur Wärmeübertragung durch Freie Konvektion in einem GWR-Wärmeübertrager.
Interner Arbeitsbericht des Instituts für Thermodynamik der HSBwM München, 1982

[3] U. Grigull, H. Sandner:
Wärmeleitung
Springer-Verlag, Berlin, Heidelberg, New York, 1979

[4] E.R.G. Eckert:
Einführung in den Wärme- und Stoffaustausch, 3. Auflage
Springer-Verlag, Berlin, Heidelberg, New York, 1966

Half Model Testing Applied to Wings above and below Stall

H. Bippes, M. Turk
Deutsche Forschungs- und Versuchsanstalt für Luft- und Raumfahrt e.V.
Bunsenstraße 10, D-3400 Göttingen

Summary

For an experimental study of the flow on a rectangular wing with rear separation the half-model technique was applied. Below stall unsymmetrical flow conditions due to the displacement effect of the interferent boundary layers on the wing-reflection wall intersection can be minimized by boundary layer suction. Beyond stall the interacting vortex sheets, caused by separation of the three-dimensional boundary layers on the wing and the reflection wall, form a vortex system which is different from that developing on the corresponding full model. The topological structure of this vortex system and its effect on the flow symmetry are discussed.

Introduction

An approximate calculation method for the flow on aircraft wings including viscosity effects and rear separation is being developed at the DFVLR. In order to provide experimental data for proper mathematical modeling as well as for testing the validity of the calculation method a series of experiments has been started in a "3m x 3m Low Speed Wind Tunnel". The model consists of a rectangular wing with an aspect ratio of 3.1. In order to supply data for wings with larger aspect ratios at the same Reynolds number ($2{,}1 \times 10^6$) the wing was also tested as half model. For this purpose it was mounted on a plane plate, as reflection wall which allowed boundary layer suction within a circular area surrounding the profile section.

Because of the large spanwise length of the model compared to its chordwise length the violation of the symmetrical flow conditions due to interferent boundary layers on the wing-reflection wall intersection was thought to be of small effect. The experimental results indicate that this is true only for small angles of incidence. In this case a small loss of lift is measured in the corner region. As soon as rear separation takes place there is a distinct increase of lift near the reflection wall which influences the total lift coefficient remarkably.

As the half model technique is often used in aerodynamic testing it seems to be worth-while to investigate in detail the features of secondary flows on the wing-reflection wall intersection. These secondary flows lead to errors in the symmetrical flow conditions which considerably exceed the displacement effect of the interferent boundary layers on the wing and the reflection wall.

Experimental arrangement

The experiments were carried out in the "3m x 3m Low Speed Wind Tunnel" of the DFVLR in Göttingen, the model was a rectangular wing with a NACA 4415 airfoil section (Fig. 1). In the test

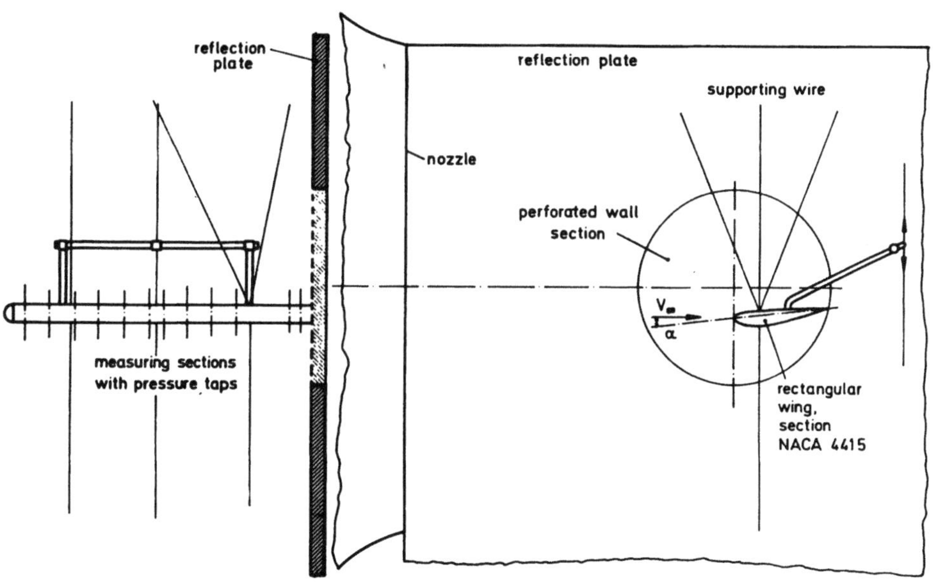

Fig. 1 Experimental set-up.

section it was suspended by tautly tensioned wires of 3 mm diameter. Cables to the measuring probes and pressure ducts were lead to the model through two struts, which are also used for arbitrarily variing the angle of incidence [1]. For the use as half model the wing was placed with one end against a plane wall having a rounded leading edge (Fig. 1). A foam-rubber seal in the joint prevented pressure compensation between the upper and lower sides of the wing. A perforated circular section on the reflection wall of 1.2 m diameter allowed boundary layer suction.

The attachment of the flow was monitored visually by a series of tufts and by the visualization of the limiting streamlines with the oil-film technique. The deflection of the oncoming flow through the displacement effect of the wing plate configuration is established by velocity field measurements with a directional probe. Pressure measurements on the wing surface close to the sidewall indicate the pressure loss due to boundary layer effects in the wing reflection wall area. Lift and drag were determined by integrating the wall pressure. For this purpose the model was equipped with 44 pressure taps along the contour of each of 12 wing sections.

Blockage effects of the wing reflection wall configuration

Blockage effects of the wing reflection plate configuration and boundary layer separation near the leading edge of the plate were seen as essential sources for unsymmetrical flow conditions. The attachment of the flow from the leading edge of the side plate was controlled by tufts. The direction of the oncoming flow, measured with the directional probe is shown in Fig. 2 for a high angle of incidence, i.e. an extreme case. The measurements were taken in the plane of the leading edge and the flow direction at various distances in front of the model. If the velocity field induced by the wing is taken into account the deflection of the oncoming flow is certainly less than $\beta = 1°$. At zero angle of attack the deflection is close to zero.

Secondary flow in the wing reflection wall intersection

As it is well known from single roughness elements the boundary layer flow approaching the stagnation zone of an obstacle separates and forms a rolling up vortex sheet, so that the

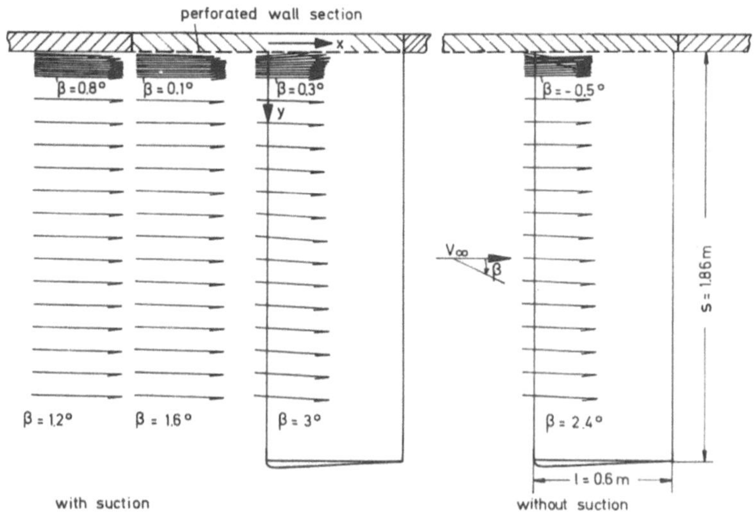

Fig. 2 Velocity distribution in the plane of the flow direction and the leading edge. β = deflection of the oncoming flow, α = 23°, Re = 2,1 x 10^6.

obstacle is surrounded by a vortex in a horse-shoe like manner. In the oil flow pattern of Fig. 3 the separation line (S) is identified by the envelope of the limiting streamlines and indi-indicates the spatial extent of the secondary flow. The effect of the secondary flow on the local lift and drag distribution

Fig. 3 Oil flow pattern developing on the reflection plane. S = 3-D separation line and contour of the horse-shoe vortex.

as well as on the pressure close to the side-plate may be estimated indirectly from those measured on the full model (aspect ratio Λ = 3.1) in preliminary experiments [1]. According to that the local lift assumes its maximum in the center plane (Fig. 4) as long as no rear separation takes place. On the half model

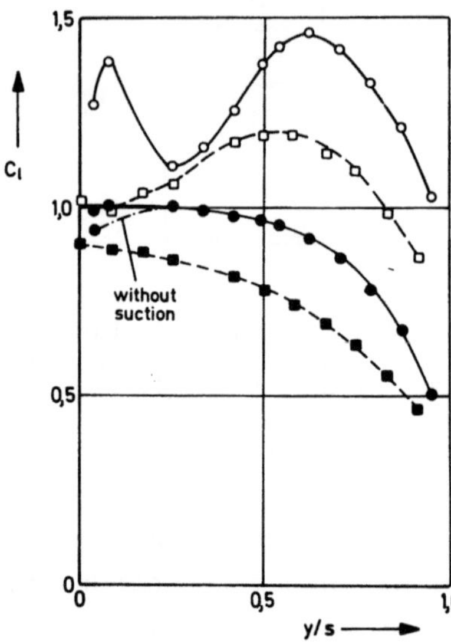

Fig. 4 Local lift distribution on the full-model (Λ = 3.1) and on the half-model (Λ = 6.2) with boundary layer suction.
full-model: ■ α = 10°; □ α = 21°; Re = 2,1 x 10^6;
half-model: ● α = 10°; ○ α = 21°;

the local lift decreases in the neighbourhood of the reflection wall, i.e. in the symmetrical section of the flow. This loss of local lift can partly be avoided by boundary layer suction so that the local lift distribution on the half-model and the full model are qualitatively in acceptable agreement. The suction rate was chosen in such a way that the lift loss was minimized.

At higher angles of incidence the interference of the horse-shoe vortex and the separated flow on the wing largely prevent symmetrical flow conditions. This becomes evident by comparing the surface oil flow patterns in the center plane of the full model (Fig. 5a) and in the proximity of the reflection wall on the half model (Fig. 5b). The additional spiral node (N' in Fig. 5b) formed there reveals the existence of a further secondary vortex system. The three-dimensional separation bubble developing behind the separation line (S) apparently also closes on the side of

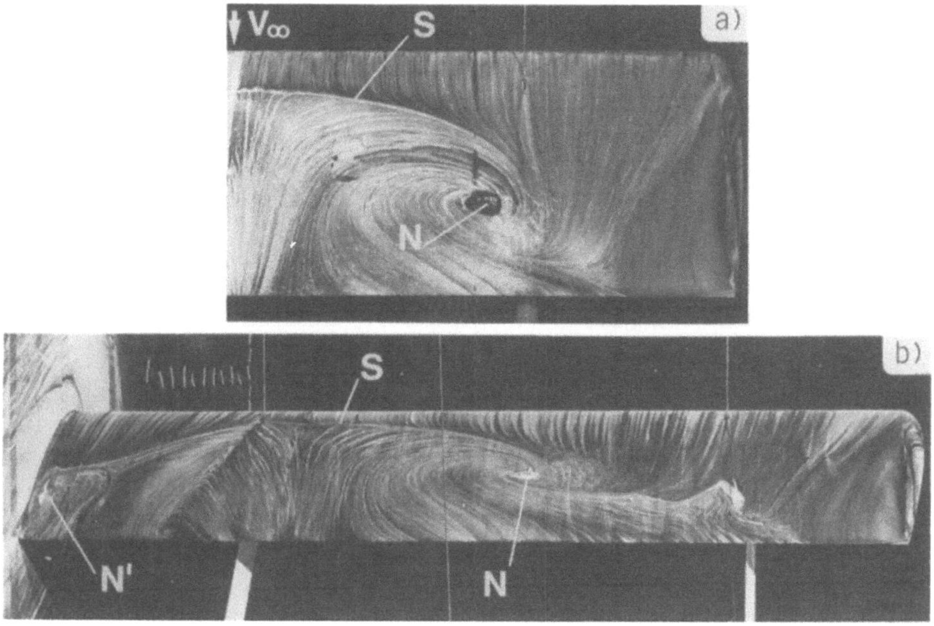

Fig. 5 Oil flow pattern on the suction side of the wing beyond stall. $\alpha = 21°$; Re = $2{,}1 \times 10^6$.
a) semi-span of the full-model
b) half-model

the reflection plane (Fig. 5b). Consequently the separation line (S) is shifted downstream. This leads to a corresponding increase of local lift (Fig. 4) whereas on the full model the local lift shows a minimum in the center plane.

In order to diminish the error in the symmetrical flow conditions different attempts have been made. Those are boundary layer suction as well as partial and complete cover of the suction holes on the perforated wall section in order to prevent pressure compensation between the upper and lower sides of the wing. But the pressure distribution in the neighbouring measuring section could not be influenced (Fig. 6) by this means. The intensity and the spatial extent of the secondary flow is too large to be affected by changing only the conditions at the wall.

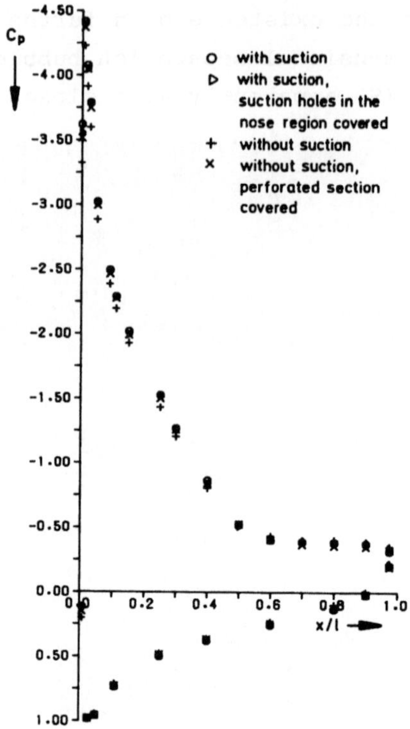

Fig. 6 Pressure distribution in the wing section at a distance y/s = 0.079 from the reflection wall. α = 21°; Re = 2,1 x 10^6.

For an exact physical explanation of this phenomenon the structure of the interacting vortex systems caused by three-dimensional separation on the reflecting wall and the wing must be known in detail. From the time averaging surface oil flow patterns a unique description of the total flow field cannot be established. An indication is given in a general study of the topological structure of separated flows by Hornung and Perry [2]. In this paper the possible vortex system on a wing-wall configuration is mapped. It is shown that the formation of an additional spiral node near the reflection wall is topologically necessary (Fig. 7).

Following Hornung and Perry's physical model the measured and visualized features become conceivable. It gives us an idea to what extent unsymmetrical flow conditions will be caused. According to that the largest shift of the separation line on the wing would arise if the center plane of the three-dimensional separation bubble on the corresponding full model is situated in

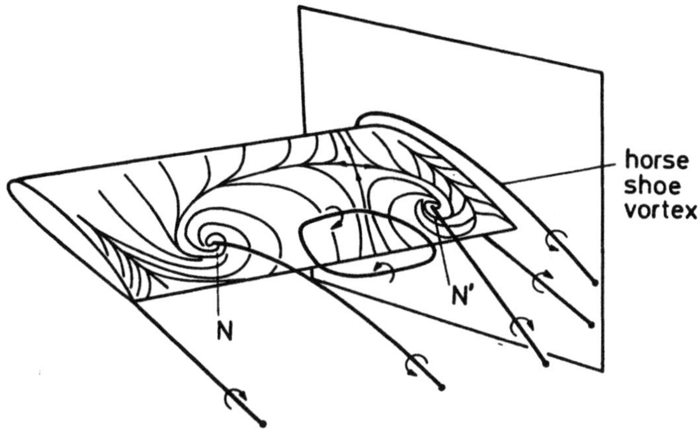

Fig. 7 Schematic of the vortex system caused by 3-D separation on the half-model given by Hornung and Perry [2]. N = spiral node; N' = additional spiral node.

the symmetric section and on the other hand the shift would be smallest if the three-dimensional separation on the corresponding full model forms a spiral node in the center plane. The latter case is possible when more than one separation bubble is formed on the wing (Fig. 8). As shown by Winkelmann and Barlow [3] the number of three-dimensional separation bubbles depends on the angle of attack, the aspect ratio and, presumably, the Reynolds number.

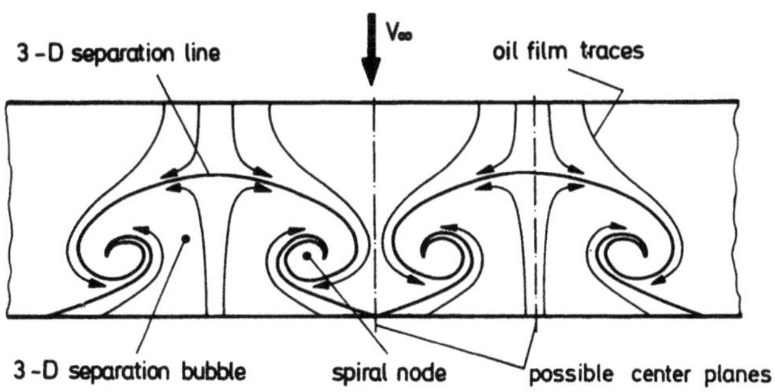

Fig. 8 Schematic of the possible locations of the separation bubbles with respect to the center plane of the full-model.

Conclusions

The main cause for the failure to establish symmetrical flow conditions in the described half model test is not a simple displacement effect of the interferent boundary layers developing on the reflection wall and on the wing but the three-dimensional separation in front of the model. For an estimate of the effect of the interacting secondary flow induced by the separated vortex sheet the flow conditions on the model must be taken into account. On the airfoil symmetrical flow conditions can be approximated satisfactorily with the use of boundary layer suction only below stall. In the presence of a three-dimensionally separated flow region the horse-shoe vortex forces the formation of an additional spiral node near the reflection wall. The size of the error in the mirror image of the flow depends on the location of this nodal point with respect to the nodal points of the imbedded vortex system on the corresponding full model. As long as the dependence of the structure of the separated flow on the aspect ratio, the Reynolds number and the angle of attack is not well understood the validity of the results of half model tests is uncertain. In the even more complicated case of wing fuselage configurations this necessitates special considerations. The consequence of this argument would be the idea to prevent boundary layer suction on the reflection wall by intensive local suction in the region of the strong adverse pressure gradient in front of the model. Such an attempt has not yet been made.

References

[1] Bippes, H.; Jacob, K.; Turk, M.: Experimentelle Untersuchung zur abgelösten Strömung an einem Rechteckflügel. DFVLR-FB 81-12 (1981).

[2] Hornung, H.; Perry, A.E.: Streamsurface Bifurcation, Vortex Skeletons and Separations. DFVLR-IB 222-82 A 25 (1982).

[3] Winkelmann, A.; Barlow, B.: Flowfield Model for a Rectangular Planform Wing beyond Stall. AIAA-Journal, Vol. 18, No. 8, August 1980, pp. 1006-1008.

On the Turbulence-Modeling Requirements of Three-Dimensional Boundary-Layer Flows

Tuncer Cebeci and K. C. Chang

Douglas Aircraft Company
Long Beach, California 90846

Summary

Appropriate three-dimensional equations have been solved, in finite-difference form, and with boundary conditions corresponding to the infinite swept wing of van den Berg and Elsenaar and the full three-dimensional data of East and Hoxey. In the former case, results were obtained with an algebraic eddy-viscosity formulation and a two-equation model which allows for transport of turbulence kinetic energy and dissipation rate. The results show that both models yield similar mean-flow characteristics, provided the same wall boundary conditions are employed, and that these deviate from the measurements with increasing adverse pressure gradient. As with previous investigations of two-dimensional flows, the procedure used to generate the initial turbulence energy profile can significantly influence the calculated results. The calculations of the fully three-dimensional flow made use of the algebraic eddy-viscosity formulation and, in keeping with the previous results for two-dimensional flows and the swept wing, the agreement with measurements is excellent until the separation region is approached.

Introduction

In a previous paper, Cebeci and Meier [1] considered the relative merits of three turbulence models for the calculation of the flows around geometric configurations of relevance to airplane and missile design. Calculated results were obtained with the three turbulence models and were compared to each other and to the boundary-layer measurements selected at the 1968 Stanford Conference [2]. The results demonstrated that the velocity and skin-friction behavior of the attached turbulent flows around aircraft and missiles can be represented to engineering accuracy by boundary-layer equations with an algebraic eddy-viscosity hypothesis. The use of higher-order models is difficult to justify and the existence of possible "coherent" structures does not appear to influence this conclusion.

The present paper extends the work of [1] to three-dimensional flows and considers the relative merits of turbulence models for such flows. The two turbulence models, corresponding to the algebraic eddy-viscosity formulation of Cebeci and Smith [3] and the transport model of Hanjalic and Launder [4] involving differential equations for turbulent kinetic energy and dissipation rate are described in forms appropriate to the three-dimensional boundary layers in the following section together with the governing mean-flow equations. The results are presented in the third section with descriptions of the two test cases, the numerical approach and the comparisons with experimental data. The paper ends with a general discussion of the results and their implications for future calculations of boundary-layer flows about three-dimensional bodies and of turbulence-model requirements.

Equations and Models

Due to space limitations, the equations and the models are not presented here; the reader is referred to Cebeci and Huang [5] for details of the equations and their boundary conditions.

Comparison Between Calculated Results with Two Turbulence Models and Experiments

We use the same numerical method with both turbulence models. This is a second-order, two-point finite-difference method developed by Keller [6] and extensively used by Cebeci and his associates for two- and three-dimensional boundary-layer flows with a wide range of boundary conditions. Details of this previous work have been described, for example, by Bradshaw et al. [7].

The first test case considered here corresponds to the measurements obtained by van den Berg and Elsenaar [8] in a three-dimensional boundary layer in an adverse pressure gradient flow under infinite swept-wing conditions. This experimental configuration appears to achieve yawed wing conditions on the test plate except for a mild adverse pressure gradient which should not be very important. We should note, however, that the test configuration corresponds to the flow in a yawed diffuser with adjustable blockage at the exit and not to a swept wing. We know from tests with two-dimensional and axisymmetric configurations that separation in a diffuser can be eliminated by introducing a high resistance element (blockage) into the flow although the precise mechanism by which this is accomplished is not presently known. Since the flow around the blocking elements is separated and can cause oscillatory wakes, it may be argued that the upstream travelling pressure pulses somehow alter the boundary-layer behavior. For this reason, the calculated results near separation should be interpreted with caution and the differences between predictions and data should not wholly be blamed on the turbulence model.

The calculations with the Cebeci-Smith (CS) eddy-viscosity formulation were made with two separate wall boundary conditions. The first used the no-slip boundary conditions, namely,

$$y = 0, \qquad u = v = w = 0 \tag{1a}$$

$$y = \delta, \qquad u = u_e, \qquad w = w_e \tag{1b}$$

and the other used the law-of-the-wall boundary conditions,

$$u_0 = u_\tau \beta_x \left(\frac{1}{\kappa} \ln y_0^+ + c\right), \qquad w_0 = u_\tau \beta_z \left(\frac{1}{\kappa} \ln y_0^+ + c\right), \tag{2a}$$

$$v_0 = \beta_x^{-1/2} \left\{\frac{u}{u_\tau} \frac{du_\tau}{dx} y - \frac{\nu}{2} \left[\beta_z \frac{d\theta}{dx} + \sin\theta \frac{d}{dx} (\tan\beta_w) \int_0^{y_0^+} \phi_1(y^+) dy^+\right\}\right. \tag{2b}$$

Here θ denotes the angle between the external streamline and the x-coordinate line and β_w denotes the limiting cross-flow angle; the parameters β_x and β_z are [5]:

$$\beta_x = \cos\theta - \tan\beta_w \sin\theta, \quad \beta_z = \sin\theta + \tan\beta_w \cos\theta \tag{3}$$

The initial velocity profiles were obtained at $x = 0.52$m by first generating the streamwise profile according to Coles' formula and the cross-flow profile according to Mager's formula and by resolving them into x and z-components.

Figure 1 shows the effect of the different "wall" conditions on the distribution of total local skin-friction coefficient c_{ft}, wall cross-flow angle β_w, displacement and momentum thicknesses, δ^* and θ, respectively. Except for the cross-flow angle β_w, we see that the two sets of boundary conditions yield similar results when the adverse pressure gradient is weak. When the pressure gradient becomes stronger and the flow approaches separation, however, the effect of the boundary conditions become more pronounced. As in two-dimensional flow studies, the results obtained by the log-law boundary conditions yield better results than those obtained

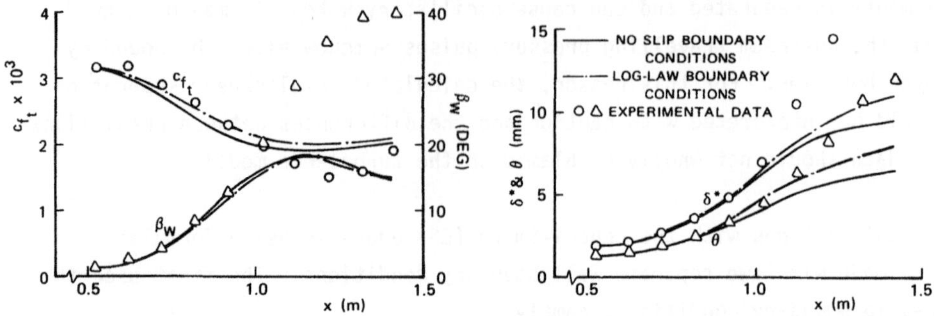

Fig. 1. Effect of different "wall" boundary conditions to the predictions of Cebeci-Smith model for the infinite swept wing data of van den Berg and Elsenaar.

with no-slip boundary conditions. This indicates that the wall functions used in the Cebeci-Smith model become increasingly inappropriate as separation is approached. It is to be expected that the law-of-the-wall assumptions will also break down but, apparently, slightly closer to separation.

Figures 2, 3 and 4 show results obtained with the K-ϕ model. From figure 2, it is clear that the influence of the turbulence model is negligibly small, even as separation is approached, provided the same wall boundary conditions are incorporated. On the other hand, and as shown by figure 3, the procedure used to generate the initial profile of kinetic energy can have a significant effect on the results as shown by figure 3. These results are consistent with those previously presented in [1] for

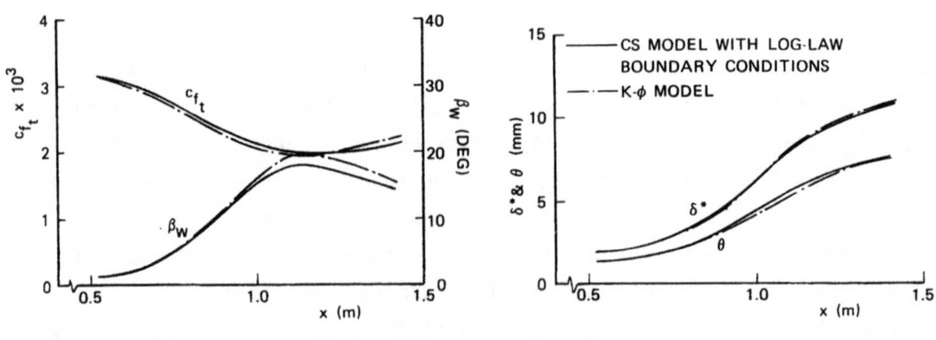

Fig. 2. Comparison between the predictions of the CS model and the K-ϕ model for the infinite swept wing.

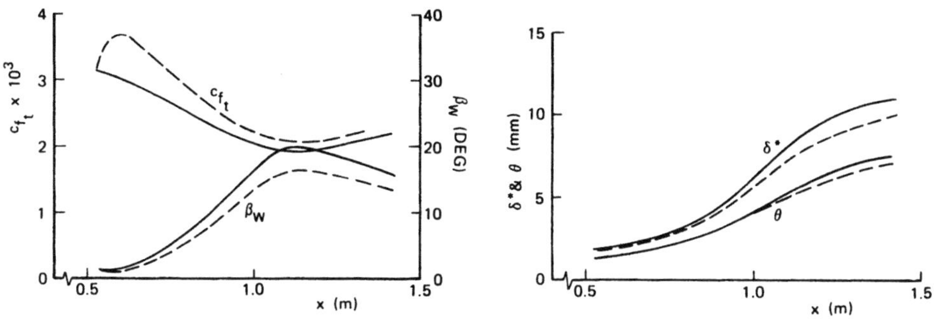

Fig. 3. Effect of initial conditions on the K-ϕ model. Solid lines correspond to initial stress profile generated from local similarity [5] and dashed lines to those from mixing length model with ℓ = 0.4y.

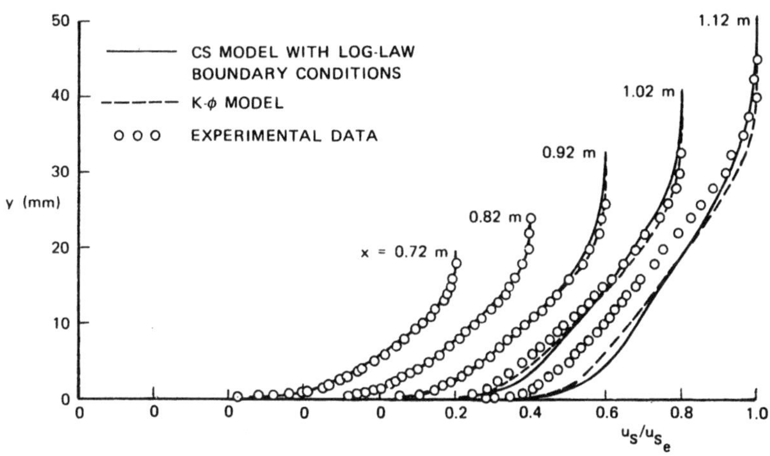

Fig. 4. Comparison of the calculated velocity profiles with experimental data of van den Berg and Elsenaar.

two-dimensional boundary layers and again it appears that the local similarity approach to the initial K-profile yields results closer to experiment and to those obtained with the algebraic eddy-viscosity formulation. As can be seen from figure 4, both models lead to velocity profiles which are essentially identical to the measurements up to the 0.92m station: further downstream, and as a consequence of the increasing adverse pressure gradient, the calculations with both models deviate from the measured results.

The data of East and Hoxey [9], the second test case considered here, were obtained in the flow formed by placing an obstruction in a thick two-dimensional boundary layer. The strong pressure gradient imposed by the obstruction, fig. 5, caused the boundary layer to become three-dimensional and to separate. The measurements were made in the three-dimensional boundary layer upstream of and including the three-dimensional separation.

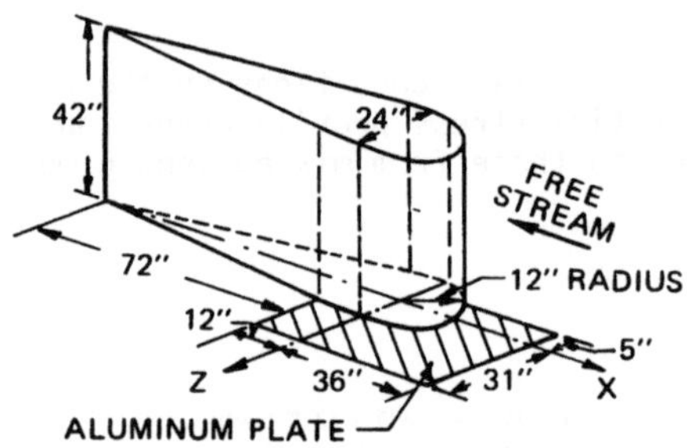

Fig. 5. Schematic drawing of East and Hoxey's test setup.

Three-dimensional boundary-layer calculations require that initial conditions in the (x,y) and (z,y) planes be given. The initial conditions along the (z,y) plane at $x = x_0$ can be obtained by inputting the measured velocity profiles. Since the z = 0 line (see fig. 6) is a line of symmetry, the initial conditions in the (x,y) plane can easily be obtained by solving the so-called attachment-line equations for a given external velocity distribution and for a specified velocity profile at $x = x_0 =$ 30". However, since the flow separates on the attachment line, an alternative procedure is required and the present calculations for obvious reasons were started on a plane parallel to the attachment-line plane as shown on figure 6. This procedure required a special numerical method to account for the zones of influence and dependence and that developed by Cebeci and Stewartson [7], and referred to as the Characteristic Box scheme, was used to obtain the results shown in figures 7 to 9. These calculations were started by specifying the initial condi- tions, in terms of measured data starting at x_0 = 30", z = 9" up to x = 21".

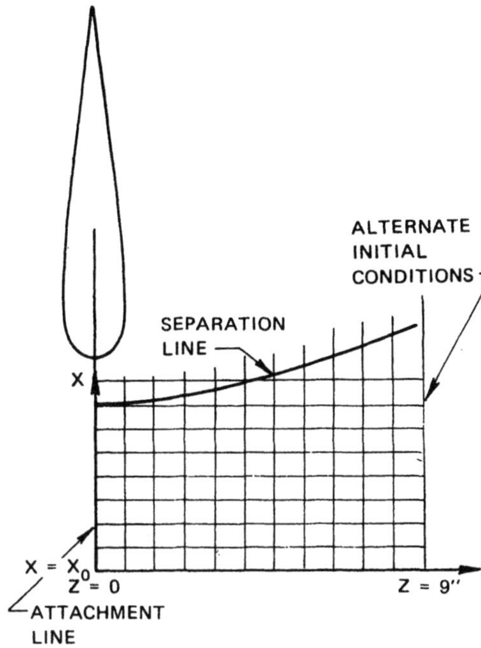

Fig. 6. Notation for East and Hoxey data.

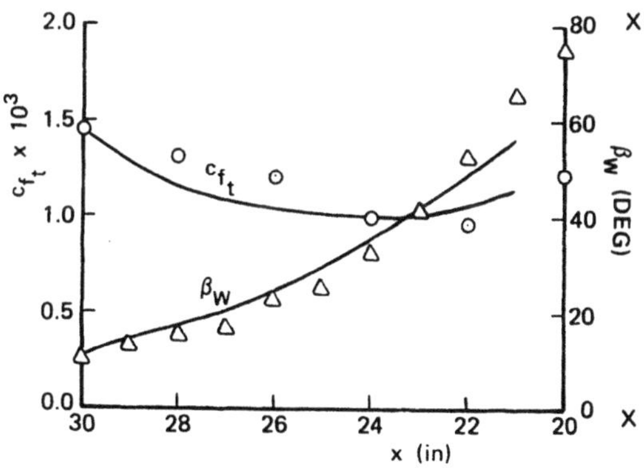

Fig. 7. Comparison of calculated results (solid line) with the experimental data (symbols) of East and Hoxey along z = 6".

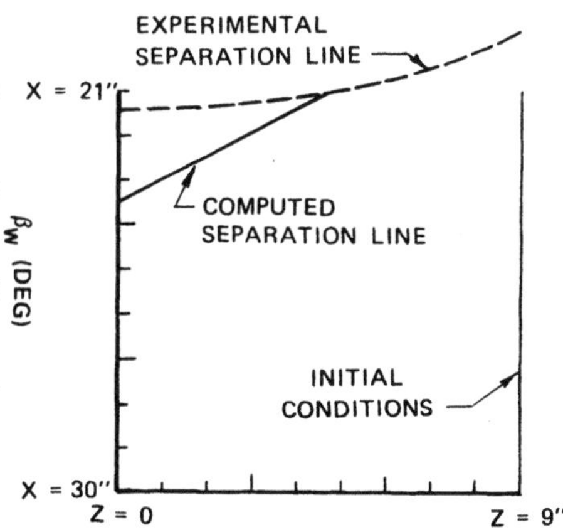

Fig. 8. Comparison between experimental and computed separation lines for the data of East and Hoxey. Note that the initial conditions at z = 9" were available for x up to 21".

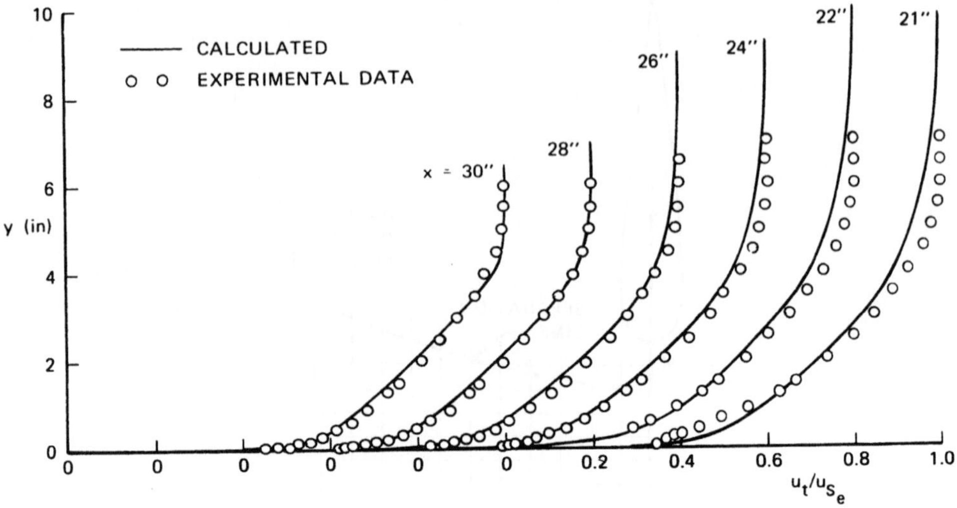

Fig. 9. Comparison of calculated velocity profiles with the experimental data of East and Hoxey along z = 6".

Results are presented on figures 7 to 9 and were obtained only with the algebraic eddy-viscosity formulation. The above discussion of the flow of reference 9, and related calculations, make clear the negligible differences resulting from the choice of turbulence model in contrast to differences between calculations and measurements close to separation and to those resulting from initial-condition uncertainties. With this fully three-dimensional flow, the separation line is calculated upstream of the measured value, figure 8, and comparison of measured and calculated mean-velocity profiles, figure 9, again show excellent agreement until the separation region is approached.

Concluding Remarks

It is clear from the result of the previous section that an infinite swept-wing flow and a fully three-dimensional flow can be represented by the present calculation method within measurement error except in the region close to separation. The choice of turbulence model is unimportant although it appears that a logarithmic law-of-the-wall boundary condition is slightly to be preferred. These conclusions are consistent with those resulting from the previous investigation of two-dimensional flows [1] and the use of the algebraic eddy-viscosity formulation is probably to be recommended on grounds of simplicity and economy.

Acknowledgment

This work was performed under the sponsorship of the Independent Research and Development Program of the McDonnell Douglas Corporation.

References

1. Cebeci, T. and Meier, H.U.: Modeling Requirements for the Calculation of the Turbulent Flow Around AIrfoils, Wings and Bodies of Revolution. AGARD Conference on Turbulent Boundary Layers; Experiments, Theory and Modeling, Den Haag, the Netherlands, 26-26, 1979.

2. Coles, D. and Hirst, E.A.: Computation of Turbulent Boundary Layers - 1968, AFOSR-IFD-Stanford Conference, Vol. 2, Thermoscience Division, Stanford University, Stanford, 1969.

3. Cebeci, T. and Smith, A.M.O.: Analysis of Turbulent Boundary Layers, Academic Press, New York, 1974.

4. Hanjalic, K. and Launder, B.E.: Sensitizing the Dissipation Equation to Irrotational Strains. J. Fluid Engineering, Trans. ASME, Vol. 102, Mar. 1980.

5. Cebeci, T. and Huang, T.T.: Description of Two Turbulence Models Used in the Finite-Difference Calculation of Three-Dimensional Boundary Layers. Proceedings of Berlin Workshop on Three-Dimensional Boundary Layers, Apr. 1982.

6. Keller, H.B.: A New Difference Scheme for Parabolic Problems. In Numerical Solution of Partial-Differential Equations. Bramble, J. (ed.), Vol. II, Academic Press, New York, 1970.

7. Bradshaw, P., Cebeci, T. and Whitelaw, J.H.: Engineering Calculation Methods for Turbulent Flows, Academic Press, London, 1981.

8. Van den Berg, B., and Elsenaar, A.: Measurements in a Three-Dimensional Incompressible Turbulent Boundary Layer in an Adverse Pressure Gradient Under Infinite Swept-Wing Conditions. NLR TR 72092U, 1972.

9. East, L.F. and Hoxey, R.P.: Low-Speed Three-Dimensional Turbulent Boundary-Layer Data, Pt. 1, Royal Aircraft Establishment, Farnborough, England, TR 69041, Mar. 1969.

Topological Analysis of Computed Three-Dimensional Viscous Flow Fields

GEORGE S. DEIWERT

NASA Ames Research Center
Moffett Field, California

Summary

Computed solutions of the time-dependent, Reynolds-averaged Navier-Stokes equations for three-dimensional flows having thin shear layers are analyzed, using topological concepts. Specific examples include the transonic flow over a body of revolution with conical afterbody at moderate angles of incidence to the free stream. Experimental flow-visualization techniques are simulated graphically to visualize the computed flow. Scalar and vector fluid dynamic properties, such as pressure, shear stress, and vorticity on the body surface, are presented as topological maps, and their relationship to one another in terms of orientation and singular points is discussed. The extrapolation from these surface topologies toward the understanding of external flow-field behavior is discussed and demonstrated.

Introduction

Computational fluid dynamics has progressed in recent years such that with present numerical algorithms and modern computational facilities it is becoming feasible to simulate complex viscous flows over three-dimensional configurations of practical interest. Typical solutions are obtained in the form of very large data sets, of the order of millions of words, that define the flow-field properties at a large number of discrete points on and about the body of interest. To visualize flow-field features, techniques devised and used by experimentalists can be emulated, and graphic displays can be generated that yield comparable information. Experimental visualization techniques include surface-oil-streak patterns, shadowgraphs, Schlieren photographs, interferograms, holograms, tufts, dye, smoke, vapor screens, and hydrogen bubbles. In addition, flow fields are probed for quantitative information with tubes, wires, films, and light beams. Similar quantitative information can be extracted from computed data bases in a straightforward manner.

Particular information concerning complex three-dimensional flow features is found in the flow-field topology on or very near the surface. This topology is the foundation upon which the external flow rests, and an understanding of the surface features can provide substantial insight into the behavior of the rest of the flow field. The analysis by Squire [1] shows that surface oil-flow visualizations closely approximate the skin-friction lines and limiting streamlines. Some considerable analyses by Legendre [2], Lighthill [3], Perry and Fairlie [4], Hunt et al. [5], and Tobak and Peake [6] provide a mathematical basis that permits detailed descriptions of three-dimensional separation and reattachment patterns based on surface vector patterns and their singular points. By extracting surface or near-surface vector and scalar topologies from discretized computed three-dimensional flow fields, the computational fluid dynamicist can use these mathematical concepts to analyze regions of separation and reattachment and to relate computed patterns to experimental observations.

To illustrate some of the topological principles for computed data bases and to relate the behavior of various vector and scalar topological patterns to each other, a series of computed results for flow over a conical afterbody at moderate angles of incidence to the free stream is considered.

The thin-shear-layer approximation to the time-dependent, three-dimensional, compressible, Reynolds-averaged Navier-Stokes equations was solved numerically, using the implicit finite-difference method of Beam and Warming [7]. The computations correspond to a series of experiments by Shrewsbury [8] for flow over a pointed cylindrical forebody of caliber nine followed by a 15° conical afterbody and a two-thirds-forebody-diameter cylindrical sting. The flow conditions correspond to a free-stream Mach number of 0.9 and a Reynolds number based on forebody diameter of nearly 3×10^6. Steady-state solutions were computed for angles of incidence to the free stream of 0°, 2°, 4°, 6°, 8°, and 12°. Quantitative comparisons of computed and experimentally measured surface pressure data for angles of incidence up to 6° have been reported by Deiwert [9] and good agreement was found. Computed results for the 8° and 12° cases will be used as illustrative examples here.

Discussion

To relate various surface properties, the incompressible Navier-Stokes equations are written for the surface where all velocity components are zero and $\omega_\zeta = 0$:

$$p_\xi = \mu u_{\zeta\zeta} = \mu(\omega_\eta)_\zeta$$

and

$$p_\eta = \mu v_{\zeta\zeta} = \mu(\omega_\xi)_\zeta \quad ,$$

where ξ and η are coordinates in the plane of the body surface and ζ is normal to the surface. The above relations tell us that the tangential vorticity is aligned in the direction of surface isobars. For a small distance ζ from the surface, the velocity vector field in the plane parallel to the surface is given by

$$\vec{V} = (\vec{\omega}_w \times \vec{n})\zeta = \vec{\tau}_w \zeta/\mu$$

where $\vec{\omega}_w$ is the vorticity vector at the surface and $\vec{n}$ is the unit surface normal vector. This relation tells us that $\vec{\tau}_w$ (surface shear) is orthogonal to $\vec{\omega}_w$ (tangential vorticity) and hence to the direction of the surface isobars.

Discretized solutions obtained from finite-difference equations do not provide detailed descriptions of behavior at singular points. Small-perturbation analyses of the differential equations in the vicinity of singular points can provide more detail and enhance our understanding of the flow behavior in these critical regions, which occur when the flow separates and reattaches. Two types of separation typically occur in three-dimensional flows: one when the components of tangential vorticity approach zero simultaneously, and the other when there is a convergence of surface streamlines. In the latter type of separation we can see from continuity that for flow through a rectangular stream tube whose base b, is the distance between two adjacent limiting streamlines and whose variable height is ζ ,

$$|\vec{\omega}_w| \zeta^2 b/2 = \text{constant} .$$

Thus, when surface streamlines converge, b decreases and there is a corresponding increase in ζ and a "piling up" of stream tubes that results in three-dimensional separation in the form of a vortex sheet. In the former type separation ($\omega_w \to 0$), small-perturbation analysis suggests that in the vicinity of a singular point there is also a pressure extremum which can be defined by $p_\xi = p_\eta = 0$, where ξ and η are mutually orthogonal.

For the present cases, limiting streamlines were determined by tracing particle paths parallel to the body surface (neglecting the surface normal velocity component) at a very small distance from the surface ($\zeta/d = 0.00008$). These lines are identical to skin-friction lines and, according to Squire [1], should closely approximate oil-streak patterns. Shown in Fig. 1 are side views of limiting streamline patterns over the complete configuration at angles of incidence of 8° (Fig. 1a) and 12° (Fig. 1b). In both cases the streamlines on the cylindrical body strongly converge near the generator, 45° from the lee generator, indicating a three-dimensional separation in the form of a vortex sheet.

Immediately aft of the cylinderical-forebody/conical-afterbody junction, the streamlines turn abruptly and completely circumscribe the afterbody. This again is an indication of three-dimensional separation and will be described in more detail subsequently. Just aft of the conical-afterbody/cylindrical-sting junction, the flow reattaches along a line circumscribing the sting, and the streamlines proceed to converge downstream along a lee generator of the sting.

Shown in Fig. 2 are surface topologies of several fluid dynamic properties for the afterbody at 8° incidence to the free stream. Included are lee, lateral, and windward views showing lines of surface shear (solid) with lines for $p_\xi = 0$ and $p_\eta = 0$ (dashed) superposed, where ξ and η are the axial and circumferential directions, respectively. Just aft of the forebody/afterbody junction there is a line of separation that completely circumscribes the afterbody. This line begins at a separation saddle point on the lee generator and terminates at a separation node on the windward generator. Immediately upstream of this separation line is a line of $p_\xi = 0$ (dashed) which intersects $p_\eta = 0$ lines coincident with the lee and windward generators and $p_\eta = 0$ lines a few degrees to either side of each generator. All saddle and node points are denoted by symbols and are defined in the figure captions. On the lee side, the pressure extrema are of the form saddle-node-saddle and correspond to the separation saddle point indicated by the limiting streamlines on the lee generator. On the windward side, the pressure extrema are of the form node-saddle-node and correspond to the separation node indicated by the limiting streamlines on the wind generator. Aft of the afterbody/sting junction there is a line of reattachment completely circumscribing the sting. There are reattachment nodes on both the lee and windward generators and a reattachment saddle 60° from the lee generator, as indicated by the limiting streamlines. Immediately downstream of the line of reattachment is the dashed line of $p_\xi = 0$, which intersects $p_\eta = 0$ lines coincident with the lee and windward generators and the $p_\eta = 0$ line at 83° from the lee generator. These points are node, node, and saddle, respectively, and correspond to the reattachment singular points directly. Between the line of separation and the line of reattachment, the flow over the surface is reversed and a separation bubble exists. The directions of the limiting streamlines under the bubble indicate counterclockwise circulation when seen in the lateral view. Downstream of reattachment there is a strong convergence of streamlines about 55° from the lee generator, indicating separation of the second kind.

Shown in Fig. 3 is a lateral perspective of the afterbody at 8° incidence. Figure 3a is a surface pressure map, in which C_p is shown as a function of axial distance (ξ) and circumferential angle (ϕ) from the lee generator. Surface pressure contours (dashed) are mapped on the afterbody in Fig. 3b and are superposed over the limiting streamlines to illustrate regions of orthogonality. The sonic surface envelope over the afterbody is shown in Fig. 3c, with skin-friction lines drawn on the afterbody surface. It is the termination of the supersonic flow inside the sonic surface by means of a three-dimensional shock front that induces flow separation on the afterbody near the forebody/afterbody junction. Included also in Fig. 3c is a dashed line for $(\vec{\omega}_w \cdot \vec{n}) = 0$, which is aligned with the vortex sheet that separates from the sting. There are also zero vorticity lines near both the lee and windward generators that cross these generators at the singular points indicated in the limiting streamline patterns.

As the angle of incidence is increased, the complexity of the surface topology can increase. Shown in Fig. 4 are lee, lateral, and windward views for the afterbody at 12° incidence. As observed in the 8° solution, there is a line of separation aft of the forebody/afterbody junction and a line of reattachment aft of the afterbody/sting junction. Both lines completely circumscribe the body. Beginning at the intersection of the separation line and the lee generator and progressing along the separation line to the windward generator, there is a series of singular points as follows: node-saddle-node-saddle-node-saddle at ϕ = 0°, 22°, 67°, 102°, 160°, and 180°, respectively. These points are preceded in the streamwise direction by pressure extrema at ϕ = 0°, 30°, 145°, and 180°, which are node-saddle-node-saddle points, respectively. It is of particular interest to note that there exist more singular points on the line of separation than there are pressure extrema in the vicinity. The "extra" points appear to be the separation node and saddle at ϕ = 67° and 102°, respectively. In addition, the flow pattern under the separation bubble is broken into several distinct counterrotating zones, with the most prominent discriminator being a line connecting the separation saddle at ϕ = 102° and the reattachment saddle at ϕ = 75°. Along this line there is strong convergence of the limiting streamlines, indicating a prominent bubble bifurcation by means of a vortex sheet. The complex three-dimensional flow behavior in the separation bubble obviates a simpler surface analysis and a one-to-one correspondence between pressure extrema and singular points in the surface pattern.

Along the line of reattachment there are nodal points on both the lee and windward generators and a saddle 75° from the lee. These are accompanied by pressure extrema slightly downstream with nodal points on both the lee and windward generators and a saddle-node-saddle cluster at $\phi = 65°$, 75°, and 97°, respectively, near the reattachment saddle. Again, as in the 8° case, convergence of streamlines on the sting near the $\phi = 70°$ generator indicates the separation of a vortex sheet.

Shown in Fig. 5 are perspective views of the 12° solution. Figure 5a shows the surface pressure map, $C_p(\xi,\phi)$. Dashed surface pressure contours are superposed on the surface streamline pattern in Fig. 5b. The sonic surface and surface streamlines are included in Fig. 5c, with a dashed line for $(\vec{\omega}_w \cdot \vec{n}) = 0$ superposed. This zero vorticity line is coaligned with the location of converged surface streamlines. Other zero vorticity lines appear near the lee and windward generators, crossing these lines at the separation and reattachment saddles and nodes.

Concluding Remarks

In the examples discussed above, we have seen that in regions of three-dimensional separation where the surface vorticity goes to zero ($\vec{\omega}_w \cdot \vec{n} \to 0$), the singular points defining the lines of separation are preceded by accompanying pressure extrema of like character (i.e., node:node and saddle:saddle) and singular points defining lines of reattachment precede accompanying pressure extrema of like character (note that clusters of singular points behave as a singular point so that a node-saddle-node behaves as a node and a saddle-node-saddle behaves as a saddle). For the 8° case, there is a direct correspondence between the pressure extrema and the singular points of separation and reattachment. For the 12° case, where the flow pattern is more complex, more singular points were observed along the separation line than were suggested by accompanying pressure extrema. In both cases (8° and 12°) singular points are indicated both by the limiting streamlines and by zero surface vorticity. For the separated vortex sheet developing from the reattachment saddle there is an accompanying line of zero surface vorticity co-aligned with the converging limiting streamlines.

References

1. Squire, L. C.: The motion of a thin oil sheet under the boundary layer on a body. RAE Aero 2636, Feb. 1960.
2. Legendre, R.: Separation de l'ecoulement laminaire tridimensionnel. Rech. Aero. 54 (1956) 3-8.
3. Lighthill, M. J.: Attachment and separation in three-dimensional flow. In Laminar Boundary Layers, L. Rosenhead, II (ed.). Oxford U. Press, 1963.
4. Perry, A. E. and Fairlie, B. D.: Critical points in flow patterns. In Advances in Geophysics, New York: Academic Press, 1974.
5. Hunt, J. C. R.; Abell, C. J.; Peterka, J. A.; and Woo, H.: Kinematical studies of the flows around free or surface-mounted obstacles: applying topology to flow visualization. J. Fluid Mech. 86 (1978) 179-200.
6. Tobak, M. and Peak, D. J.: Topology of three-dimensional separated flows. Ann. Rev. Fluid Mech. I4 (1982) 61-85.
7. Beam, R.; and Warming, R. F.: An implicit finite-difference algorithm for hyperbolic systems in conservation-law-form. J. Comp. Phys., 22, Sept. (1976).
8. Shrewsbury, G. D.: Effect of boattail juncture shape on pressure drag coefficients of isolated afterbodies. NASA TM X-1517, 1968.
9. Deiwert, G. S.: Numerical simulation of three-dimensional boattail afterbody flowfields. AIAA J., 19 (5), May (1981).

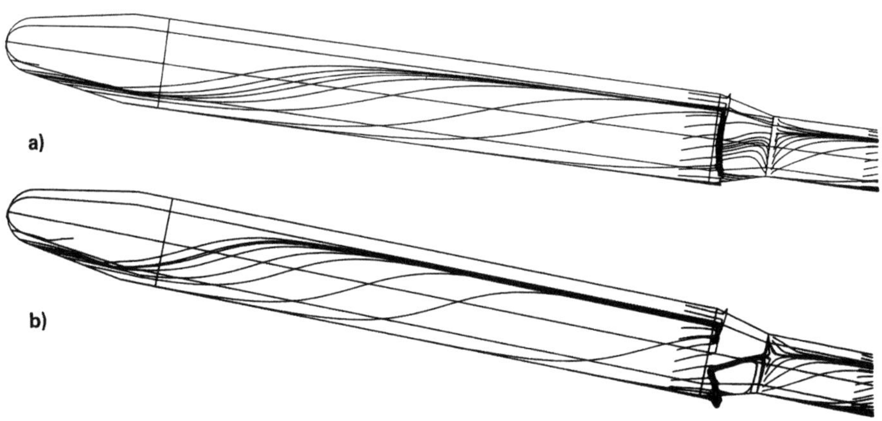

Fig. 1. Limiting streamline pattern on complete afterbody model at incidence (α) to the free stream (lateral view) $M_\infty = 0.9$, $Re_d = 3 \times 10^6$: (a) $\alpha = 8°$; (b) $\alpha = 12°$.

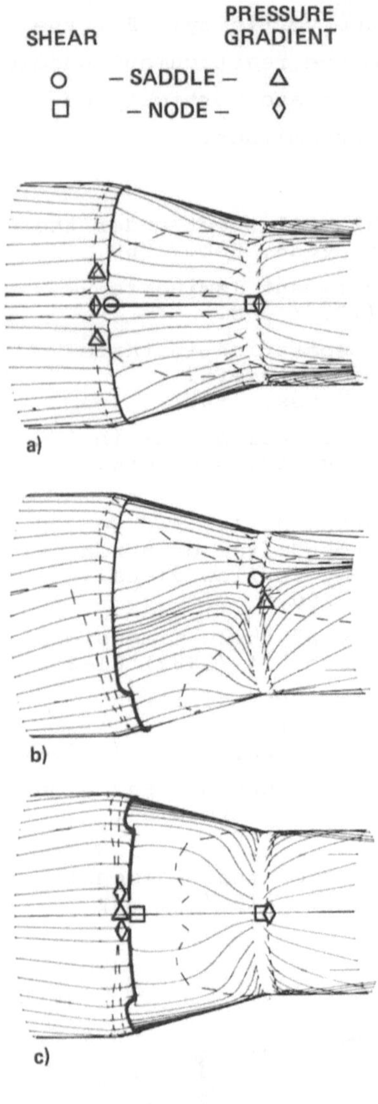

Fig. 2. Surface shear (—) and zero pressure gradient (--) topology for conical afterbody at 8° incidence, $M_\infty = 0.9$: (a) lee; (b) lateral; (c) windward side.

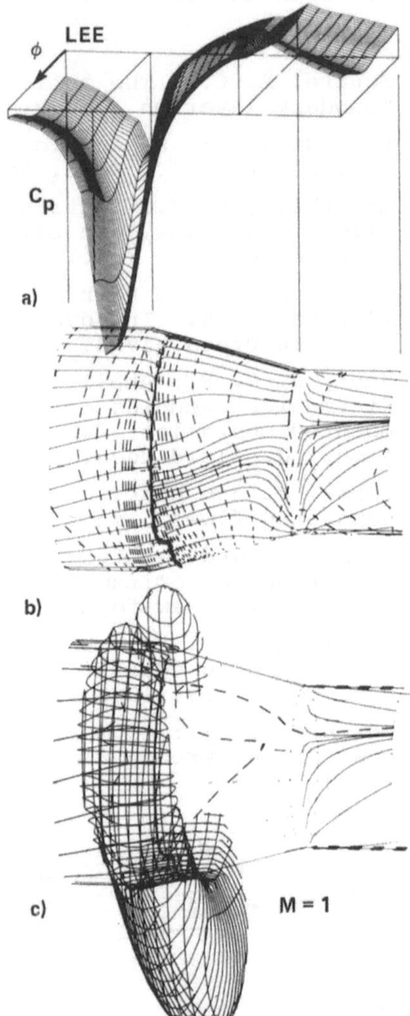

Fig. 3. Lateral perspective of flow features for conical afterbody at 8° incidence, $M_\infty = 0.9$: (a) surface pressure; (b) isobars (--) and surface shear (—); (c) sonic surface, surface shear (—) and zero vorticity ($\vec{\omega}_w \cdot \vec{n} = 0$) (--).

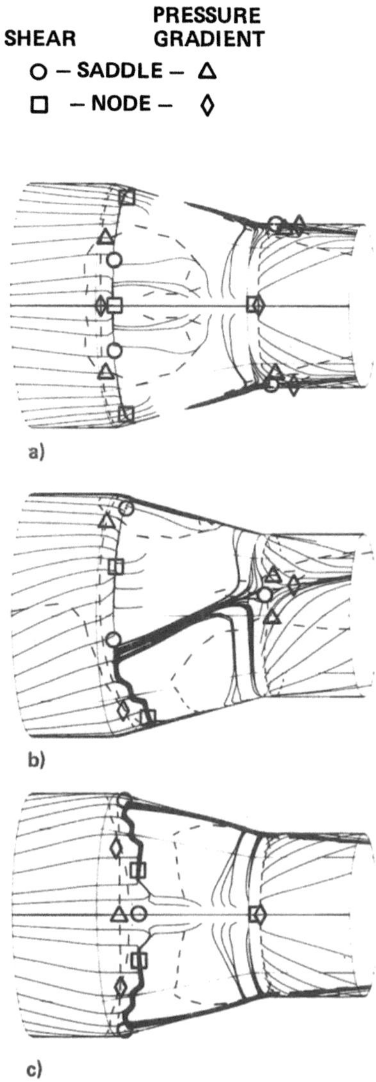

Fig. 4. Surface shear (—) and zero pressure gradient topology (--), for conical afterbody at 12° incidence, $M_\infty = 0.9$: (a) lee; (b) lateral; (c) windward side.

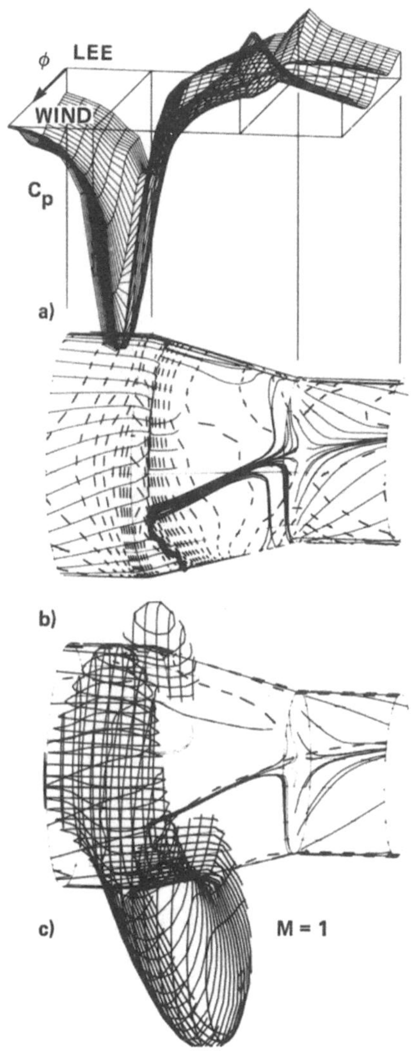

Fig. 5. Lateral perspective of flow features for conical afterbody at 12° incidence, $M_\infty = 0.9$: (a) surface pressure (b) isobars (--) and surface shear (—); (c) sonic surface, surface shear (—) and zero vorticity ($\vec{\omega}_W \cdot \vec{n} = 0$) (-·).

Instability and Coherent Structures in Jet Flames

H. EICKHOFF

Institut für Antriebstechnik
der DFVLR Köln

Summary

Coherent structures are of importance for mixing and reaction in jet diffusion flames. Using a Toepler-schlieren set up with a nano-light as the light-source, a double vortex structure, a high frequency oscillation and a low frequency oscillation separated by the reaction front, is observed. At higher Reynolds numbers the vortices break down to threedimensional turbulence retaining some degree of coherence. Compared with constant density jets exhibiting one vortex system only, jet flames are more stable and the vortices are by far more dominant. The effect of combustion on hydrodynamic instability is discussed by the aid of the inviscid disturbance differential equation.

Introduction

Although of great importance for mixing and reaction the knowledge of orderly flow structures in jet diffusion flames is very limited. Low frequency oscillations of the order of 10 - 20 Hz have been identified as burning interfaces [1,2]. Grant and Jones [1] following Kimura [3] in referring to linear stability theory pointed out that several unstable modes may exist in diffusion flames. But they were able to observe only low frequencies.

On the other hand, Yule et al [4,5] observed a double vortex structure in burning jets of propane and propane/air mixture, an inner high frequency and an outer low frequency oscillation. The low frequencies were attributed to Kelvin - Helmholtz instabilities, but the high ones were assumed to be combustion driven acoustic resonance of the nozzle. The reacting interfaces were attributed to the inner vortex system. Obviously even a qualitatively correct understanding of the vortex formation in jet flames and the location of the reactive layer is still lacking .

In the present investigation the flow field is studied by flow visualization. The oscillations are qualitatively analysed by the aid of the disturbance equation for variable density shear flows.

Experimental observations

Before starting detailed local measurements it is useful to study qualitatively the basic physical phenomena of vortex formation in flames. This can be done very effectively by flow visualization. In flames due to the steep density gradients the schlieren technique is an appropriate method. For the present investigation a Toepler-schlieren set up was used. It turned out that a very short exposure time is necessary in order to visualize clearly those high frequency vortices occuring in the interior of jet flames. Therefore a nano-light having a spark time of $2 \cdot 10^{-8}$ seconds was used as the light source.
By appropriate focusing it is possible to get both vortex systems visualized by their schlieren picture on the same photograph together with the reaction zone visualized by its' self-luminosity. Fig. 1 clearly shows this for three different downstream positions. The exposure time of 1/30 seconds of the luminous reaction zone, being located between an outer large scale and an inner small scale vortex system, is large compared with the spark time of $2 \cdot 10^{-8}$ seconds for the schlieren pictures. It is determined by the time necessary to trigger the nano-light. Due to the relatively large exposure time an orderly structure of the luminous flame zone cannot be identified. However, from published data of frequencies of temperature and ionisation fluctuations [1,2] it can be concluded that the reaction zone is fluctuating with frequencies in the same range as the outer vortex system does (10 - 20 Hz).

Frequencies of the outer and inner vortex systems in the present investigations have been determined from high speed cine-photography utilizing up to 6000 frames per second. The inner vortices move at a speed which is approximately equal to the flow velocity. Its' frequency, depending on the flow velocity is several orders of magnitude larger than that of the outer vortices.

Fig. 2a shows the rolling up of initial disturbances and formation of the inner vortices which increase in size by pairing

downstream. This process of vortex pairing is irregular at lower Reynolds numbers. Depending on the Reynolds number, two, three or even four vortices are pairing more or less completely before they start a new pairing process. The vortex structures including the onset of three dimensional disturbances with increasing Reynolds number, Fig. 2b and c, look very similar to those which have been observed in plane shear layers [6].
The outer diameter of the inner ring vortices increases linearly with the distance x from the orifice:

$$b(x) \simeq 0{,}15\ x + d_o,$$

and the wave length of the vortices is:

$$\lambda \simeq 0{,}25\ x.$$

Defining a dimensionless frequency by the Strouhal-number one obtains:

$$St = f\ .\ d_o/u_o \simeq 4\ d_o/x,$$

with the convection velocity of the vortices being approximately equal to the initial flow velictiy u_o and with the wave length given above. This relation is valid for both a contoured nozzle and a sharp edged orifice. The frequency of these inner vortices is several orders of magnitude larger than that of the outer ones which oscillate at about 15 Hz.

Analysis of the instability

The instability of free shear layers including variable density effects has been discussed in some detail by Michalke [7]. In addition to gravity, which is important for meteorological problems Michalke considered inertia effects caused by variable density. The governing disturbance equation is derived by introducing small disturbances into the equation of motion and mass conservation. With the assumptions of incompressible, nonviscous flow and neglecting buoyancy one obtains for parallel plane flow:

$$\frac{d^2\phi}{dy} + \frac{d\rho/dy}{\rho}\frac{d\phi}{dy} - \{\frac{\frac{d}{dy}(\rho\, du/dy)}{(u-c)} + \alpha^2\}\,\phi = 0. \tag{1}$$

In this equation ρ and u are the density and velocity in the shear layer, ϕ is the velocity disturbance and α the wavenumber; $c = c_r + ic_i$, where c_r is the phase velocity and c_i the amplification rate of the disturbance.
For round jets the assumption of plane flow is valid, because the initial shear layer, to which the above equation applies, is small compared with the nozzle diameter. The influence of buoyancy on stability and vortex formation was tested by comparison of flames burning vertically and horizontally. Besides a slide deformation no essential changes of the vortex system were observed. It can be concluded therefore, that gravity exerts no essential influence on the stability of jet diffusion flames.

Compared with the Rayleigh-equation valid for constant density flows two modifications occur in equation (1), which are due to variable density. One term including $d\phi/dy$ and a more important one resulting from a modification of the vorticity term:

(2) $$K(y) = \frac{1}{\rho}\frac{d}{dy}\left(\rho\frac{du}{dy}\right).$$

For constant density flow K(y) is proportinal to the gradient of vorticity:

(3) $$K(y) \sim d\Omega/dy,$$

while for variable density:

(4) $$K(y) \sim \Omega\left(\frac{d\Omega/dy}{\Omega} + \frac{d\rho/dy}{\rho}\right) = \Omega\frac{d}{dy}\left[\ell n\,(\Omega\rho)\right].$$

It has been shown for constant density flow that the instability increases with the maximum value of the vorticity at constant shear layer thickness [8] and that the number of possible unstable modes is equal to the number of zeros of K(y) [9].
Profiles of velocity and density typical of the initial region of a jet diffusion flame are shown in Fig. 2. The reaction zone of low density $\rho_{f\ell}$ is embedded between the fuel gas and the air.

From the u- and ρ-profiles an $\Omega\rho$-profile may be deduced and is shown in Fig. 3 below.
The effect of combustion is to increase stability due to the decrease of density. Because K(y) becomes zero at three different positions, three unstable modes are possible. Out of these three unstable modes two modes, which are connected with the maxima of the $\Omega\rho$-profile are observed. These disturbances are rolling up to form the two vortex systems. A third unstable mode belonging to the minimum of $\Omega\rho$ probably does not exist due to the damping effect of the high viscosity in the reaction zone.
The less important effect of the $d\rho/dy$-term in eq. (1) is to damp the instability for $d\rho/dy < 0$ and vice versa.

low frequency
vortex system

luminous
reaction zone

high frequency
vortex system

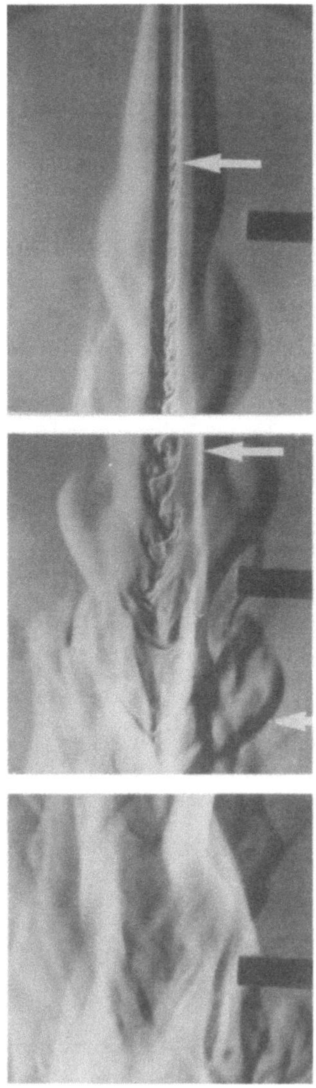

Fig. 1. Schlieren photographs of a propane diffusion flame at three different downstream positions; $d_o = 5$ mm, $u_o = 7{,}5$ m/s;

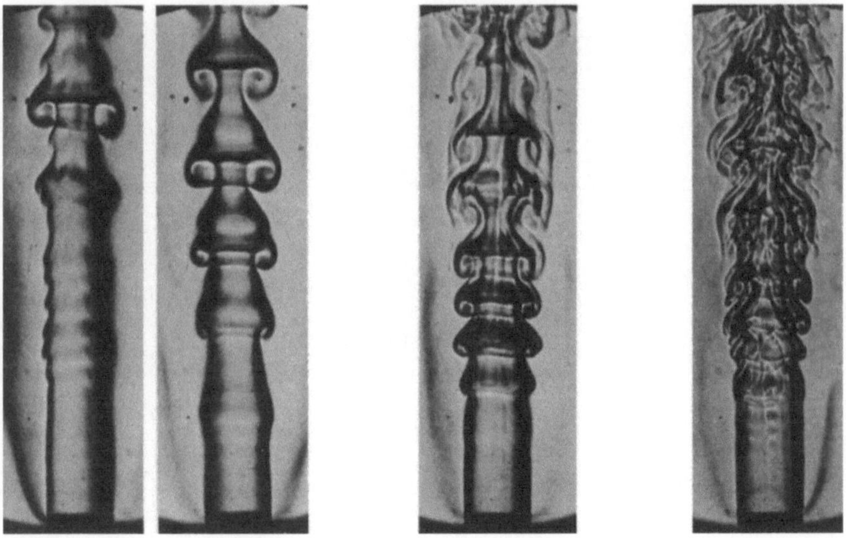

a) b) c)

Fig. 2. Schlieren photographs of propane diffusion flames at three different nozzle exit velocities u_o

a) u_o = 3,2 m/s, b) u_o = 7,4 m/s, c) u_o = 9,5 m/s; nozzle diameter d_o = 10 mm, exposure time $2 \cdot 10^{-8}$s.

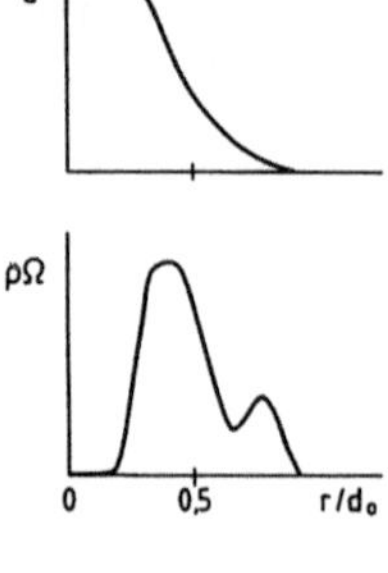

Fig. 3. Qualitative representation of radial profiles of density ρ, velocity u and ρΩ in the initial region of jet diffusion flames

References

1. Grant, A.J.; Jones, J.M.: Low frequency diffusion flame oscillations. Combustion and Flame 25, (1975) 153 - 160

2. Durão, D.F.G.; Whitelaw, J.H.: Instantaneous velocity and temperature measurements in oscillating diffusion flames. Proc. R. Soc. Lond. A. 338 (1974) 479 - 501

3. Kimura, I.: Stability of laminar-jet flames. Tenth Symposium (International) on Combustion (1965) 1295 - 1300

4. Yule, A.I.; Chigier, N.A.; Ralph, S.; Boulderstone, R.; Ventura, J.: Combustion-transition interaction in a jet flame. AIAA Paper 80-0077. 18th Aerospace Science Meeting, Pasadena, Cal. (1980)

5. Chigier, N.A.; Yule, A.J.: The physical structure of turbulent flames, A.I.A.A. paper 79-0217, 17th Aerospace Science Meeting, New Orleans (1979)

6. Roshko, A.: Structure of turbulent shear flow: a new look. A.I.A.A. paper 76-78. A.I.A.A. 14th Aerospace Science Meeting, Washington D.C. (1976)

7. Michalke, A.: Der Einfluß variabler Dichte auf die Instabilität einer freien Scherschicht. Ingenieur-Archiv 40 (1971) 29 - 39

8. Michalke, A.; Schade, H.: Zur Stabilität von freien Grenzschichten. Ingenieur-Archiv 33. Bd, 1. Heft, (1963) 1 - 23

9. Howard, L.N.: The number of unstable modes in hydrodynamic stability problems. J. Méchanique 3 (1964) 433

Finite-Difference Solutions for Laminar Boundary Layer Flows with Separation

E. ELSHOLZ+, E.U. SCHADE* and F. THIELE*

+Unternehmensbereich Transportflugzeuge, Messerschmitt-Bölkow-Blohm GmbH, Bremen
*Hermann-Föttinger-Institut für Thermo- und Fluiddynamik, Technische Universität Berlin

Summary

Both attached and separated laminar, two-dimensional, incompressible boundary layer flows are investigated by means of an inverse procedure with prescribed displacement thickness. Based on the variational principle the numerical solution is performed by an accurate finite-difference scheme of Hermitian type. Stable results are obtained provided that the FLARE approximation is applied within the vicinity of zero wall shear stress. Comparison of the numerical results with experimental data demonstrates the reliability of the procedure proposed.

Introduction

Locally bounded separation regions have a strong influence on the development of viscous shear layers. If these phenomena occur in high Reynolds number flows investigations can be performed on the basis of the boundary layer equations. For this purpose it is essential that the separation bubble is of small extend and does not burst leading to a large dead water area. Examples of such a bubble type separation region are laminar separation bubbles at the leading edge of an aircraft wing with transition to turbulence inside the bubble as well as turbulent separation which may take place in connection with shock-boundary layer interaction. However, turbulence modelling for reverse flow regions is somewhat doubtful up to now. Especially, there is a severe lack of transitional criteria in this regime. So the present investigations are restricted to laminar separated flows.

First numerical results on separated boundary layers have been presented by Catherall and Mangler [1] using a downstream-marching procedure. With the displacement thickness prescribed they found a smooth velocity distribution at the outer edge. No irregularities occured from Goldstein's singularity passing the separation or reattachment point. Further investigations have shown that no singular behaviour arises using the inverse formulation with regularly prescribed distributions of displacement thickness [2, 3], wall shear stress [4, 5] or normal velocity at the boundary layer outer edge [6].

In this paper an inverse boundary layer procedure with prescribed displacement thickness is applied which makes use of the variational principle (see e.g. [7]). The resulting differential equations are then solved by an accurate finite-difference method of Hermitian type [8]. Various low speed flows with separation are predicted. The numerical results are compared with experimental data [9, 10, 11] as well as theoretical predictions of Horton [3].

Basic Equations

Introducing the usual boundary layer transformation

$$d\xi = dx$$
$$d\eta = \left(\frac{u_e}{\nu\xi}\right)^{1/2} dy \tag{1}$$

and the stream function

$$\psi(x,y) = (u_e \, \nu \, \xi)^{1/2} f(\xi,\eta) \tag{2}$$

the boundary layer equations for two-dimensional (2-D), laminar incompressible flow can be written in the form [12]

$$f''' + m_1 f f'' + m_2\left[1 - \gamma(f')^2\right] + \frac{\eta}{2}(\gamma - 1) f' f'' =$$
$$= \xi\left(\gamma f' \frac{\partial f'}{\partial \xi} - f'' \frac{\partial f}{\partial \xi}\right) . \tag{3}$$

Here, the parameters are defined by

$$m_1 = (1 + m_2)/2 \; , \qquad m_2 = \xi/u_e \cdot du_e/d\xi \; .$$

For $\gamma = 1$ eq. (3) is the usual stream function equation whereas $\gamma = 0$ represents the FLARE approximation [13] as discussed below. The boundary conditions are

$$\begin{aligned} \eta = 0: \quad & f(0) = 0\,, \quad f'(0) = 0 \\ \eta = \eta_e: \quad & f'(\eta_e) = 1\,. \end{aligned} \tag{4}$$

In the direct problem with given pressure gradient parameter $m_2(\xi)$ eq. (3) is solved with respect to the boundary conditions (4). However, in the inverse formulation the velocity u_e at the outer edge is unknown and has to be determined such that the displacement thickness

$$\delta^*(\xi) = \frac{\xi}{\sqrt{Re_\xi}} \left[\eta_e - f(\eta_e)\right] \tag{5}$$

is varified. $Re_\xi = u_e\xi/\nu$ is the local Reynolds number.

Following the variational principle the function

$$\phi(u_e) = \delta^*(\xi) - \delta_0^*(\xi) \tag{6}$$

with δ_0^* prescribed has to be satisfied by proper choice of $u_e(\xi)$. This requires an iterative process. Here, Newton's method

$$u_e^{\nu+1} = u_e^{\nu} - \frac{\phi(u_e^{\nu})}{\partial\phi(u_e^{\nu})/\partial u_e} \qquad \nu = 0,1,2\ldots \tag{7}$$

is applied where the derivative is

$$\frac{\partial\phi(u_e)}{\partial u_e} = -\frac{\xi}{\sqrt{Re_\xi}}\frac{\partial f_e}{\partial u_e} - \frac{\delta^*}{2u_e}\,. \tag{8}$$

As the derivative $\partial f_e/\partial u_e$ is yet unknown we solve the corresponding variational equation numerically. This is presented in the next section.

Numerical Procedure

For the numerical solution of eq. (3) all ξ-derivatives are approximated by two-point backward finite-differences

$$f''' + \alpha_1 f f'' - \gamma\alpha_2 (f')^2 + Ef'f'' = \alpha(f_i f'' - \gamma f_i' f') - m_2 \quad . \qquad (9)$$

Linearization by the Newton-Raphson method yields the ordinary differential equation at the station ξ_{i+1}

$$\begin{aligned} f''' &+ (\alpha_1 \bar{f} + E\overline{f'} - \alpha f_i) f'' + (-2\gamma\alpha_2 \bar{f}' + \alpha\gamma f_i' + E\overline{f''}) f' \\ &+ \alpha_1 \overline{f''} f = - m_2 + \alpha_1 \overline{f f''} - \gamma\alpha_2 \overline{(f')^2} + E\,\overline{f'}\,\overline{f''} \; . \end{aligned} \qquad (10)$$

In eqs. (9) and (10) the abbreviations

$$\alpha = \xi_{i+1}/\Delta\xi \qquad E = \eta(\gamma - 1)/2$$

$$\alpha_1 = m_1 + \alpha \qquad \alpha_2 = m_2 + \alpha \qquad m_2 = \alpha[u_e - (u_e)_i]/u_e$$

are used. The bars in eq. (10) denote the values of the previous iteration.

In order to derive the variational equation we introduce $F = \partial f/\partial u_e$, $F' = \partial f'/\partial u_e$, etc. Differentiation of eq. (10) with respect to u_e results in the linear equation

$$\begin{aligned} F''' &+ (\alpha_1 f + Ef' - \alpha f_i) F'' + (-2\gamma\alpha_2 f' + \alpha\gamma f_i' + Ef'') F' \\ &+ \alpha_1 f'' F = \frac{\partial m_2}{\partial u_e} \left[-1 - \frac{1}{2} f f'' + \gamma (f')^2\right] . \end{aligned} \qquad (11)$$

This variational equation has to be solved subjected to the boundary conditions

$$\begin{aligned} \eta = 0 : &\qquad F(0) = 0 \,, \qquad F'(0) = 0 \\ \eta = \eta_e : &\qquad F'(\eta_e) = 0 \end{aligned} \qquad (12)$$

The linear third-order differential equations (10) and (11) are numerically solved by means of the accurate finite-difference method of Hermitian type. The representation of the Hermitian formulas is somewhat lengthy, details may be found in ref. [8].

The overall iteration process starts with the solution of eq. (10) for a given value u_e. From eq. (11) $\partial f_e/\partial u_e = F_e$ is obtained so that an improved value u_e can be calculated by Newton's method (7). This process is repeated until two consequtive values u_e differ less than a small error ε.

The procedure developed so far works properly within the attached as well as separated flow regime. Numerical characteristics turn out to be

$$\partial\phi/\partial u_e < 0 \quad \text{if} \quad f_w'' > 0 \qquad \text{(attached)} \quad ,$$

$$\partial\phi/\partial u_e > 0 \quad \text{if} \quad f_w'' < 0 \qquad \text{(separated)} \quad .$$

Near separation or reattachment where f_w'' tends to zero the discontinuity is removed by applying the FLARE approximation. This approach neglects the convective term $u\ \partial u/\partial x$ when $u < 0$. However, rather small step sizes $\Delta\xi$ and underrelaxation are necessary to get a converged solution. Once passed through the regions specified by very small wall shear stress underrelaxation and FLARE approximation can be dropped without affecting the stability and accuracy of the procedure.

Results and Conclusions

Three different test cases with sufficient experimental data available are investigated. These are

case 1: Separation over an elliptical cylinder,

case 2: Short separation bubble,

case 3: Short separation bubble at sweep angle.

The first experiment was carried out by Schubauer [9] on an elliptical cylinder, axis ratio 2.96 : 1, with the major axis

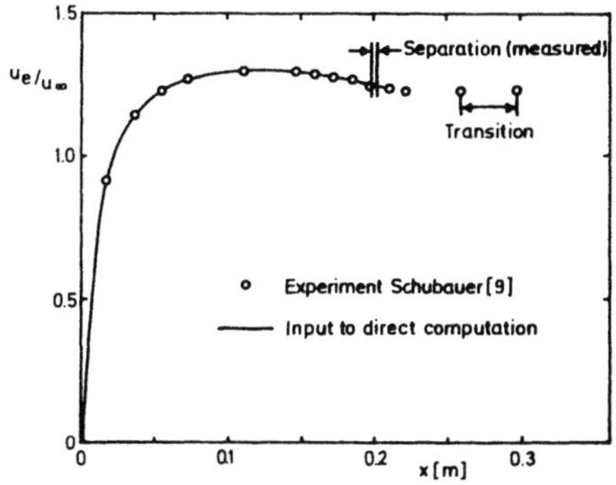

Fig. 1. External velocity distribution, case 1

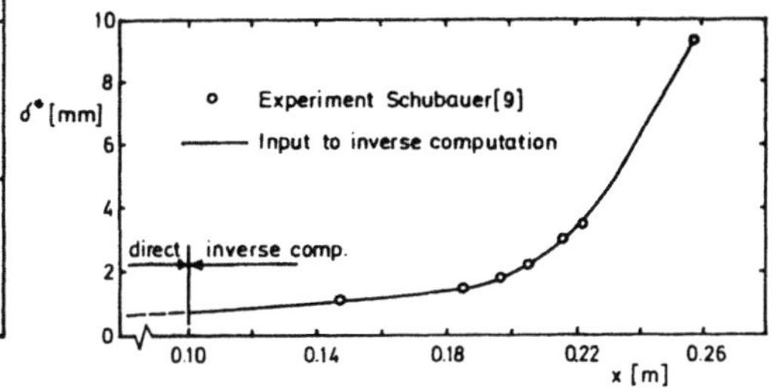

Fig. 2. Displacement thickness distribution, case 1

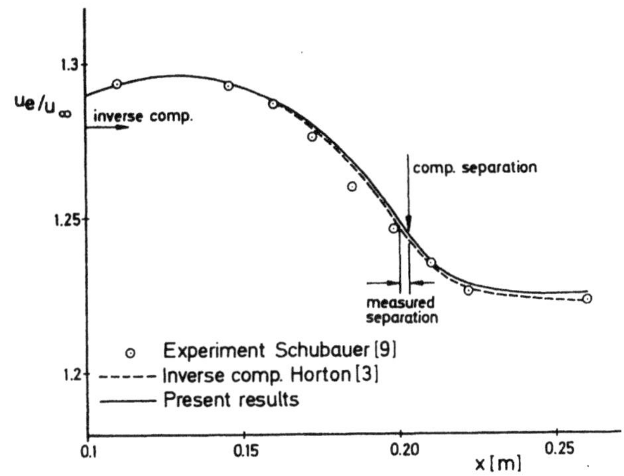

Fig. 3. Computed and measured edge velocity, case 1

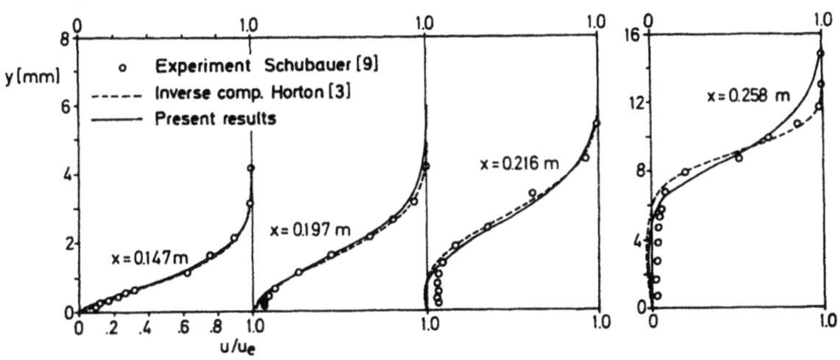

Fig. 4. Computed and measured velocity profiles, case 1

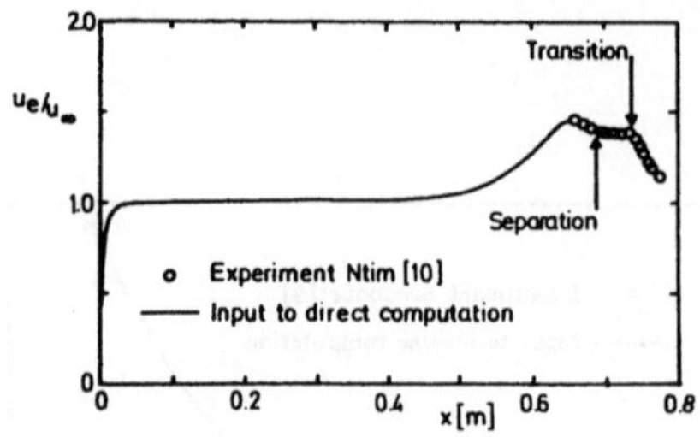

Fig. 5. External velocity distribution, case 2

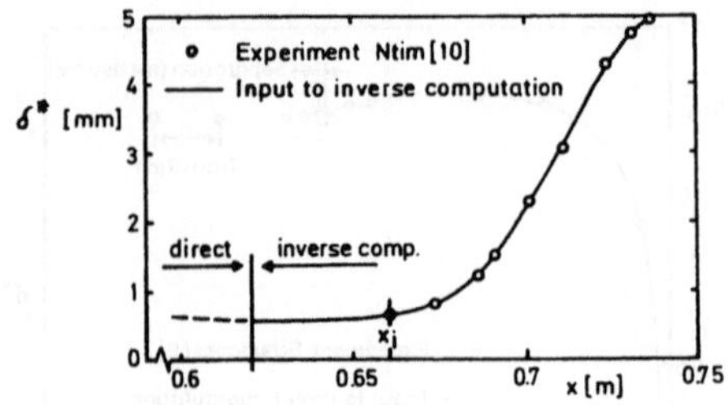

Fig. 6. Displacement thickness distribution, case 2

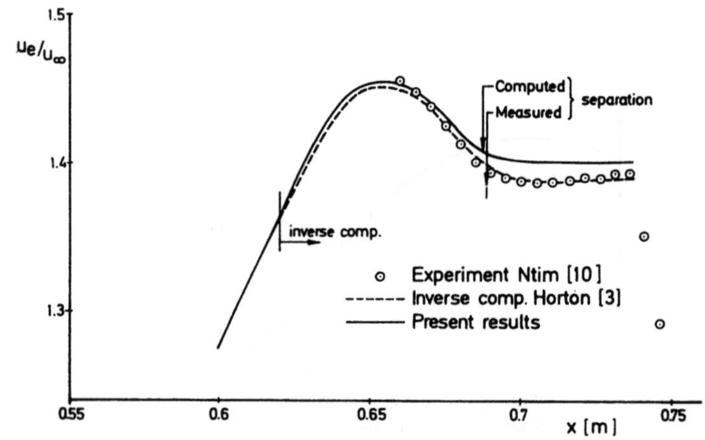

Fig. 7. Computed and measured edge velocity, case 2

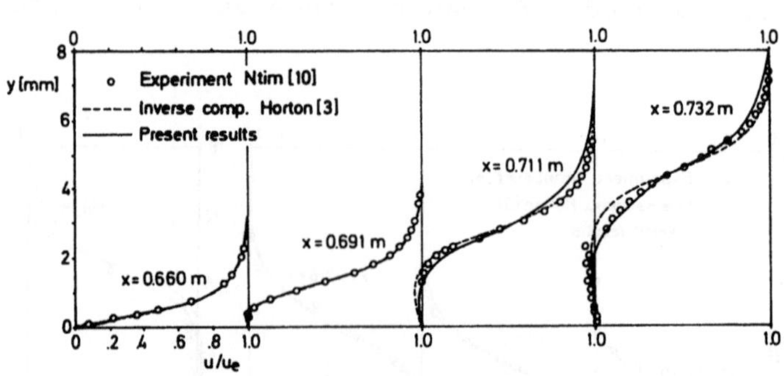

Fig. 8. Computed and measured velocity profiles, case 2

aligning with the free stream direction. Although there are some known deficiencies of the profile data near the model surface, this experiment has been chosen because of its status as a standard test case for laminar boundary layer calculation methods. Figs. 1 and 2 show the measured external velocity and displacement thickness distribution, respectively. The latter is taken from ref. [3]. The transition has not been measured directly but is believed to take place at $0.258 \leq x[m] \leq 0.294$. So the present laminar results cannot succeed further downstream than $x_{tr} = 0.258[m]$. Computations have been carried out in the direct mode up to $x = 0.1[m]$, switching there to the inverse mode with δ^* prescribed. The separation point is predicted at $x = 0.2031[m]$ which differs only slightly from the experimental value of $0.2012 \pm 0.002[m]$ observed by Schubauer. The outer edge velocity u_e/u_∞ compares very well to the measured distribution in Fig. 3. The computational results of ref. [3] are given by the dashed lines. Various velocity profiles are shown in Fig. 4 for the attached as well as for the separated flow regime. The profiles agree well with the experimental data. In the near wall region the experiment is doubtful due to the cooling effect on the large diameter hot wire used. Some discrepancies at $x = 0.258[m]$ are obvious. This might be caused by transition to turbulence.

Test case 2 is the short separation bubble by Ntim [10] (Series II, $u_\infty = 9.36[m/s]$). The experiment was performed within a separation region on a flat surface. The separation was induced by a porous circular cylinder which has been located parallel to the plate but perpendicular to the free stream direction. Figs. 5 and 6 show the external velocity distribution and the displacement thickness, respectively. Ntim observed the separation point at $x = 0.6883[m]$. From the shape of the velocity profiles one can conclude that transition occurs at $x = 0.7315$ $[m]$ as it is indicated in Fig. 8. The present results are obtained by switching from direct to inverse computation at $x = 0.62[m]$. The predicted outer edge velocity u_e/u_∞ is presented in Fig. 7. In the region of adverse pressure gradient the computed results differ slightly from the experiment.

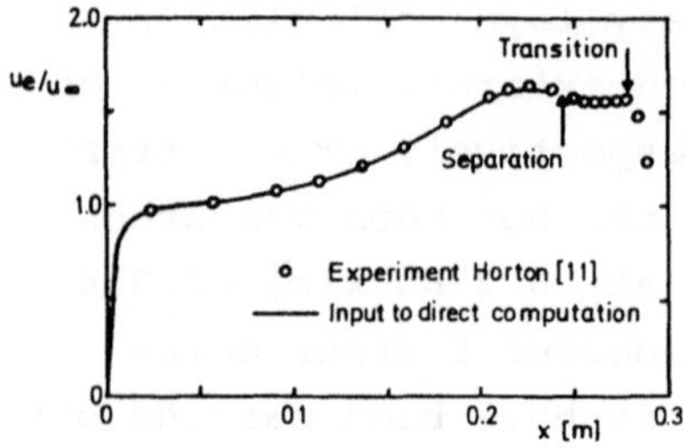

Fig. 9. External velocity distribution, case 3

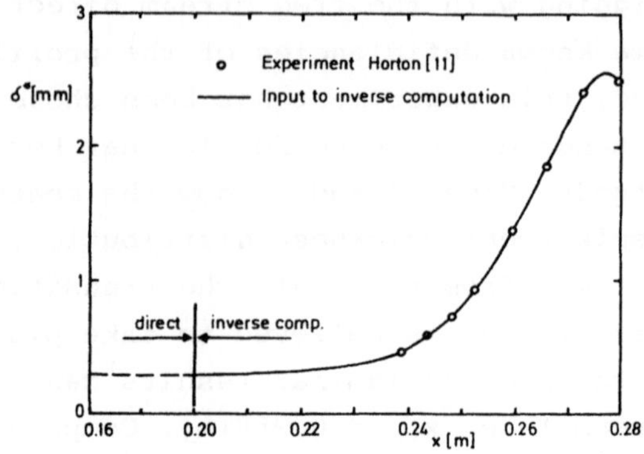

Fig. 10. Displacement thickness distribution, case 3

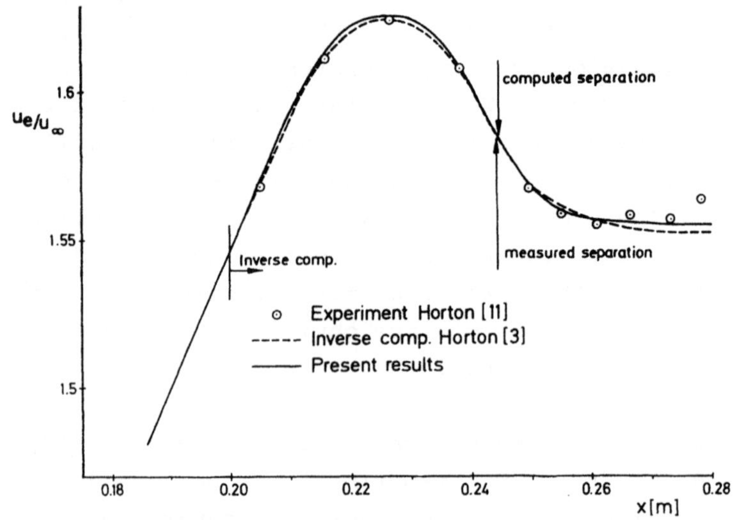

Fig. 11. Computed and measured edge velocity, case 3

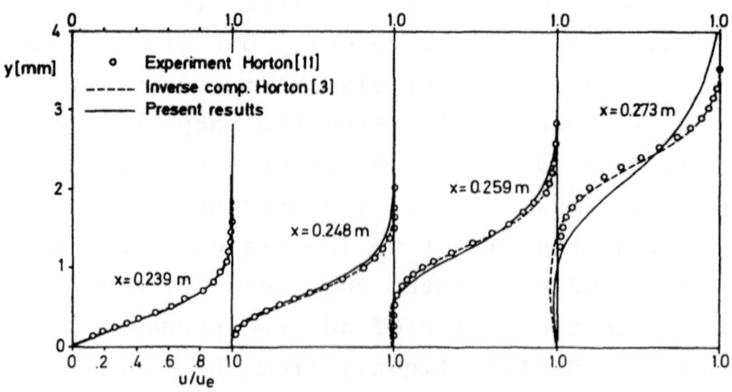

Fig. 12. Computed and measured velocity profiles, case 3

However, separation is encountered at $x = 0.685[m]$ which is close to the experimental value. The velocity profiles given in Fig. 8 are in reasonable agreement with the experiment. The numerical results of Horton [3] are presented in Figs. 7 and 8 by the dashed lines.

Test case 3 is the swept short separation bubble by Horton [11] with $u_\infty = 15.82[m/s]$ and an experimental arrangement similar to that of ref. [10]. Here, both flat plate and cylinder were swept at an angle of 26.5 deg with respect to the free stream direction. However, the flow conditions were close to that of infinite span so that the chordwise flow components are dealt with, only. The external velocity and displacement thickness distributions are shown in Fig. 9 and 10. Separation and transition were observed at $x = 0.2444[m]$ and $0.2796[m]$, respectively. The present results with the switch point from direct to inverse computation at $x = 0.20[m]$ are given in Figs. 11 and 12. Good agreement with the experiment is achieved for the outer edge velocity distribution in Fig. 11. Separation is predicted at $x = 0.2450[m]$ which is very accurate, too. But the velocity profiles obtained indicate some discrepancies within the separated flow region. This is thought to be caused mainly by the present 2-D computation with no infinite swept span condition such as constant cross flow.

The results presented prove the reliability of the numerical method proposed. In general, the prediction of the external velocity distribution agrees well with the experimental data. The location of the separation point is computed in agreement, too. However, for week 3-D flows such as represented by test case 3, deviations from the measured velocity profiles are evident. This is due to the strictly two-dimensional formulation of the boundary layer equations.

Finally it should be noted that no downstream-upstream iterations [14] were performed, as the flow usually becomes transitional prior to reattachment. As the computations have been carried out up to the transition point there is no initial condition available to restart the subsequent upstream iteration.

References

1. Catherall, D.; Mangler, K.W.: The integration of the two-dimensional laminar boundary-layer equations past the point of vanishing skin friction. J. of Fluid Mech. 26 (1966) 163-182.

2. Carter, J.E.: Inverse solutions for laminar boundary-layer flows with separation and reattachment. NASA TR R-447 (1975).

3. Horton, H.P.: Comparisons between inverse boundary-layer calculations and detailed measurements in laminar separated flows. J. Aeronaut. Q. (1981) 169-187.

4. Klineberg, J.M.; Steger, J.L.: On laminar boundary-layer separation. AIAA Paper No. 74-94 (1974).

5. Horton, H.P.: Separating laminar boundary layers with prescribed wall shear. AIAA J. 12 (1974) 1772-1774.

6. Klineberg, J.M.; Steger, J.L.: Calculation of separated flows at subsonic and transonic speeds. Lecture Notes in Physics 19, Berlin/New York: Springer-Verlag 1973.

7. Cebeci, T.: An inverse boundary-layer method for compressible laminar and turbulent boundary layers. J. of Aircraft 13 (1976) 709-717.

8. Thiele, F.: Accurate numerical solutions of boundary layer flows by the finite-difference method of Hermitian type. J. Computational Phys. 27 (1978) 138-159.

9. Schubauer, G.B.: Air flow in a separating laminar boundary layer. NACA Report 527 (1935).

10. Ntim, B.A.: A theoretical and experimental investigation of separation bubbles. Ph.D. Thesis, Queen Mary College, University of London (1969).

11. Horton, H.P.: Laminar separation bubbles in two- and three-dimensional incompressible flow. Ph.D. Thesis, Queen Mary College, University of London (1968).

12. Cebeci, T.; Bradshaw, P.: Momentum transfer in boundary layers. Washington, D.C.: McGraw-Hill/Hemisphere Publishing Co. 1977.

13. Reyhner, T.A.; Flügge-Lotz, I.: The interaction of a shock wave with a laminar boundary layer. Int. J. of Non-Linear Mech. 3 (1968) 173-199.

14. Williams, P.G.: A reverse flow computation in the theory of self-induced separation. Lecture Notes in Physics 35 (1975) 445-451.

Turbulent Flow Simulation: A Large Eddy Simulator's Viewpoint

Joel H. Ferziger

Department of Mechanical Engineering
Stanford University
Stanford, California, USA

Abstract

The state of knowledge of turbulent flows is reviewed very briefly. An argument is posed that, for the near term, the best hope for engineering simulation of turbulent flows lies in the use of phenomenological models tuned for particular flow classes. Input from experiments and higher-level simulations will aid in model construction. Finally, the state of the art in higher-level simulations is reviewed, and a number of difficulties that are impeding progress are pointed out.

1. Introduction: What is Turbulence?

A request to contribute to a commemorative volume offers the author an opportunity to be more philosophical than usual. It is appropriate in a volume honoring Dr. Walz to look both backward and forward in the area of prediction of turbulent flows--a technical area particularly suited to philosophical approach. Indeed, turbulence is one of the few areas of physical science in which there are "schools of opinion" of the kind more commonly encountered in the humanistic fields. In the author's opinion, each school has something to offer and no opinion (including mine) should be accepted without question. Accordingly, I would like to offer some opinions as to what is valuable in each point of view, where we now stand, and possible directions of future research. I do not claim objectivity, particularly with respect to higher level simulations of turbulent flow, an area in which I have been engaged for the past decade. The opinions offered are my own although they are strongly influenced by those of my colleagues, and Profs. S. J. Kline and W. C. Reynolds in particular. Also, I am guilty of some deliberate stereotyping to make particular points.

The phenomenon we call turbulence (arguments still rage over its precise definition) seems to be quite chaotic. This influences one to seek statistical approaches. Were turbulence indeed nothing but random noise, the problem would have been solved by statistical methods a long time ago.

That these methods are only partly successful indicates that there is more to turbulence than randomness.

Within the past decade, there has been an increasing emphasis on the importance of coherent structures in turbulent flows (again there is disagreement on definitions); it is also worth noting that the origins of the coherent structures viewpoint goes back at least to the 1950's. An extreme coherent structure viewpoint might envision a turbulent flow as a nearly periodic repetition of deterministic events which produce most of the important behavior; superimposed on these, we find random noise which is just a nuisance and probably a remnant of old structures.

There is undoubtedly much that is correct in both viewpoints. We already saw that the limited success of purely statistical approaches means that something else must be involved and the coherent structures picture offers an alternative. On the other hand, the failure to find coherent structures in some flows means that, at the very least, the randomness is also quite important. In fact, turbulence seems to be one of those unfortunate problems in which the key parameter is of order unity. We might regard the statistical and coherent structures viewpoints as occupying the two ends of a scale measuring an order parameter; actual turbulent flows lie somewhere near the middle of the scale. Neither an expansion in the order parameter nor its reciprocal is likely to be of great value in the region where most flows are to be found.

Several people, Lumley (1981) among them, have proposed approaches which seem to offer the possibility of finding coherent structures in an unbiased manner. The author and a number of his colleagues have recently reviewed this approach and find the approaches suggested to be more subjective than they appear to be on the surface. To render the method feasible in an actual flow, one has to make guesses as to what to seek; these are almost certainly somewhat subjective. Despite this, this approach offers promise and deserves further consideration. One should be warned to be prepared for an enormous computational task in almost any of these methods.

Where does this leave us? We do not currently have an accepted paradigm in turbulence. For the near future, predictions of turbulent flows for engineering application will need to rely on phenomenological models and whatever insights we can bring to bear on their construction. The remainder of this paper is devoted to a brief review of the prospects of this approach and the contributions that higher-level simulations might make to it. What of the longer-range future? The author is a perpetual optimist, but it is not clear at present which directions will prove fruitful. It is still sensible to follow a number of routes. Prophesies have very short half lives, so I will not offer very many.

2. Practical Simulations of Turbulent Flows

Many design applications require prediction of turbulent flows. The designer usually requires only a few overall parameters of the flow such as force and moment coefficients; the optimum method is the one which produces the required data with satisfactory accuracy at minimum cost. In practice, considerable extra data are always computed; the task of the code designer is to minimize it.

A few years ago, Kline et al. (1978) proposed a taxonomy of methods of computing turbulent flows; this scheme, with one addition and several refinements were used as the basis for classification of methods at the 1981 AFOSR-HTTM-Stanford Conference (Kline et al., 1981). Of the five levels in the original scheme, three can be called practical methods, while the last two are research methods at the current time. There are many subcategories within each level and it is possible for a single code to embody methods from more than one level. We shall make a few remarks about the first levels in this section and give further details of the last two in the following one.

Each method at a given level has a broader range of applicability than a method on the preceding level. This increased flexibility requires more extensive data sets (both a broader range of flows and more detailed data for each flow) to fix the model and an increase in the amount of computation required to apply the method to a given flow.

On the lowest level in the scheme stand correlations. These have been and will remain workhorse methods for the bulk of engineering for systems in which the level of fluid dynamics technology is relatively low; occasionally, they are used in more sophisticated applications. The stock of correlations will be added to at a modest pace and correlations will always be preferred where they are applicable.

On the second level of the scheme one finds the integral methods. These are well developed for a number of important flow classes and "fine tuning" has rendered them accurate within their limits. Integral methods are used extensively in high-technology areas. Their greatest limitation is the need for extensive reworking to add new effects.

Moving up, on the third level we find methods based on the time averaged Navier-Stokes equations. This level is particularly rich in the number of methods it contains. It includes design methods and is the object of a broad research and development effort. I shall not describe all of the methods here (see the taxonomy papers) but will be content with a few general statements.

Two points of view are commonly adopted in constructing models for the time averaged Navier-Stokes equations. The first is purely phenomenological. A plausible functional relationship is proposed, and the constants are obtained by data fitting. These models can be regarded as correlations at a higher level and have the advantage of allowing use of whatever insights can be brought to bear on the problem. The essential difference between this approach and the lower levels lies in the level of detail at which the correlations are made.

The second point of view uses exact equations as the starting point and attempts to derive models by systematic approximation. This approach is appealing and appears far more scientific than the preceding one. It is based on an idea from the kinetic theory of gases that high order statistical quantities "lock on" to low order ones after a short time and the flow is completely determined thereafter by the lower-order statistics. This argument is correct, even rigorous, in kinetic theory, but there is considerable experimental evidence to invalidate it in turbulence. In many flows the low order statistics come to "equilibrium" long before the higher order ones. This argument does not imply that models derived in this way are not valid; they may be as accurate as phenomenological models. It does mean that a model based on more equations is not <u>necessarily</u> more accurate or of broader applicability than simpler models. The proof of the model is in the testing.

The most popular single model at Level 3 at the current time is the $k-\varepsilon$ model. It is capable of predicting a large number of flows with reasonable accuracy with a single set of constants. For environmental flows, for which great accuracy is not necessary, this model is quite adequate. For high technology applications in which higher accuracy may be demanded, it is not yet clear whether the model will be sufficient. The author believes that the $k-\varepsilon$ model will need to be tailored to particular flows, i.e., different model constants for different flow types, if high accuracy is to be achieved.

Essentially all of the input to these models has been derived from experiment; higher level simulations are beginning to make some contribution as we will see in the next section. The contributions of both statistical theories and coherent structures ideas have been quite small. In the author's opinion, this is a great pity; a great deal of knowledge is not being put to practical use.

3. Higher-Level Simulations

Methods at Levels 4 and 5 of the taxonomy mentioned in the previous section (Levels 5 and 6 of the 1981 Conference scheme) are collectively called "higher level" methods. These include large eddy simulation, in which the small scales of the turbulence are modeled and the large scales computed, and full simulation, in which the Navier-Stokes equations are solved numerically. A review of work on cold flows in this area was made by the author (Ferziger, 1981); important applications to heat transfer have been made by the Karlsruhe group (e.g., Grötzbach (1982); Grötzbach et al. (1979)).

The ratio of the largest to the smallest scales of turbulence is proportional to $Re^{3/4}$ where Re is a Reynolds number based on turbulence quantities (Tennekes and Lumley, 1972). As the number of computational mesh points in each direction must be at least this large to assure an accurate result, full simulation is limited to low Reynolds number turbulent flows. Large-eddy simulation increases the accessible range of Reynolds numbers by an order of magnitude or more but falls short of being able to compute the majority of flows of technological interest with sufficient accuracy on currently available computers.

For the foreseeable future, higher-level simulations will not become a tool of the engineering designer; their utility must lie elsewhere. The most appropriate role for these methods is as a complement to experimental work and they have begun to fill this niche. On the one hand, higher-level simulations are beginning to provide information about the physical mechanisms that operate in turbulent flows. In contrast to experiments, simulations provide a great deal of spatial information over a limited time period. Analyses that are impossible in the laboratory can be simulated. For example, one can look at the flow both prior to and following a significant triggering event and can do this in all three spatial directions and for any flow variable.

Another important application of higher level methods is in evaluating models. Because one has data for all flow variables (including the hard-to-measure pressure fluctuations) it is possible to test the validity of models and to evaluate any parameters that appear in them directly. In this way, higher-level simulations play an important role in the development of models for the lower levels.

Let us review some achievements. A great deal of attention has been put on the homogeneous flows, especially by the author and his colleagues. These were chosen because they are the simplest turbulent flows and are frequently used to set model constants. However, one must be careful because that homogeneous flows are structurally different from their inhomogeneous counterparts; this is a subject which is currently under investigation. All of the homogeneous flows which have been measured in the laboratory have been simulated with satisfactory results by higher-level methods. By computing quantities which are difficult to measure experimentally it has been possible to study those model terms which depend on these quantities; a number of significant results of this kind have been produced (Shirani et al., 1981, Feiereisen et al., 1981). Higher-level methods have been used to explain the effects of rotation on homogeneous turbulence. This work lead to a modification to the $k-\varepsilon$ model which accounts for rotational effects (Bardina et al., 1982).

Free shear flows have also received a great deal of attention. Laboratory free shear flows normally develop spatially and measurements either provide time average data or flow visualizations. A major difficulty in higher level simulation of these flows is that the conditions at the inflow boundary of the considered domain are not known accurately and no suitable approximation has yet been found; there is a similar, but less serious, problem at the outflow boundary. The lack of a suitable inlet condition has led simulators to consider the time developing versions of these flows, i.e., the flows that develop when fluid streams of different velocities are suddenly brought into contact. The two kinds of flows are assumed connected by Taylor's hypothesis but as these flows develop rapidly this approximation is not well justified.

Another major difficulty in simulating free shear flows is that the length scales increase rapidly with time and quickly outgrow the computational grid. Taken together with the experimental observation that large coherent structures are important in these flows, this means that only a small portion of a typical computer run is representative of the physical flow. These problems await resolution, but considerable progress has been made in the simulation of free shear flows despite them.

Mansour et al. (1978) showed that much of the behavior of the mixing layer could be simulated in terms of vortex pairing. Metcalfe and Riley (1979) simulated the fully developed mixing layer and the wake of a self-propelled body with satisifactory results; the latter develop more slowly than the mixing layer. Cain et al. (1981) recently simulated transition region of the mixing layer and showed that it is possible to reconcile the coherent structure and randomness points of view.

The last flows to be reviewed here are the wall bounded flows. The principal difficulties are again related to length scales. Wall-bounded flows have critical Reynolds numbers below which they are not turbulent. Much of the generation of new turbulent motions in these flows occurs in small structures that lie close to the wall A consequence of these observations is that extremely large numbers of grid points are required to simulate turbulent wall bounded flows at even the lowest Reynolds numbers. Consequently, all simulations of the fully developed channel flow to date have utilized large eddy simulation; full simulation has been applied only to the relatively early stages of transition from laminar to turbulent flow.

Even large-eddy simulation of these flows requires a large number of points. To avoid this difficulty, Deardorff (1970) did not simulate the region close to the wall but instead imposed an artificial boundary condition in the logarithmic or buffer region. This approach was adopted and modified by Schumann (1979) and his colleagues applied to the simulation of flow and heat transfer in channels and annuli. The artificial boundary conditions appear to work well, despite the fact that they do not accurately represent the actual conditions at the boundary (Kim, private communication). Artificial boundary conditions of this kind will need to be employed if we are to simulate flows at high Reynolds numbers. The development of improved boundary conditions of this type is therefore a priority item in higher level simulations.

Moin and Kim (1982) have simulated the channel flow with the exact, no-slip, boundary conditions at the wall. Grotzbach (1982) and colleagues used no slip conditions in predicting natural convection flows at low to moderate Rayleigh numbers. These results have gone a long way toward establishing the credibility of higher level simulations and have produced both new physical insights and important information for turbulence modelers. Recently, system rotation and wall transpiration were added to the channel-flow simulations, with excellent results. Curved channel simulations are in progress.

Orszag (1980) and colleagues have made a major contribution in the area of transition in wall bounded flows. The ability to include nonlinear effects, allowed them to discover an important mechanism of the instability of these flows. A very interesting calculation of transition in a time developing boundary layer was made by Wray (unpublished); he was able to simulate a number of the features seen in experimental work.

4. Summary

This section sums up some of the ideas put forward in the earlier sections. Despite a century of effort, the area of turbulent flow remains as much an art as a science. Many avenues of research are being followed, but the interactions among them are, unfortunately, not as strong as they ought to be. Interaction of the subdisciplines and unification of the field deserve to be important goals of future work.

For near-term engineering applications, phenomenological methods, based on correlations obtained from experimental data and, to a lesser extent, higher-level computations, will provide the basis for engineering design work. These correlations, however, will be made at increasingly sophisticated levels.

Higher-level simulations are beginning to complement laboratory measurements and will contribute in many of the ways that experiments do. A number of problems in higher-level simulation methods need to be solved in order to broaden the range of flows that can be simulated. There will also be a continuing need for more computer resources of every kind.

Acknowledgments

I wish to thank the organizers of this volume for having invited me to contribute this paper. I am also heavily indebted to several of my colleagues--so many that I am bound to leave some out. I shall therefore mention only a few who have had the greatest influence on my thinking; these include Profs. W. C. Reynolds and S. J. Kline and Drs. P. Moin, J. Kim, A. Leonard, and M. Rubesin.

This work was sponsored in part by NASA-Ames Research Center under Grant NCC 2-15.

References

Bardina, J., J. H. Ferziger, and R. S. Rogallo, "Effect of Rotation on Isotropic Turbulence: Computation and Modeling," submitted to J. Fluid Mech.

Cain, A. B., W. C. Reynolds, and J. H. Ferziger, "Simulation of the Transition and Early Turbulence Regions of a Mixing Layer," Rept. TF-14, Dept. of Mech. Engrg., Stanford University, 1981.

Deardorff, J. W., "A Numerical Study of Three-Dimensional Turbulent Channel Flow at Large Reynolds Numbers," J. Fluid Mech., 41, 452 (1970).

Feiereisen, W. J., W. C. Reynolds, and J. H. Ferziger, "Numerical Simulation of a Compressible Homogeneous Turbulent Shear Flow," Rept. TF-13, Dept. of Mech. Engrg., Stanford University, 1981.

Ferziger, J. H., "Higher-Level Simulations of Turbulent Flows," Rept. TF-16, Dept. of Mech. Engrg., Stanford University, 1981.

Grötzbach, G., "Direct Numerical Simulation of Laminar and Turbulent Bénard Convection," Kernforschungszentrum Karlsruhe, 1979.

Grötzbach, G., "Direct Numerical Simulation of Turbulent Momentum and Heat Transfer in an Internally Heated Fluid Layer," Proc. 7th Intl. Heat Transfer Conference, Hemisphere, Washington, 1982.

Kline, S. J., J. H. Ferziger, and J. P. Johnston, "Calculation of Turbulent Shear Flows: Status and Ten-Year Outlook," ASME J. Fluids Engrg., 100, 3 (1978).

Kline, S. J., and B. J. Cantwell, Proc. 1980-81 Stanford-AFOSR-HTTM Conf. on Complex Turbulent Flows, Dept. of Mech. Engrg., Stanford University, 1981.

Lumley, J. L., "Coherent Structures in Turbulence," in Transition and Turbulence, R. E. Meyer, ed., Academic Press, New York, 1981.

Mansour, N. N., J. H. Ferziger, and W. C. Reynolds, "Large Eddy Simulation of a Turbulent Mixing Layer," Rept. TF-11, Dept. of Mech. Engrg., Stanford University, 1978.

Metcalfe, R. W., and J. J. Riley, "Direct Numerical Simulations of Turbulent Shear Flows," Proc. 7th Intl. Conf. Numerical Methods in Fluid Dynamics, Springer, 1981.

Moin, P., and J. Kim, "Numerical Investigation of Turbulent Channel Flow," J. Fluid Mech., to be published.

Orszag, S. A., and A. Patera, "Numerical Simulation of Transition in Channel Flow," Proc. 7th Intl. Conf. on Numerical Methods in Fluid Dynamics, Springer, 1981.

Schumann, U., G. Grötzbach, and L. Kleiser, "Direct Numerical Simulation of Turbulence," in Prediction Methods for Turbulent Flows, W. Kollmann, ed., Hemisphere, Washington, 1980.

Shirani, E., J. H. Ferziger, and W. C. Reynolds, "Mixing of a Passive Scalar in Isotropic and Sheared Homogeneous Turbulence," Rept. TF-15, Dept. of Mech. Engrg., Stanford University, 1981.

Tennekes, H., and J. J. Lumley, A First Course in Turbulence, MIT Press, 1972.

Reduced Turbulence Production by Increased Elongational Viscosity

B. GAMPERT, P. WAGNER

Fachgebiet Strömungslehre, Universität Essen-Gesamthochschule, Essen, West-Germany

Summary

Experimental results concerned with drag reduction by polymer additives in fully developed turbulent pipe flow are presented. Polystyrene samples with low dispersity dissolved in toluene were investigated. The Toms phenomenon is related to a reduction in bursting rate thus diminishing the production of turbulence. A Deborah number is defined as quantitative expression corresponding to the mechanism of drag reduction described.

Introduction and Experimental Procedures

By adding minute quantities of long chain macromolecules like polyacrylamide (PAA), polyethyleneoxide (PEO) or polystyrene (PS) to the turbulent pipe flow of e.g. pure water or toluene the pressure loss along the pipe is reduced when the flow rate is kept constant or the flow rate is increased when the pressure loss is kept constant. This effect of frictional drag reduction by polymer additives in turbulent flow first was observed by Toms [1] and therefore often is called the Toms phenomenon. Although the Toms phenomenon has received considerable attention in recent years the mechanism by which drag reduction occurs is still a debated topic.

In Fig. 1 the pipe flow drag coefficient λ being given by

$$\lambda \equiv \frac{2\ \Delta p\ d}{\rho_{solut}\ u_m^2\ L} \tag{1}$$

is shown as function of the Reynolds number Re. In Eq. (1) Δp is the pressure loss along the distance L and ρ_{solut} is the density of the solution. The Re number is based on the pipe diameter d, the mean flow velocity u_m and the solution's kinematic viscosity ν_{solut}.

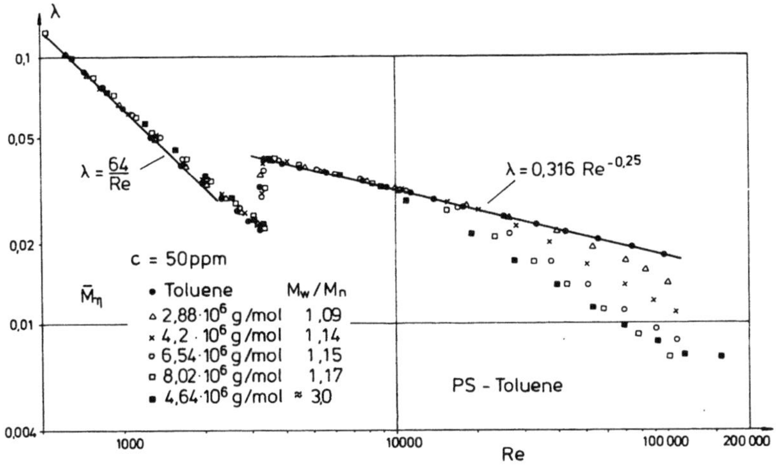

Fig.1. Friction coefficient λ as function of Reynolds number Re for polystyrene samples solved in toluene

The experimental results shown, we obtained for PS samples with different viscosity average molecular weights $\bar{M}_\eta$ dissolved in toluene. The PS concentration for these measurements was 5o ppm. The PS molecules only were drag reducing efficient if the wall shear stress was larger than a critical value τ_w^* [2]. For given pipe geometry a critical Re number Re* corresponds to τ_w^*. It can be seen from Fig. 1 that Re* depends on the average molecular weight of the PS sample being lower for larger molecules in accordance with the fact that for given Re number largger molecules are more effective than shorter ones.

Further Re* depends on the molecular weight distribution characterized by the polydispersity $\delta = M_w/M_n$ (M_w = weight average molecular weight, M_n = number average molecular weight). The PS probe with $\delta \approx 3.o$ obviously contained many very long molecules which caused the high drag reduction obtained with this sample.

The Toms phenomenon is thought to result from flow induced stretching of macromolecules [3]. Only for Re numbers sufficiently large molecular stretching is strong enough and drag reduction appears. For higher Re numbers stretching of the molecules is magnified and the drag reducing efficiency of the molecules is increased. But finally stretching of the molecules becomes too intensive and breakage of the molecular chain appears. The shorter molecules produced by this effect called mechanical degradation obviously are less drag reducing effective as can be seen from the enhanced λ-values.

Although drag reduction is heightened for larger τ_W values it cannot be concluded that stretching of polymer molecules necessarrily is caused by shear forces. It was shown in [4] that if a molecule is subjected to a pure shearing action as illustrated in Fig.2 the molecule will rotate and a given segment will first be pulled out toward 1 then pushed in from 2 and so on. As the shear rate is enlarged molecular rotation and distortion increase simultaneously. When the molecule rotates faster it has less time for stretching. Thus for very high shear rates the molecule approaches a certain limiting amount of distortion which is far too small to cause the chain to break.

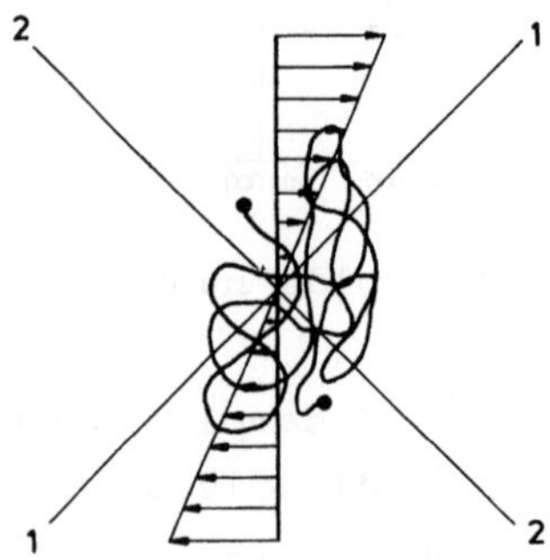

Fig.2. Macromolecule in pure shear field

It follows that mechanical degradation cannot originate from pure shear forces and the question arises by what kind of mechanism stretching of macromolecules actually is generated.

The preceding discussion and the results presented in Fig.1 show that the molecular weight is an important parameter in drag reduction experiments. If we want to characterize polymer samples effectively by its average molecular weight it is necessary to use samples with narrow molecular weight distributions.

In [5], [6] laboratory polymerized highly clean PAA samples of comparativelv low polydispersity dissolved in water have been investigated.

The only polymers of high molecular weight readily available with narrow molecular weight distributions are the polystyrenes. Thus in addition to our experiments with PAA the present paper presents some of the results obtained for polystyrene standards dissolved in toluene. Many of the physico-chemical problems one encounters working with water soluable polymers like PAA and PEO do not appear for PS.

The samples with weight average molecular mass M_w = (2.88, 4.48, 6.77 and 8.42)$\cdot 1o^6$ g/mol and polydispersivity δ between 1.o9 and 1.17 were purchased from Toyo Soda Manufacturing Co. and the sample with viscosity average molecular mass $\overline{M}_\eta = 4.64 \cdot 1o^6$ g/mol and $\delta = 2.5 \div 3.o$ was purchased from Chemische Werke Hüls (CWH).
Polymer solutions of concentrations between 1o ppm and 5o ppm were obtained by dilution of a o.1 % PS-solution which was prepared first and then homogenized by being gently shaken for several days.
The measurements were performed with a pipe flow system. The nitrogen pressure in a container drove the test liquid through a o.oo61 m diameter horizontal pipe. To prevent degradation caused by pumping a single pass rather than a recirculating system was employed. The flow rate was obtained by weighing the fluid which passed through the pipe in a measured interval of time. Two pressure taps were placed at 1.5 and 2.o m from the pipe entrance, that is in the region of fully developed flow. Viscosity measurements for PS-solutions were made in an Ubbelohde-viscosimeter.

Normal Stresses and Elongational Viscosity

By adding polymers in very dilute concentrations to a pure fluid the shear viscosity is changed only slightly while the elongational viscosity is enlarged by orders of magnitude. This is illustrated by the fact that in those laminar flows which exhibit high elongational rates as e.g. porous media flows [7] the fluid drag is strongly increased by polymer additives.

In [8] results of studies of dilute solutions (5o to 1oo ppm) of high molecular weight polystyrene dissolved in polychlorinated biphenyl (PCB) were reported. Flow birefringence was used in order to measure the extent of macromolecular distortion in a wide range of two dimensional flow types which were produced in a four roll mill. The four roll mill consisted of four cylinders, the centers of which formed a square. The flow type was determined by the relative rates and directions of rotation of rolls.

Flows ranging from purely extensional flows to flows which approached a simple shearing flow were considered. The results showed very clearly that when the flow field is purely extensional orientation and deformation of the macromolecules was strongest. With increasing shear stresses and decreasing elongational stresses in the flow field orientation and stretching of PS macromolecules were diminished.

Investigating the Toms phenomenon and assuming that interaction between macromolecules and flow field mainly takes place via elongational stresses the question arises where such elongational flow fields appear in turbulent shear flow.

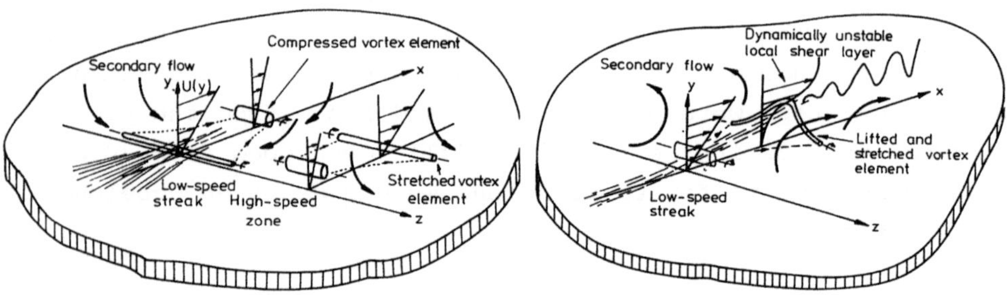

Fig.3a. The mechanics of streak formation

Fig.3b. The mechanics of streak breakup

The process of turbulence production near the wall in turbulent shear flow is shown in Fig. 3 from Ref. [9]. When flow carrying vortex elements is moving towards the wall vortex lines are stretched. They are compressed in regions of outflow. The velocity component u is reduced by vortex line compression and increased by vortex stretching leading to the formation of wall-layer streaks of low and high velocity values.

The turbulence production occurs through a sequence of events consisting of formation, lift up, oscillatory growth and 'break up' of the low-speed streaks followed by a larger scale inrush or sweep. In [1o] it was shown that 7o % of the turbulent kinetic energy is produced during these bursts.

In [11] it was suggested for the first time that the Toms phenomenon is related to a reduction in bursting rate. This hypothesis was confirmed in [12] by the observation that in drag-reducing solutions the amount of bursting and the production of turbulence actually were diminished leading to a decrease in dissipation and lower friction factors.

The drag reduction cannot occur through the interaction of a fluid eddy with an individual polymer molecule. This follows from the large disparity in size which exists between the smallest scales of turbulent fluid motion and the largest dimension of the polymer molecule. Instead the mechanism of drag reduction has to be associated with a continuum property of the solution.

From Fig. 3 it can be seen that vortex stretching and thus elongational strains are present during the formation of the streaky structure and the bursting of low momentum streaks. This observation and the two above mentioned facts a) that fluid flow-macromolecule interaction mainly takes place via extensional stress and b) that elongational viscosity is increased by addition of drag reducing molecules can be summed up by the statement that drag reduction by polymer additives in turbulent flow results from an increased resistance towards elongation diminishing the number of bursts and thus the production of kinetic energy of turbulence.

Support to this explanation is given by the observation that drag reduction only appears if polymer molecules are present in the region near the wall.

Deborah Number Definition and Discussion of Experimental Results

Interaction between flow field and macromolecules only takes place when the molecular relaxation time t_M related to the characteristic flow time t_F is sufficiently large. Otherwise the polymer molecule cannot become drag reducing effective. Our aim is to define a Deborah number $De = t_M/t_F$ as a quantitative expression related to the described mechanism of drag reduction.

In [13] a relaxation time t_M characterizing the elongation of macromolecules in solution was determined assuming that the polymer molecules behave in accordance with the model of finitely extendable nonlinear elastic dumbbells (FENE-dumbbells). For highly diluted solutions assuming isolated macromolecules the expression

$$t_M \approx \frac{[\eta]\ \eta_s\ M}{A\ k\ T} \tag{2}$$

was obtained, with $[\eta]$ being the intrinsic viscosity, η_s being the shear viscosity of the solvent, M being the molecular weight of the macromolecule, A and k being Avogadro's and Boltzmann's number respectively. In drag reduction studies concentrations generally are very low. Thus Eq. (2) is applicable.

The characteristic 'flow time' t_F we represent by the time t_B for one bursting period. In [9], [12] for t_B the relationship

$$t_B \sim 1/u_\tau^2 \quad , \quad u_\tau^2 \equiv \tau_w/\rho_{solut} \tag{3}$$

was found. In order to obtain t_F appropriately dimensionalized we multiply $1/u_\tau^2$ by the kinematic viscosity of the solution ν_{solut} which yields

$$t_F = K_1\ \nu_{solut}\ /\ u_\tau^2 \tag{4}$$

From Eqs. (2) and (4) we obtain for the Deborah number with $\nu_{solut} = \nu_s\ (1+c[\eta])$, ν_s and c being the kinematic viscosity of the solvent and the concentration of the polymer solution respectively

$$De = K\ \frac{[\eta]\ \eta_s\ M\ u_\tau^2}{A\ k\ T\ (1+c[\eta])\nu_s} \tag{5}$$

For the following discussion we don't need the exact value of K=const but assume K=1.

According to the definition of the De number De $\equiv t_M/t_F$ drag reduction is expected to increase for higher De numbers.

The expression for the De number given in Eq. (5) qualitatively is in accordance with experimental results. From Fig. 1 and Eq. (5) it can be seen that De number and drag reduction in common are enlarged for higher molecular weight and higher shear stress.

As Eq. (2) was derived for non-interacting molecules in Eq. (5) the concentration c only has influence on the De number as far as the solution's viscosity ν_{solut} is increased for higher concentration of the polymer. Besides of this comparatively small effect the drag reducing efficiency of the macromolecules is concentration independent. Thus from Eq. (5) it follows that the critical shear stress τ_w^*, too principally is concentration independent in highly diluted solutions. But for drag reducing efficient macromolecules it depends on the concentration whether drag reduction is measureable.

Previous results [5] have shown that high concentration values cannot compensate lack in molecular weight.

As the Deborah number refers to the single macromolecule the drag reduction DR is devided by (c · [η]), which is a quantity proportional to the volume consumed by polymer molecules. This yields

$$DR^* = \frac{\lambda_T - \lambda_{solut}}{\lambda_T} \cdot \frac{1}{c \cdot [\eta]} \qquad (6)$$

with λ_T and λ_{solut} being the drag coefficient of pure toluene and PS solution respectively. In Fig. 4 and 5 DR^* is plotted versus the De number.

In Fig. 4 the polymer concentration is constant. For PS samples with very different molecular weights but only slightly varying dispersity a single curve for DR^* as function for De has been obtained.

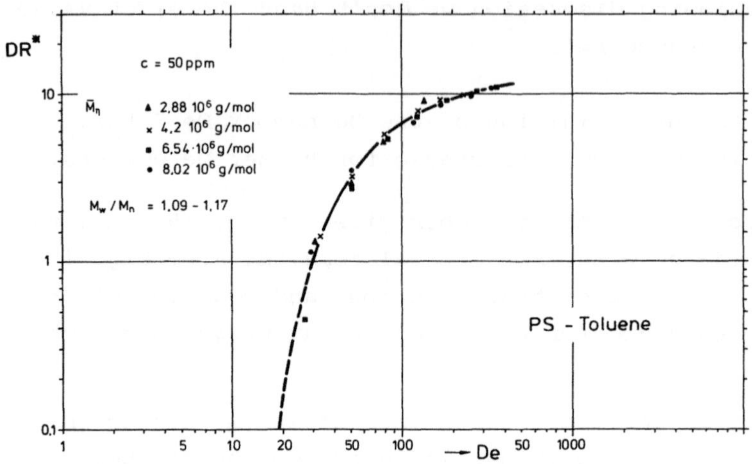

Fig.4. Drag reduction DR^* as function of De number for PS samples with different average molecular weights

As the De number describes the fluid-polymer interaction the De number für the onset of drag reduction is an universal quantity as far as the presuppositions of the calculations presented in [13] are fulfilled. This can be seen in Figs. 4 and 5.

One parameter determining the fluid-polymer interaction that was not taken into account in Eq. (2) was the dispersity δ. Thus the influence of this quantity is seperated and is shown in Fig. 5 where results are presented for PS samples with various concentrations and molecular weights and two different dispersities δ. The onset De number does not depend on concentration but is changed towards lower De numbers for higher dispersity.

For larger De numbers the single curve obtained for given dispersity δ is splitted and the concentration c appears as an additional parameter. One reason is that the increase of drag reduction is less than linear for higher De numbers. Then the fact that DR is devided by c in order to obtain DR^* has no counterpart in a related enhancement of drag reduction for larger concentration values. It can be seen that for higher De numbers DR^* values are lowest for highest concentrations.

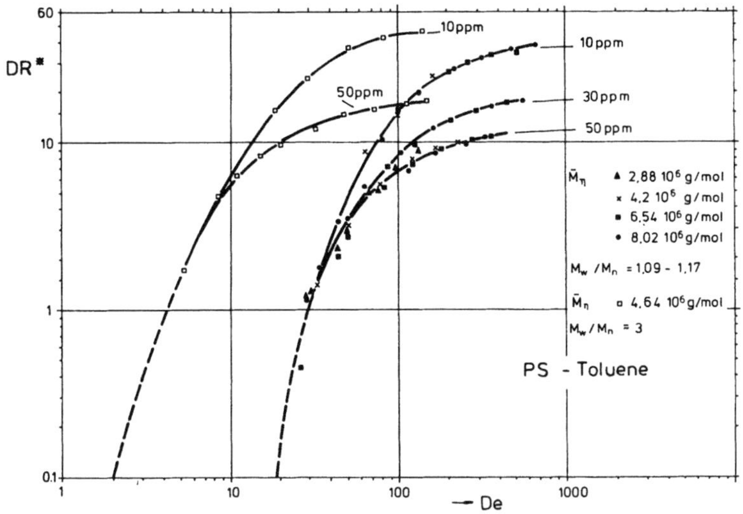

Fig.5. Drag reduction DR* as function of De number for PS samples with different polydispersity. Influence of concentration and mean molecular weight are shown.

References

1. Toms, B.A.; Some Observations on the Flow of Linear Polymer Solutions Through Streight Tubes at Large Reynolds Numbers. In Proc. First Intern. Congr. on Rheology, Vol. II (1948) 135-141.

2. Virk, P.S.; Drag Reduction Fundamentals.AIChE Journal, Vol. 21, No. 4 (1975) 625-565.

3. Lumley, J.L.; Drag Reduction in Turbulent Flow by Polymer Additives.J. Polymer Sci., Macromolecular Reviews, Vol. 7 (1973) 263-29o.

4. Bueche, F.; Mechanical Degradation of High Polymers. J.Appl. Polymer Sci., Vol. IV, No. 1o (196o) 1o1-1o6.

5. Gampert, B.; Wagner, P.; Turbulent flow with polymer additives. Archives of Mechanics, No. 4 (1982).

6. Gampert, B.; Wagner, P.; Experimentelle Untersuchung turbulenter Rohrströmungen hochreiner wässriger Polyacrylamid-Lösungen. Will be published in Rheol. Acta.

7. Durst, F.; Haas, R.; Dehnströmungen mit verdünnten Polymerlösungen: Ein theoretisches Modell und seine experimentelle Verifikation. Rheol. Acta 2o (1981) 179-192.

8. Fuller, G.G.; Leal, L.G.; Flow birefringence of dilute polymer solutions in two-dimensional flows. Rheol. Acta 19 (198o) 58o-6oo.

9. Kline, S.; Reynolds, W.C.; Schraub, F.A.; Rundstadler, P.W.; The structure of turbulent boundary layers. J. Fluid Mech.,vol. 3o, part 4 (1967) 741-773.

1o. Kim, H.T.; Kline, S.J.; Reynolds, W.C.; The production of turbulence near a smooth wall in a turbulent boundary layer. J. Fluid Mech., vol. 5o, part 1 (1971) 133-16o.

11. Gadd, G.E.; Turbulence damping and drag reduction produced by certain additives in water. Nature, 2o6 (1965) 463-467.

12. Donohue, G.L.; Tiederman, W.G.; Reischman, M.M.; Flow visualization of the near-wall region in a drag-reducing channel flow. J. Fluid Mech., vol. 56, part 3 (1972) 559-575.

13. Bird, R.B.; Hassager, O.; Armstrong, R.C.; Curtiss, C.F.; Dynamics of Polymeric Liquids. Vol. II. New York: J. Wiley & Sons (1977).

Some Aspects to the Structure of Separated Flows

Dieter Geropp

Universität -GH- Siegen

Abstract

First the different phenomena of separated flow, especially of dead water respectively wake, are discussed in the non-steady view. The time average picture of dead water behind bodies with different contours is illustrated on the basis of pressure and LDA-measurements. Typical dàta obtained on pressure distributions, vortex structures and extensions of dead waters can serve as a basis for the development of a dead water model.

Introduction

If the flow of a viscous fluid along a wall is too strong retarded or is led to a sharp edge the flow will separate. Between the undisturbed external flow and the wall a turbulent region is formed with different eddies designated as dead water or more general as wake. This region is characterized by the severest energy dissipation in flow fields and is the reason for severe pressure losses or the drag. Furthermore heat and mass transfer conditions in this region are changed fundamentally. Consequently the phenomenon of flow separation is very important in many fields of application e.g. in the aerodynamics, naval hydrodynamics, traffic engineering, in the field of turbomachines, in structural aerodynamics and chemical engineering, even in the medicine (vortex development behind stenoses and cardiac valves).

Until now the calculation of flows with large separated regions using the Navier Stokes equations is impossible, especially for low viscosity or high Reynolds number flows. The turbulent boundary layer on the body but also the turbulent motions in dead water require additional physical informations in the form of empirical laws to help solve the NS-equations.

Corresponding calculation methods (e.g.Grashof /1/, Jacob /2/) are already existing for drag and lift determination of airfoils. These are based on a model proposed by Walz /3/ in which the dead water is simulated by a fictive displacement body. But this model has proven to be too far from reality so that the calculation results are often quite inexact.

Therefore the present review on older and recent investigations should contribute to a better understanding of the physics of dead waters and certainly to the improvement of dead water models required for theoretical approaches.

The study is restricted to two-dimensional incompressible flows.

Patterns of flow separation

Figure 1 shows some typical flow patterns with separation, which can occur in fluid-machines and other technical equipment. The examples of an airfoil and a Carnot diffusor show that a separated flow (S: point of separation, R: point of reattachement) can reattach under specific circumstances e.g. following a laminar-turbulent transition or a changed pressure distribution. Generally such "separation bubbles" have only little influence on the whole stream field in contrast to separated flows without reattachement which, will be dealt with in the following.

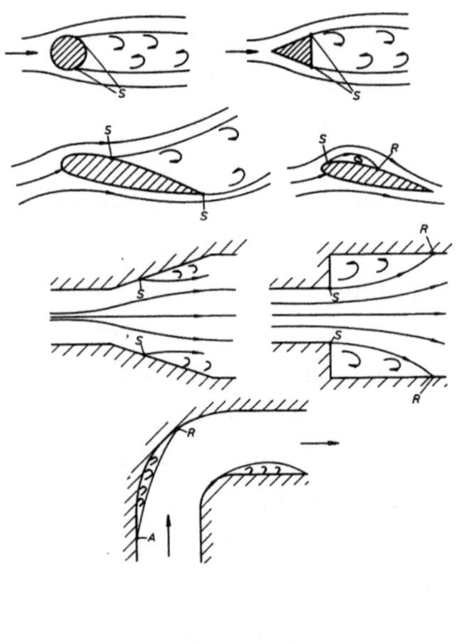

Fig. 1 Forms of separated flows

The beginning of separation or of dead water is given by two types of separation points or separation lines:
a) the separation point on a continously curved wall

(e.g. circular cylinder), where the position of S is varying being mainly influenced by the pressure gradient along the wall as well as by the transition to turbulence.

b) the separation point on a sharp edge, where the position of S ist fixed and independent of any flow parameter.

In both cases regions with recirculating flow occur near the wall behind the separation point. The so formed dividing or shear layers are unstable; already minor pressure fluctuations cause them to transform into small eddies. Similar to a starting airfoil these eddies combine to form large vortices separating periodically and moving downstream with a velocity somewhat lower than the free stream. Finally they break up by molecular and turbulent dissipation.

The processes of vortex shedding and decay are dependent upon the Reynolds number. Figure 2 shows the fundamental states of the dead water and wake in the case of a circular cylinder according to Morkovin /4/. The single instant pictures in Figure 2 show the separation to start at Re > 5 with a clearly bounded laminar dead water consisting of a stationary vortex pair fixed to the cylinder

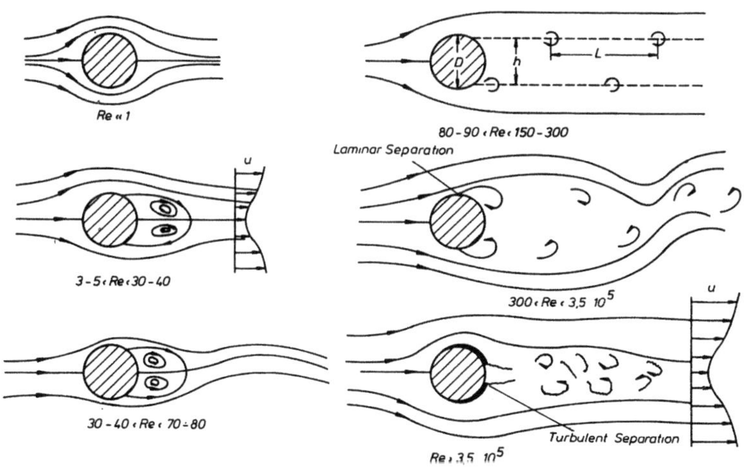

Fig. 2   Influence of Re-number on the wake

The dead water shows for the first time a so called free stagnation point FS, in which the total head is appreciable lower than in the frontal stagnation point. Above Re > 40 first indications of instable oscillations are detected in the streamline from FS. Around Re ≈ 150 the vortices shed periodically and form the so called Kármán vortex street. Flow is still laminar but unsteady. In the range $Re < 3.5 \cdot 10^5$ the attached flow is laminar. The vortices shed periodically, but they are fully turbulent and do not form any stable configuration. Above $Re > 3.5 \cdot 10^5$ the attached flow is already turbulent resulting in a delayed separation. Therefore the dead water has a smaller cross-sectional dimension than with the lower Reynolds number flow and it is characterized by turbulent stochastic vortex motions. But even in this range one may find, against the prevalent opinion in the literature, stable Kármán vortex streets as confirmed by Figure 3. The satellite photo analyzed by Wenderoth /5/ shows a Kármán vortex street at $Re \approx 10^{10}$ leeward of a mountain ($\bar{D}$ = 10 km average diameter) on the island Jan Mayen. The vortices are fully turbulent.

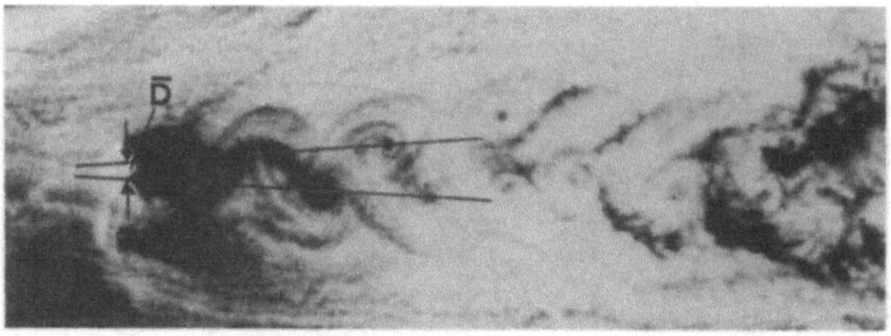

Fig. 3 Kármán vortex street leeward island Jan Mayen

These and other examples do not confirm the general validity of the known stability investigation of Kármán concerning a constant width h of the vortex street and a constant ratio h/L (width of vortex street/distance between vortices). The reason is the neglecting of the molecular and turbulent viscosity in the stability investigation.

Time averaged consideration of the dead water

The considerations are so far mainly based on instantaneous pictures of separated flows. Leaving out the range Re < 40 and considering the unsteady, periodic, mostly high frequency motions of the wake it appears reasonable to set up a quasi-steady picture of the dead water region based on time averaged considerations. In such a manner steady vortex structures can be detected, which can serve as the basis for a theoretical dead water model.

Figure 4 shows some typical time averaged pressure distributions in the twodimensional flow field of bodies with large separation regions. The pressure coefficient c_p is plotted against the coordinates x/L respectively x/d. The pressure was measured along the dotted marked lines (for the NACA-airfoil coinciding with the suction side). The position x/L respectively x/D = 0.0 marks the onset of the dead water.

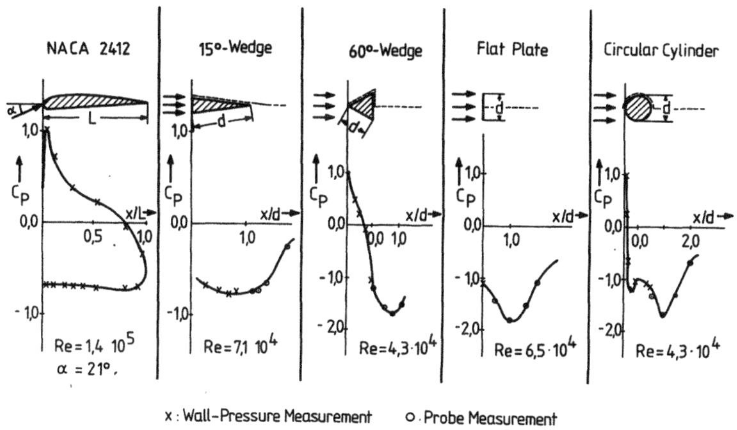

Fig. 4 Pressure coefficient c_p in the dead water

Typical for all exemples is a more or less well defined minimum of pressure in the dead water region. The formation of the minimum can be explained by assuming that the separated flow region, similar to the solid contour, acts as a fictive displacement body the maximum cross-sectional dimension of which found in the region of the pressure minimum. Figure 4 makes it evident that the assumption of a constant dead water pressure of the existing models according to /1/ or /2/ is not generally valid.

The structure and extent of various dead water regions were investigated in detail in /6/ with the help of a laser-doppler-system. A computer controlled the positioning of the measuring volume as well as the signal processing.

Figure 5 shows some examples of the measuring results of three models with different contours and angles of attack. The upper half shows the velocity vector map of the time averaged flow.

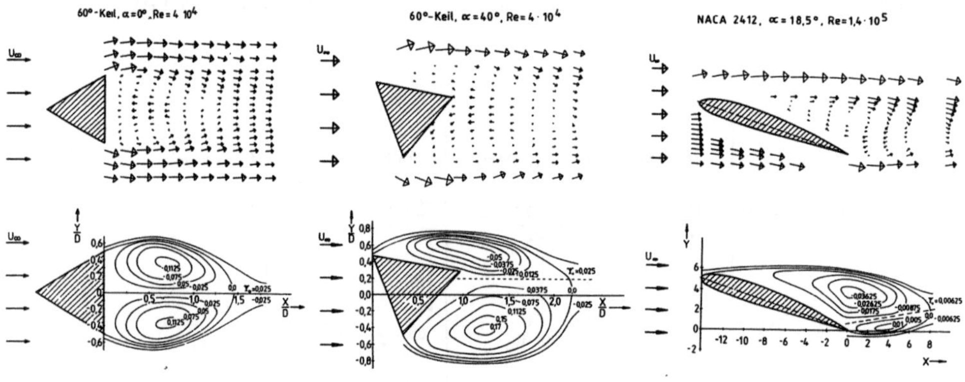

Fig. 5 LDA-measurements in dead water

The lower half represents the time averaged streamlines as calculated from the velocity profiles. ψ_o characterizes a dimensionless stream function.

It es evident that after separation the external flow is accelerated along the dead water contour and delayed later transitioning into the wake depression. The recirculating flow region, which is defined as the actual dead water is formed by two symmetric vortices. This pairwise occuring, time averaged vortex structure is characteristic for all dead waters investigated. The angle of attack changes the relative sizes and vorticity of the vortices.

It is interesting to note the distance B of the free stagnation point behind the body as a function of the Re number. It characterizes the extent of the dead water as defined by the zero-streamline. In figure 6 measuring results from /6/ and /7/ are plotted for the case of a circular cylinder. There is a distinguished maximum in the range of Re ≈ $2 \cdot 10^2$, in which it is well known that first indications of turbulence appear in the Kármán vortex street.

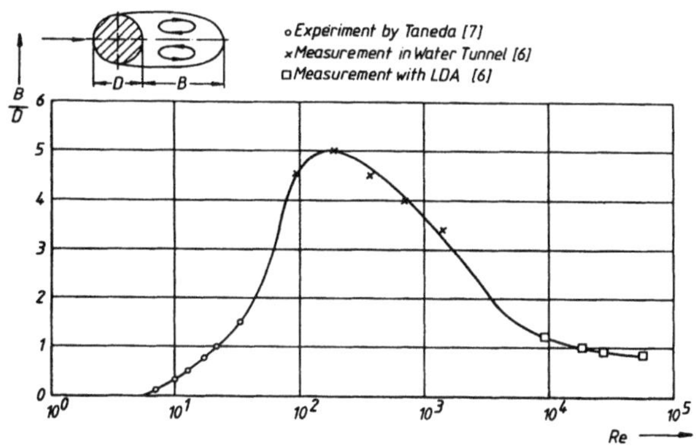

Fig. 6 Free stagnation point in dead water

With the Re-number increasing the length B decreases because of the growth of turbulence. Above Re ≈ 10^3 the value of the laminar case at Re ≈ 30 is reached. An explanation for this behaviour can be given by the solutions of the two-dimensional vorticisity transport equation, if instead of the molecular viscosity the effective eddy viscosity is introduced. This has a constant value in the dead water region.

Development of a simple dead water model

Finally a first set up of a dead water model should be considered, in which the illustrated experimental results are approximately taken into account.

As shown in fig. 7 (left), the displacement effect of the dead water can be simulated by the zero-streamline. In fig. 7 (right) the measured pressure distribution is compared with the one calculated using a panel-method. The calculated and measured pressure data agree very well directly at the wedge but the pressure minimum in the dead water is simulated with much less accuracy. The reason are dissipative effects in the dead water,which are not taken into account by the panel-method applied in potential theory.

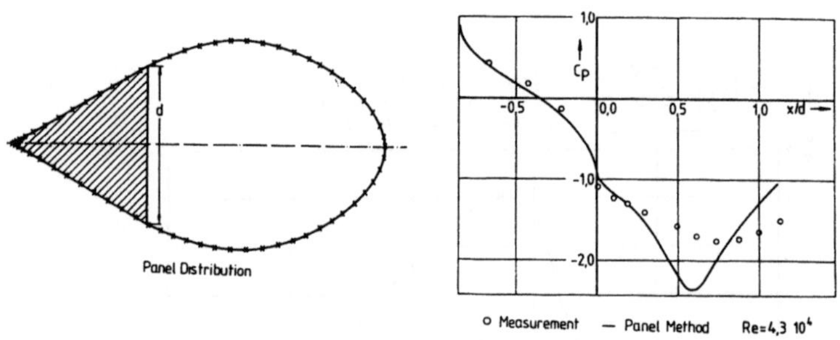

Fig. 7 Dead water simulated by the zero-streamline

Further improvement of the results are expected, if the calculation according to the panel-method is combined with a moving wall concept introduced into the boundary layer theory. Corresponding investigations are presently under way.

References

/1/ Grashof, J.: Berechnung der Druck- und Schubspannungsverteilung auf Körpern mit Totwasser in ebener inkompressibler Parallelströmung. Diss. TH Karlsruhe, 1973.

/2/ Jacob, K.: Weiterentwicklung eines Verfahrens zur Berechnung der abgelösten Profilströmung mit besonderer Berücksichtigung des Profilwiderstandes. Deutsche Luft- und Raumfahrt FB 76-36 (1976)

/3/ Walz, A.: Berechnung der Druckverteilung an Klappenprofilen mit Totwasser. Jb. d.deutschen Luftfahrtforschung I (1940) 265 - 277

/4/ Morkovin, M.V.: Flow around Circular Cylinder - A Kaleidoscope of Challenging Fluid Phenomena. Symposium on Fully Separated Flows, Housen, A.G., ed., ASME, New York (1964) 102 - 118.

/5/ Wenderoth, O.: Beitrag zur Untersuchung der Strouhal-Zahl hoher Bauwerke im natürlichen Wind bei sehr großen Reynolds-Zahlen. Diss. Karlsruhe 1979.

/6/ Leder, A.; Geropp, D.: LDA-Messungen in abgelösten Strömungen und eine einfache Berechnungsmethode für den Widerstand scharfkantiger Zylinder. ZAMM 63, Heft 4/5 (1982).

/7/ Taneda, S.: Experimental investigation of the wakes behind cylinders and plates at low Reynolds numbers. J. Phys. Soc.Japan 11 (1956) 302 - 307.

Isentropic Magnetogasdynamic Flow of a Perfect Plasma

K. Gersten, D. Glück, K. Oberländer

Institut für Thermo und Fluiddynamik
der Ruhr-Universität Bochum
D-4630 Bochum

Summary

The steady isentropic one-dimensional magnetogasdynamic (MGD) channel flow is investigated for a perfect plasma, i.e. a perfect gas of infinite electrical conductivity. A cross-field device is considered when the electric field and the transverse magnetic field are perpendicular to the main flow. Hall effects are neglected. The conservation equations for mass and momentum as well as Maxwell's equations can be reduced to a single ordinary differential equation whose solutions are independent of the streamwise coordinate. Consequently, each solution for prescribed initial Mach number Ma_1 and MGD parameter K can be used for all possible channel geometries. It is shown that even in an isentropic MGD-channel flow (MGD generator) not all of the mechanical energy can be transformed into electrical energy. The results are applied for the design of a typical MGD-energy converter (Ma_1 = 0.85 ; K = 1.09).

1. Introduction

In order to design a magnetogasdynamic (MGD) energy-conversion system the efficiency must be determined. The efficiency, sometimes called polytropic efficiency, see (1), is defined as

$$\eta_P = \frac{(\Delta h_o)_e}{(\Delta h_o)_i} \tag{1}$$

where $(\Delta h_0)_e$ is the change in total enthalpy and $(\Delta h_0)_i$ is the isentropic or ideal change in total enthalpy within the same change in pressure.

The purpose of this paper is to determine the isentropic one-dimensional magnetogasdynamic channel flow and, in particular, the change $(\Delta h_0)_i$ of total enthalpy in such an ideal MGD energy-conversion system. In the limiting case without magnetic and electric field the flow is reduced to the well-known one-dimensional isentropic flow of a perfect gas with constant total enthalpy.

2. Basic Equations

A perfect gas, i.e. an ideal gas ($p^* = \rho^* RT^*$) with constant heat capacity (c_p=const, κ=const), that has infinite electrical conductivity ($\sigma \to \infty$) is denoted as perfect plasma. The one-dimensional channel flow of a perfect plasma is considered according to Figure 1.

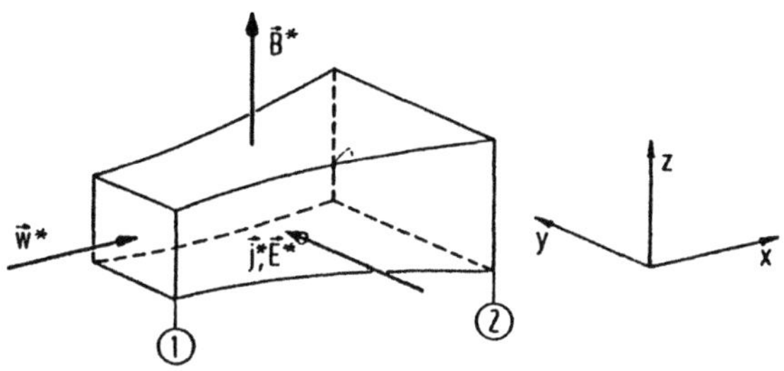

Figure 1:
Geometry of MGD channel flow (cross-field device)
(1) : entrance section
(2) : outlet section

It is a cross-field device where the electric field $\bar{E}^*$ and the electric current density $\vec{j}^*$ (y-direction) as well as the transverse magnetic field $\vec{B}^*$ (z-direction) are perpendicular to the main stream of velocity $\vec{w}^*$ (x-direction). The side walls of the channel are electrodes, while the wall on the top and bottom are assumed to be of electrically insulating material; see (1), p.408, (2),p.193. The velocity w^*, pressure p^*, and density ρ^* are assumed to be approximately constant across the cross-section. Since Hall effects are neglected, the vectors of the electric and magnetic field have only a single non-zero component in the cartesian coordinate system chosen. These components are:

$$E_y^* = E^* \ , \quad j_y^* = j^* \ , \quad B_z^* = B^* \tag{2}$$

All characteristics $w^*(x)$, $p^*(x)$, $\rho^*(x)$, $E^*(x)$, $j^*(x)$, and $B^*(x)$ may be functions of the streamwise coordinate x.

From these values further convenient flow variables can be determined:

Mach number:
$$Ma = \frac{w^*}{\sqrt{\kappa \frac{p^*}{\rho^*}}} \tag{3}$$

Total specific enthalpy:

$$h_o^* = \frac{\kappa}{\kappa - 1} \frac{p^*}{\rho^*} \left(1 + \frac{\kappa - 1}{2} Ma^2\right) \tag{4}$$

Under the assumptions of a one-dimensional flow the basic equations are:

(i) Conservation of mass: $\rho^* w^* A^* = \text{const}$ (5)

(ii) Momentum equation:

$$\rho^* w^* \frac{dw^*}{dx} + \frac{dp^*}{dx} - j^* B^* = 0 \qquad (6)$$

The product $j^* B^*$ is the body force due to the magnetic field. Body forces due to the gravity field are neglected. This equation can also be written in the form of an energy equation:

$$\frac{dh_o^*}{dx} - \frac{j^* B^*}{\rho^*} \qquad (7)$$

(iii) Isentropic relation:

$$\frac{p^*}{\rho^{*\kappa}} = \text{const} \qquad (8)$$

(iv) Faraday's law for steady flows (Maxwell's equation)

$$\text{curl } \vec{E}^* = 0 \qquad (9)$$

from which follows that $E_y^* = E^*$ is independent of x.

(v) Ampère's law (Maxwell's equation)

$$\mu_o \vec{j}^* = \text{curl } \vec{B}^* \qquad (10)$$

It is assumed that the permeabilityof the gas is identical with the permeability of vacuum μ_o. In the cross-field device considered Eq.(10) reduces to

$$j^* = - \frac{1}{\mu_o} \frac{dB^*}{dx} \qquad (11)$$

(vi) Ohm's law (Hall effects neglected)

$$\vec{j}^* = \sigma (\vec{E}^* + \vec{w}^* \times \vec{B}^*) \qquad (12)$$

Because of infinite electrical conductivity ($\sigma \to \infty$) Eq.(12) leads to

$$E^* = w^*(x) \; B^*(x) \qquad (13)$$

The five equations (5), (6), (8), (11) and (13) determine the five functions $\rho^*(x)$, $w^*(x)$, $p^*(x)$, $j^*(x)$ and $B^*(x)$, when the channel geometry $A^*(x)$ and the constant electric field E^* are prescribed. By using Eq.(11) the electric current density $j^*(x)$ can be eliminated from Eq.(6) so that Eqs.(5), (6), (8) and (13) form a set of four equations for the four unknowns $\rho^*(x)$, $w^*(x)$, $p^*(x)$ and $B^*(x)$ for given $A^*(x)$ and E^*.

The following dimensionless variables are introduced:

$$(x) = \frac{\rho^*(x)}{\rho^*_{o1}} , \quad w(x) = \frac{w^*(x)}{\sqrt{\frac{p^*_{o1}}{\rho^*_{o1}}}} \qquad (14)$$

$$p(x) = \frac{p^*(x)}{p^*_{o1}} , \quad h_o(x) = \frac{h^*_o(x)\,\rho^*_{o1}}{p^*_{o1}} \qquad (15)$$

$$A(x) = \frac{A^*(x)\,\rho^*_{o1}\sqrt{\frac{p^*_{o1}}{\rho^*_{o1}}}}{\rho^*_1 \; w^*_1 \; A^*_1} , \quad B(x) = \frac{B^*(x)\sqrt{\frac{p^*_{o1}}{\rho^*_{o1}}}}{B^*_1 \; w^*_1} \qquad (16)$$

Reference values are the values at the entrance section (Index 1) as well as the total pressure p^*_{01} and the total density ρ^*_{01} at the entrance section:

$$p^*_{o1} = p^*_1 \left(1 + \frac{\kappa - 1}{2} Ma_1^{\;2}\right)^{\frac{\kappa}{\kappa-1}} \qquad (17)$$

$$\rho^*_{o1} = \rho^*_1 \left(1 + \frac{\kappa - 1}{2} Ma_1^{\;2}\right)^{\frac{1}{\kappa-1}} \qquad (18)$$

The momentum equation, Eq. (6), can be reduced to an ordinary non-linear first-order differential equation for the function p(w):

$$\frac{dp}{dw} = \frac{K}{w^3} - w\,p^{1/\kappa} \qquad (19)$$

with the initial conditions for a given entrance Mach number Ma_1:

$$w_1 = \frac{\sqrt{\kappa}\,Ma_1}{\sqrt{1 + \frac{\kappa-1}{2} Ma_1^{\;2}}} \qquad (20)$$

$$p_1 = \left(1 + \frac{\kappa-1}{2} Ma_1^{\;2}\right)^{\frac{\kappa}{1-\kappa}} \qquad (21)$$

The constant K, defined as $K = \dfrac{E^{*2}\,\rho^*_{o1}}{\mu_o\,p^{*\,2}_{o1}}$, (22) is called MGD parameter. K is always positive and can be considered as the dimensionless electric field. From the solution $p(\kappa, w, Ma_1, K)$ of Eq.(19) the other flow characteristics can be determined:

$$\rho = p^{1/\kappa} \tag{23}$$

$$B = \frac{1}{w} \tag{24}$$

$$A = \frac{1}{\rho w} = \frac{1}{wp^{1/\kappa}} \tag{25}$$

$$\Delta h_B = -\Delta h_o = h_{o1} - h_o = h_{o1} - \left(\frac{\kappa}{\kappa-1}\frac{p}{\rho} + \frac{w^2}{2}\right) \tag{26}$$

$$Ma = \frac{w}{\sqrt{\kappa \frac{p}{\rho}}} \tag{27}$$

For MGD flows two additional Mach numbers are important: Magnetic Mach number (Alfvèn number)

$$Ma_m = \frac{w^*}{\sqrt{\frac{B^{*2}}{\mu_o \rho^*}}} = w^2\sqrt{\frac{\rho}{K}} \tag{28}$$

Combined Mach number

$$\overline{Ma} = \frac{w^*}{\sqrt{\kappa \frac{p^*}{\rho^*} + \frac{B^{*2}}{\mu_o \rho^*}}} \tag{29}$$

The three Mach numbers Ma, Ma_m and $\overline{Ma}$ are interrelated by

$$\frac{1}{\overline{Ma}^2} = \frac{1}{Ma^2} + \frac{1}{Ma_m^2} \tag{30}$$

Differentiation of Eq.(26) with respect to the velocity w leads to:

$$\frac{dh_o}{dw} = \frac{K}{w^3\ p^{1/\kappa}} \tag{31}$$

Since the right hand side of Eq.(31) is always positive, increasing velocity corresponds to increasing total specific enthalpy (pump or accelerator), while in a MGD generator the velocity decreases. It is worth mentioning that the relationships between p, w, ρ, A, B and h_B are universal and independent of the particular channel geometry, i.e., independent of the variation of these flow parameters with the streamwise

coordinate x.
In the limiting case K = 0 the solution of Eq.(22) reduces to the well-known solution for the isentropic flow of a perfect gas:

$$p = \left(1 - \frac{\kappa-1}{2\kappa} w^2\right)^{\frac{\kappa}{\kappa-1}} \tag{32}$$

or as implicit solution

$$w = \sqrt{\frac{2\kappa}{\kappa-1}\left(1 - p^{\frac{\kappa-1}{\kappa}}\right)} \tag{33}$$

In the special case $\kappa=1$, Eq.(19) becomes a linear first-order differential equation, whose solution can be given analytically.
For various initial Mach numbers Ma_1 and MGD parameters K, Eq.(19) has been solved numerically by the Runge-Kutta method.

3. Results

In Figure 2 the variables w, B, ρ, A and Δh_o are shown as functions of the pressure p for a particular example with the initial Mach number $Ma_1 = 0.85$ ($w_1 = 0.9401$, $p_1 = 0.6235$). The curves shown in the figure correspond to various MGD parameters K.
Figure 2a shows the velocity w, which is identical with the reciprocal value of the magnetic field B. The curve K=0 corresponds to the classical isentropic flow of a perfect gas, where the velocity w varies between $w_{max} = \sqrt{2\kappa/(\kappa-1)} = 2.6458$ and zero and the pressure between zero and one, respectively. The limiting curve $K\to\infty$ is horizontal, i.e. the velocity w and the magnetic field B are constant. All other curves K = const have two solutions w at each prescribed pressure p. According to Eq.(31) the upper branch of the solution corresponds to pumps or accelerators (conversion of electrical energy into mechanical energy of the flow) while the lower branch represents the MGD generators. For each given initial Mach number Ma_1 a special MGD parameter exists

$$K_{S1} = w_1^{\,4}\, p_1^{\,1/\kappa} \quad (Ma_{m1} = 1) \tag{34}$$

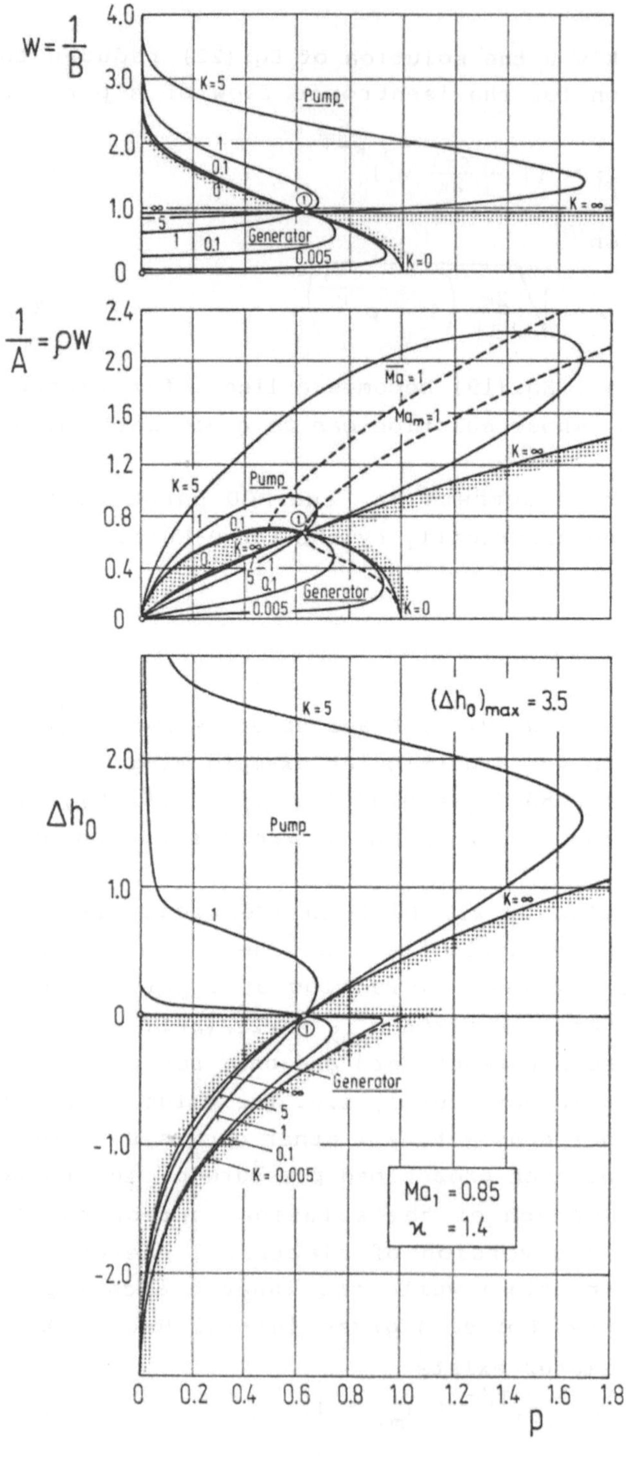

Figure 2:
Flow characteristics for a MGD channel flow of a perfect plasma ($\kappa = 1.4$, $\sigma \to \infty$). Example: $Ma_1 = 0.85$

MGD-parameter K according to Eq. (22)

a) Velocity w (Magnetic field B) as function of pressure p.

b) Product of density and velocity ρw (cross-section area A) as function of pressure p.

c) Change of specific magnetic enthalpy Δh_o as function of pressure p.

for which the curve w(p) has a vertical tangent at the entrance section. For this curve all pressures p are smaller than p_1 (in the example: $K_{S1} = 0.5574$). Pressures p higher than p_1 exist for generators and $K<K_{S1}$ as well as for pumps and $K>K_{S1}$. In the upper branch ($w>w_1$, conversion of electrical into mechanical energy) one can distinguish between pumps ($K>K_{S1}$) and accelerators ($K<K_{S1}$, $p<p_1$). The points of pressure maximum correspond to $Ma_m = 1$, i.e. in these points the velocity w* is equal to the so-called Alfvèn velocity $a_a = B^*/\sqrt{\mu_o \rho^*}$. At each curve K = const points with velocities higher than the Alfvèn velocity ($Ma_m>1$) could be called "magmnetically supersonic". The rest of the curve ($w < a_a$) corresponds to "magnetically subsonic flow". It follows that accelerators ($w > w_1$, $K < K_S$) always operate in "magnetically supersonic" flow. According to Eq.(27) the points of constant Mach number lie on curves

$$w = Ma \sqrt{\kappa}\, p^{\frac{\kappa-1}{2\kappa}} \tag{35}$$

In particular, the sections with sonic speed have no special features in MGD flows. Flow characteristics which cannot be reached with the initial Mach number Ma_1 chosen are marked by shaded areas.

Figure 2b shows the product of density and velocity, which is identical with the reciprocal value of the cross-section area. All curves K = const. are closed. The points of pressure maximum (dp/dw = 0) correspond to Ma = 1. In the points of minimum cross-section area the combined Mach number is equal to one, $\overline{Ma} = 1$. The area between the curve $\overline{Ma} = 1$ and the curve K = 0 is characterized by "magnetically supersonic" flows.

MGD generators will always have divergent channel flow (increasing A for decreasing w). For pumps the cross-section area will first decrease for increasing velocity until the combined Mach number $\overline{Ma}$ reaches unity. Beyond the point $\overline{Ma} = 1$ the cross-section will increase. For pumps usually a converging channel will be used, whereas accelerators must have a convergent-divergent channel contour.

The combined Mach number $\overline{Ma}$ has in MGD flow an equivalent role as the Mach number Ma for classical isentropic flow without magnetic and electric fields. Velocities increase in MGD flow in convergent channel for $\overline{Ma} > 1$. At the entrance section the combined Mach number $\overline{Ma}$ can have initial values Ma_1 between zero ($K \to \infty$) and Ma_1 ($K = 0$). For supersonic initial Mach number $Ma_1 > 1$ the flow is also "magnetically supersonic" ($\overline{Ma} > 1$), when the MGD parameter K is smaller than

$$K_{S2} = \frac{Ma_1^2 - 1}{Ma_1^2} K_{S1} \qquad (\overline{Ma}_1 = 1 \; , \quad Ma_1 > 1) \tag{36}$$

Ma_1	w_1	p_1	ρ_1	K_{S1}	K_{S2}
0	0	1	1	0	-
0.2	0.2357	0.9725	0.9803	0.0030	-
0.6	0.6857	0.7840	0.8404	0.1858	-
0.85	0.9401	0.6235	0.7136	0.5574	-
1.0	1.0801	0.5283	0.6339	0.8625	0
1.5	1.4739	0.2724	0.3950	1.8640	1.0356
2.0	1.7638	0.1278	0.2300	2.2264	1.6698
3.0	2.1213	0.0272	0.0762	1.5430	1.3716
4.0	2.3094	0.0066	0.0277	0.7868	0.7376
6.0	2.4792	0.0006	0.0052	0.1962	0.1908
∞	2.6458	0	0	0	0

Table 1: Characteristic Values of the MGD Channel Flow of a Perfect Plasma ($\kappa = 1.4$)

Typical values of $K_{S1}(Ma_1)$ and $K_{S2}(Ma_1)$ are given in Table 1. All possible flow conditions, depending on the parameters Ma_1 and K, are listed in Table 2.

Ma_1	K	Ma_{m1}	$\overline{Ma}_1$	Geometry: Generator	Geometry: Pump or Accelerator
<1	$0<K<K_{S1}$	>1	<1	divergent	convergent-divergent (Acc.)
	$K_{S1}<K<\infty$	<1	<1	divergent	convergent-divergent (Pump or Acc.)
>1	$0<K<K_{S2}$	>1	>1	conv.-div.	divergent (Acc.)
	$K_{S2}<K<K_{S1}$	>1	<1	divergent	convergent-divergent (Acc.)
	$K_{S1}<K<\infty$	<1	<1	divergent	convergent-divergent (Pump or Acc.)

Table 2: Possible Flow Conditions for MGD Channel Flow Depending on Parameters Ma_1 and K

Figure 2c shows the change of specific total enthalpy as function of pressure p. For pumps (accelerators) the increase in total enthalpy is in principal unlimited if K is chosen large enough. On the other side, generators can convert only a finite amount of energy. It is worth mentioning that generally not all the total specific energy can be converted into electrical energy. From Figure 2a it is obvious that even after an expansion of the flow into vacuum ($p \to 0$) a finite velocity, i.e. kinetic energy remains. The ratio of maximum increase of magnetic enthalpy Δh_{Bmax} and initial total enthalpy h_{01} is depicted as function of the parameters Ma_1 and K in Figure 3. The larger the Mach number Ma_1 the smaller is the percentage of the initial total enthalpy that is convertible into electric energy. For large parameters K an asymptotic formula can be used:

$$\frac{\Delta h_{Bmax}}{h_{o1}} = \frac{1}{1 + \frac{\kappa-1}{2\kappa} \frac{\rho_1 w_1^2}{p_1}} \left[1 + \frac{\kappa-1}{\kappa} \frac{K_{S1}}{K}\right] \quad (K \to \infty) \tag{37}$$

The dotted curves in Figure 3 corresponds to Eq.(37).

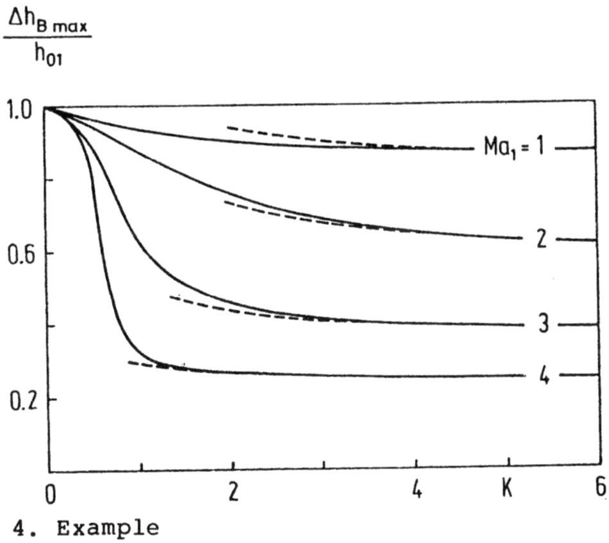

Figure 3: Maximum increase of magnetic enthalpy based on entrance total enthalpy $\Delta h_{Bm}/h_{o1}$, as function of MGD-paramater K and entrance Mach number Ma for generators of perfect plasmas (κ = 1.4 ,

4. Example

To demonstrate the application of the results, an example characterized by the parameters $Ma_1 = 0.85$ and $K = 1.09$, is considered. Figure 4 shows the particular curves $w(p)$, $A^{-1}(p)$ and $\Delta h_B(p)$ for these initial conditions. The exit cross-section is prescribed by $A_2 = 2$.

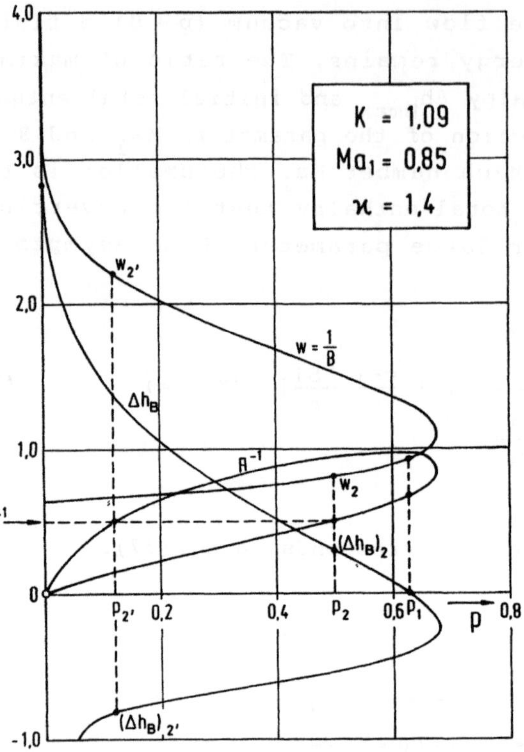

Figure 4: Application of diagrams in Figure 2 to a special example: Ma =0.85 ; K=1.09 ; A_1=1.49

Two solutions are possible. If the channel geometry is convergent-divergent with the minimum cross-section A_{min} = 1.04 , the device is an accelerator (Index 2`). The second solution refers to a generator (Index 2). The results are given in Table 3.

	Accelerator	Generator
Geometry	convergent-divergent	divergent
p_2	0.12	0.50
w_2	2.21	0.81
Δh_{B2}	-0.82	0.29
$\overline{Ma}_2$	1.63	0.42

Table 3: Example of MGD converter:
Ma_1 = 0.85
K = 1.09
A_1 = 1.4906
A_2 = 2

References

(1) G.W. Sutton, A. Sherman; Engineering Magnetohydrodynamics, Mc Graw-Hill, New York, 1965

(2) I.A. Shercliff: A Textbook of Magnetohydrodynamics, Pergamon Press, Oxford, 1965

Investigation of the Three-Dimensional Transonic Flow around an Air Intake by a Finite-Volume Method for the Euler Equations

J. GRASHOF

Dornier GmbH, Friedrichshafen, F.R.G.

Summary

The inviscid flow around a Pitot-type air intake has been calculated for a fixed freestream Mach number of 0.8 and three angles of attack (-6°, 0°, $+6^{\circ}$). The ratio of static pressures at channel exit and infinity has been held constant. The analysis was carried out for an isolated air intake, i.e. the fuselage is omitted and the splitter plate is extended to an infinite flat plate. The Euler equations written in integral-conservation form are solved by a finite volume method applying a Runge-Kutta scheme with local time-stepping. The computational mesh is composed by a set of two-dimensional grids, constructed by a complex transformation combined with the solution of geometric partial differential equations. The main features of the predicted flow fields are described and compared with experimental results for a realistic inlet-fuselage combination.

1. Introduction

The efficiency of modern aircraft depends to a large extend on the engine characteristics, which are essentially influenced by the air-intake performance. The so-called Pitot-inlets show some favourable properties, also with respect to operation and fabrication. Therefore, this type of inlets has been chosen for detailed numerical investigations. The present studies concentrate on the flow properties at the inlet-lip region, since the lip influences the overall inlet flow essentially. Hence, the fuselage and the splitter plate etc. are neglected respectively replaced by a flat plate at the inlet-fuselage junction. The predicted flow quantities of main practical interest are the static pressures and Mach numbers on the inlet-body surface, helping the designer to check if undesired events (e.g. shock waves or strong adverse pressure gradients) are present. In addition, important informations concerning flow distortions in the inlet channel can be extracted from the computational results.

2. Mesh generation

The boundary of the computational domain has to be chosen in such a manner that accurate boundary conditions can be prescribed on each segment of the boundary. Naturally, the body surface and the exit cross-section of the inlet channel are parts of the boundary. The computational domain is closed by the far-field boundary. The mesh boundaries are aligned to the body surface and to the other segments of the boundary of the computational domain. For the present study a mesh generator [1] has been developed which is superior to other

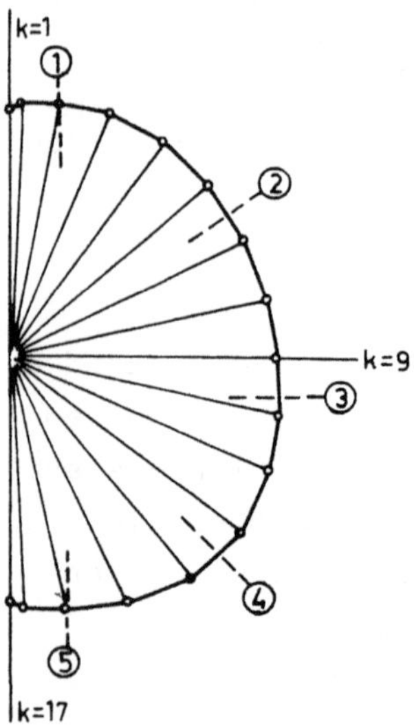

Fig. 1: Nose-point line, grid sections k and pressure hole sections 1 - 5

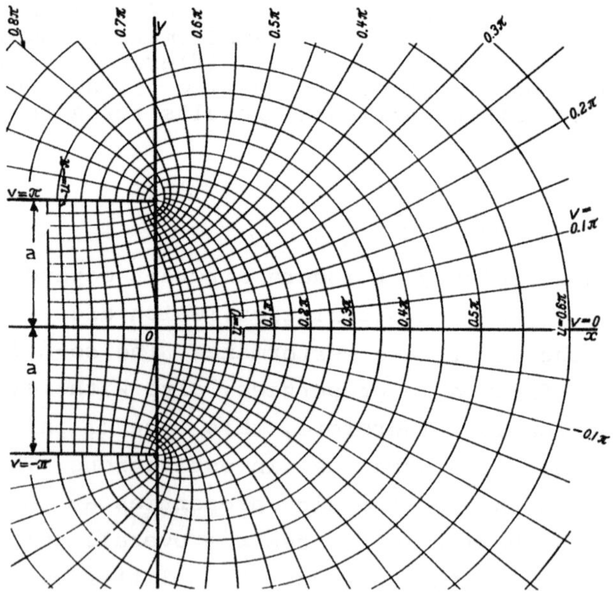

Fig. 2: Maxwell curves

possible methods [2]. The three-dimensional grid is composed by a set of two-dimensional ones, counted by index k (see fig. 1). For each section k = const a two-dimensional grid is constructed taking advantage of the complex Maxwell transformation [3]:

$$z = \frac{a}{\pi} \cdot (w+1+e^{w}) \tag{1}$$

$$z = x + i \cdot y, \quad w = u + i \cdot v \tag{2}$$

where z is the physical plane and w the transformed one. The curves u = const and v = const constitute a grid depicted in fig. 2 [3]. If the pair a slits $y = \pm a$ iş interpreted as an inlet with zero thickness the transformation (1) forms an optimal grid for flow calculations. The grid generation algorithm is based on this observation and consists of the following steps: descretize the body contour, the exit section and the x-axis in an appropriate manner and compute the related values w by an iterative inversion of equ. (1); discretize u and v (stretching) in the intervalls u (body) to u (axis) and v (body) to v (axis) and get thus the transformed grid u (i,j), v (i,j) - i,j are the mesh indices -; improve the transformed grid by solving a set of geometric Poisson-type equations in the transformed plane w; transform the solution back to the physical plane z using equ. (1). The method produces a "high-quality" grid for inlet-flow calculations. A sample is shown in fig. 3. The figure gives also an impression of a reasonably chosen far-field boundary. The algorithm should not be confounded with grid generation by conformal mapping and is much easier to handle than the latter. Since the basic idea of the present procedure is very simple and not restricted to inlet-type geometries, it can be applied to other flow problems by replacing the Maxwell transformation by another, properly selected, complex function (e.g. parabolic function for blunt-body flows).

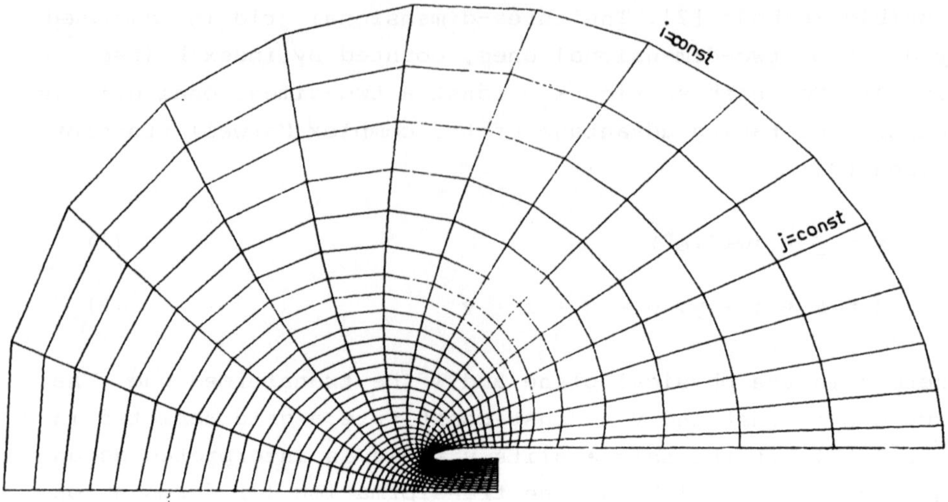

Fig. 3: Maxwell-grid for inlet section k = 9

3. Basic equations and initial/boundary conditions

The Euler equations including the energy equation may be written in brief notation and in integral-conservation form (see e.g. [4])

$$\int_V \frac{\partial}{\partial t} \vec{U}\, dV + \int_S \bar{\bar{H}}\, d\vec{S} = 0 \tag{3}$$

with

$$\vec{U} = \begin{bmatrix} \rho \\ \rho u \\ \rho v \\ \rho w \\ e \end{bmatrix} \quad , \quad \bar{\bar{H}} = \begin{bmatrix} \rho \cdot \vec{q} \\ \rho u \cdot \vec{q} + p \cdot \hat{i}_x \\ \rho v \cdot \vec{q} + p \cdot \hat{i}_y \\ \rho w \cdot \vec{q} + p \cdot \hat{i}_z \\ (e+p) \cdot \vec{q} \end{bmatrix} \tag{4}$$

Space vector: $\vec{r} = x \cdot \hat{i}_x + y \cdot \hat{i}_y + z \cdot \hat{i}_z$ (5)

Velocity vector: $\vec{q} = u \cdot \hat{i}_x + v \cdot \hat{i}_y + w \cdot \hat{i}_z$ (6)

Since a finite-volume approach is used to solve the basic flow equations (3), no coordinate transformations are necessary, i.e. all quantities are measured with respect to a cartesian basis. Nomenclature: V = control volume fixed in space, S = surface of V, dV = element of V, $d\vec{S}$ = outer area vector of surface element dS, t = time, $\vec{U}$ = formal vector, $\bar{\bar{H}}$ = formal flux tensor, $\hat{i}_{x,y,z}$ = unit vectors of the cartesian basis, ρ = density, p = static pressure, e = total energy per volume unit. - The set of basic equations is completed by an equation of state (ideal gas):

$$p = (\gamma-1) \cdot [e - \frac{1}{2} \rho \cdot (u^2+v^2+w^2)] \qquad (7)$$

Due to the integral-conservation form no special treatment of discontinuities is needed, e.g. shocks are "captured".

Initial conditions: The computation is started from scratch, i.e. in the whole computational domain the undisturbed flow quantities are prescribed.

Boundary conditions: Improper treatment of boundary conditions can lead to serious errors and perhaps instability [5]. The fundamental idea how to formulate properly posed boundary conditions is based on the analysis of disturbance propagations at the boundaries, i.e. on the distinction between incoming and outgoing characteristics. Numerical "error waves" should be absorbed at a boundary or they should travel out of the computational domain instead of being reflected. These principles are taken into account in the present method. Instead of going into details, one special condition is outlined: At the channel exit (subsonic flow) parallel stream is assumed and the static pressure p_{ex} is prescribed. The flow quantities are then calculated by solving locally the flow equations (3).

4. Numerical solution of the basic equations

In this article no details can be presented, instead the algorithm is characterized by some "key words". The set of integro-differential equations (3) is integrated in time by an "explicit, four-stage, two-level" Runge-Kutta scheme including "filtering" and "enthalpy forcing term" ([5], [6]). The convergence to a steady-state is accelerated by the use of "local time-stepping" and by the mentioned forcing term, which is proportional to the difference between the local total enthalpy and its feestream value. The latter implementation is justified by the condition that the total enthalpy must become constant everywhere at the final steady-state of the computation. The "filtering" (Shuman-type filter for the pressure) removes preferably high-frequency numerical disturbances and thus accelerates the convergence. The scheme is stable for CFL-numbers up to large values (an appropriate value is CFL = 2.8).

5. Results and discussion

For the inlet depicted in fig. 4, where the computational grid on the body surface and the grid in the symmetry plane are shown, a series of flow calculations has been performed. Some results will be presented below. The grid consists of 17 two-dimensional sections k = 1 to 17 (see fig. 1). Static pressures have been measured at stations 1 to 5. Since the finite-volume approach is used, the computed quantities are stored in the volume centers, i.e. the results for section k belong to the centers of volumes build up by section k and section k+1.

For all computed cases the freestream Mach number is $M_o = 0.8$ and the ratio of static pressures at the channel exit and at infinity is $p_{ex}/p_o = 1.327$.

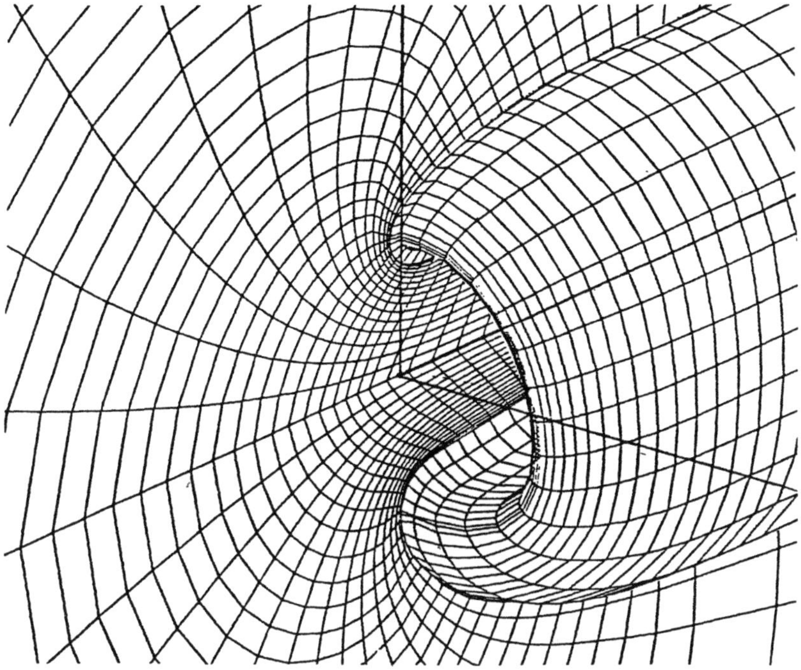

Fig. 4: Grid on inlet body and symmetry plane

The theoretical distributions of wall pressures and Mach numbers at the sections k = 3 (top), k = 9 (side) and k = 15 (bottom) are shown in fig. 5.a) to fig. 5.c) for the angle of attack $\alpha = 0^{\circ}$. The quantity EPSHL signifies the mass-flow ratio with the high-light area as reference plane; L is a reference length computed from the high-light area. In addition, related experimental pressure values are given. The discrepancy between predicted and measured pressures will be discussed later. At the side section (k = 9) a local supersonic flow is observed which is recompressed without any indication of a shock. In order to study the main influence of incidence on the sectional flow characteristics, the predicted maximum Mach numbers (suction-peak values on the external inlet-surface) are plotted on fig. 6. The influence is weak for the side-sections and well pronounced near the inlet-fuselage junctions. The curves show the trends expected.

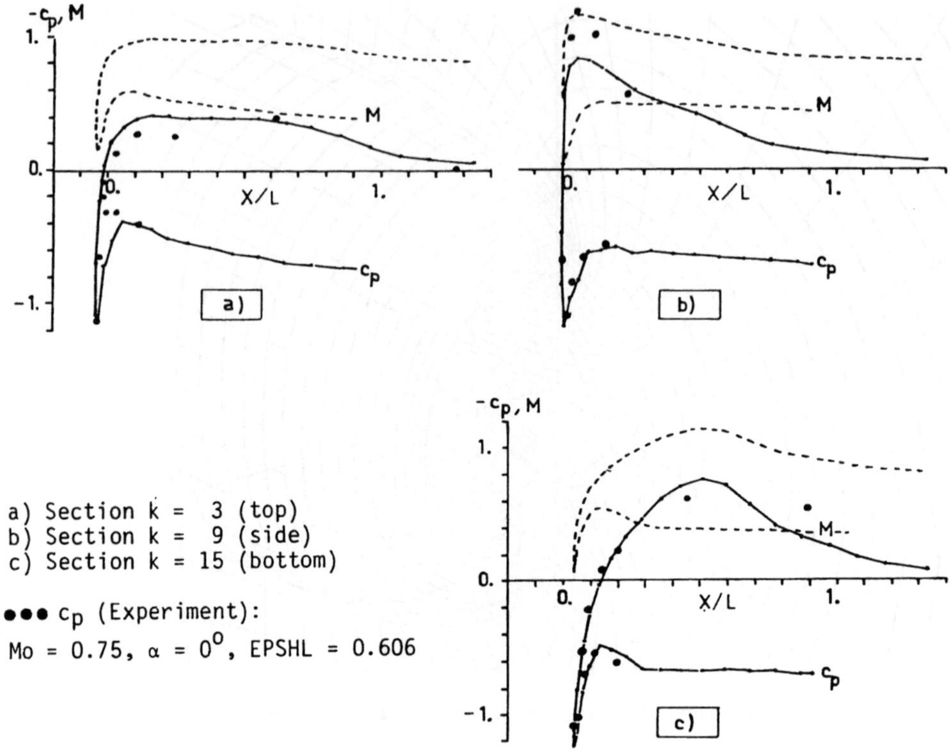

Fig. 5: Distributions of Mach number and pressure along three sections; M_o = 0,80, $\alpha = 0^o$, p_{ex}/p_o = 1.327, EPSHL = 0.6022

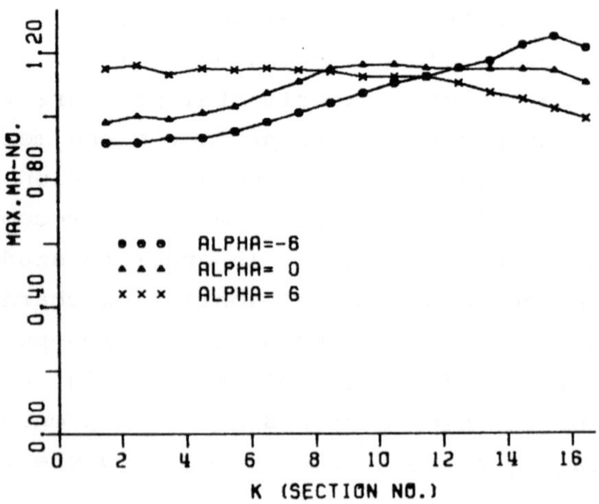

Fig. 6: Max. Mach numbers on external lip surface vs. k

The theoretical and experimental results (fig. 5) are not exactly comparable, since the values for M_o and EPSHL are slightly different and the pressure-hole sections do not coincide with the related volume centers of the computational grid. Nevertheless, this can't cause the discrepancies observed in fig. 5 (note, that in all cases the experimental pressure distributions are predicted fairly well in the channel region and near the inlet nose). The underprediction of the suction peaks for the side-sections can be explained by the fuselage of the wind-tunnel model influencing the flow in front of the inlet. The fore-body has a conical shape, approximately, and distorts the freestream to a divergent flow. Hence, in contrast to the numerical analysis, the inlet body experiences an inhomogeneous oncoming flow similar to a yaw situation, i.e. at the side-sections a lateral incidence is induced by the fore-body. A qualitative evidence of this explanation will be given below: The evaluation of a potential-flow calculation of the wing-body combination indicates, that the fore-body displaces the stream lines by about 5 degrees in lateral direction at locations corresponding to the position k = 9 of the inlet. Hence, the suction-peak difference mentioned above should vanish in the comparison of the theoretical pressure distributions with related experimental data measured at a yaw angle of $\beta = -5^o$. Unfortunately, no experiments for $\beta = -5^o$ are available. Therefore, an experimental result for $\beta = +5^o$ is taken, i.e. the yaw is just in the opposite direction, and it is expected that the suction-peak differences at the side-sections will be twice as large as for the case $\beta = 0$. This conjecture is indeed substantiated which will be shown for the case $\alpha = 6^o$, see fig. 7.a) to fig. 7.c).

So it is proven, that the mentioned pressure-peak discrepancies between theory and experiment are caused by the fore-body of the wind-tunnel model inducing a yaw-effect upon the air-intake; obviously, this can't be modelled by the present calculation for an isolated inlet. The present studies will be continued taking into account the real shape of the fore-body and the fuselage.

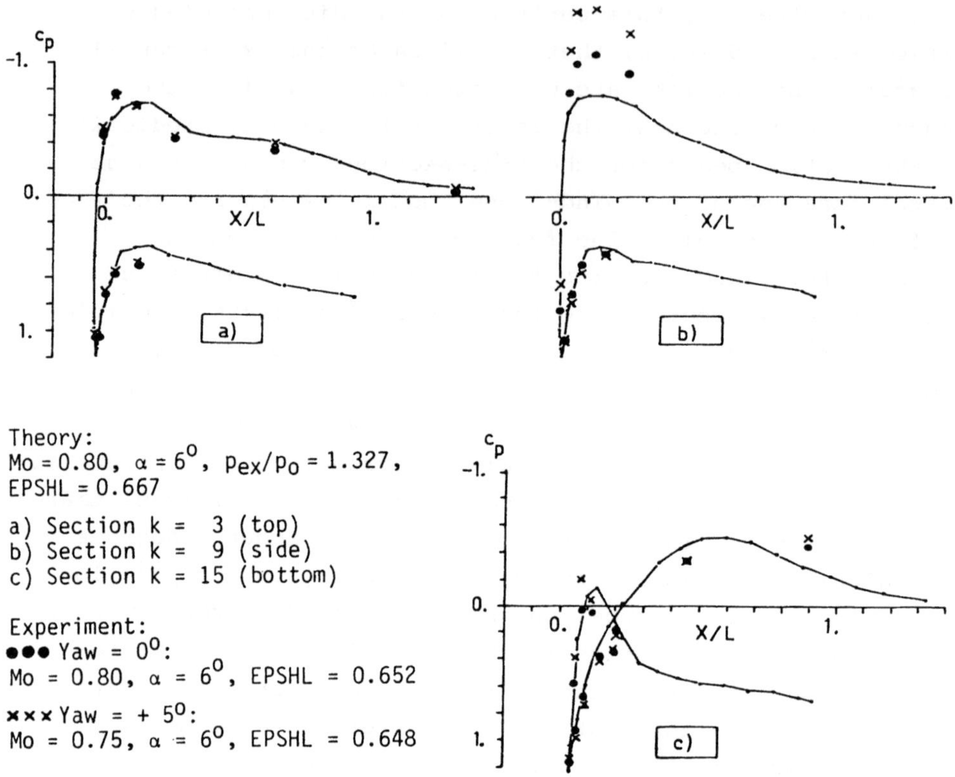

Fig. 7: Comparison of predicted and measured pressure distributions and influence of yaw

The inlet body has been designed under the constraint that the flow should be shockless in the design situation. Inspection of the theoretical results shows that this goal has been achieved, i.e. the supersonic regions depicted in fig. 8 for three angles of attack are not terminated by a shock.

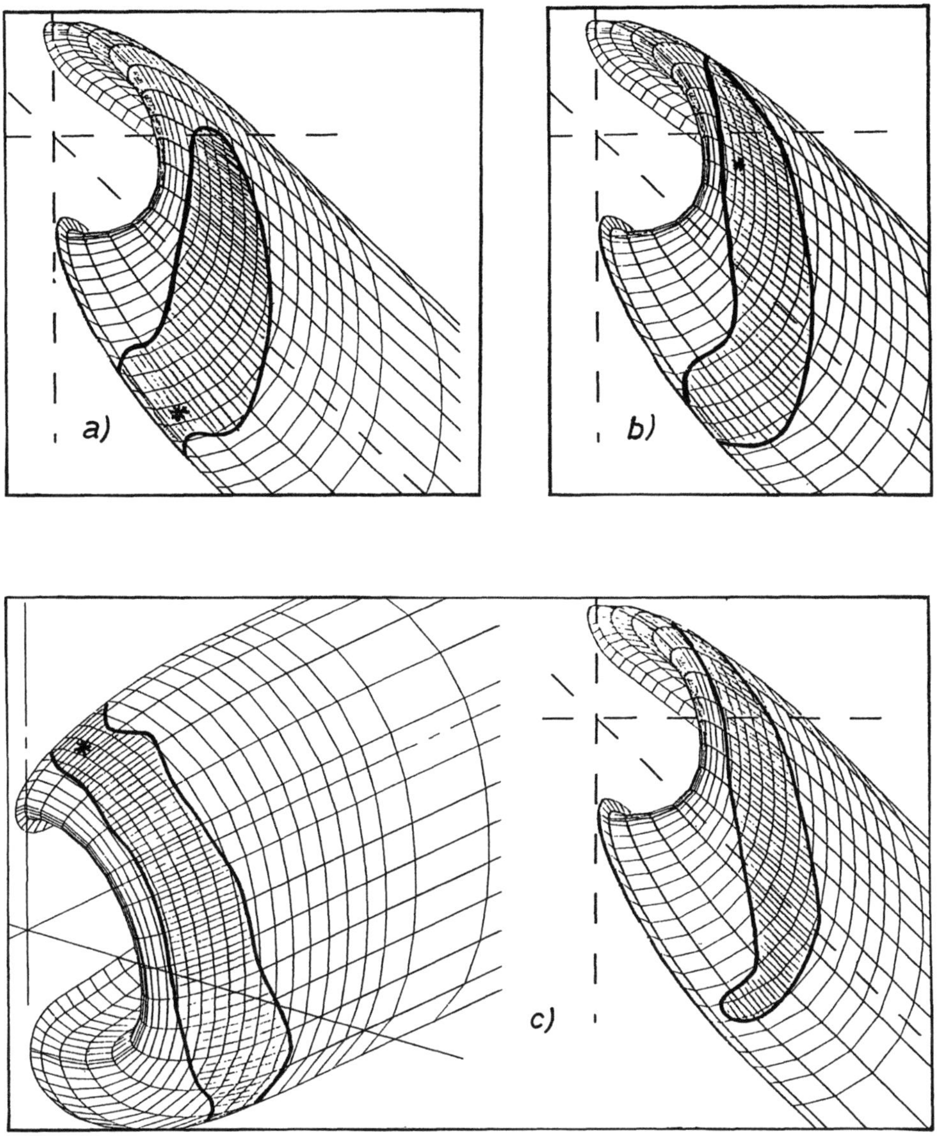

Fig.8: Predicted supersonic flow regions on inlet surface
(* Location with max. Mach number M_{max})
a) $\alpha = -6^{\circ}, M_{max} = 1.247$; b) $\alpha = 0^{\circ}, M_{max} = 1.158$; c) $\alpha = 6^{\circ}, M_{max} = 1.163$

6. References

[1] Grashof, J.: Inlet mesh generation by using a Maxwell transformation. Dornier-Note, 1982.

[2] Leicher, S.; Fritz, W.; Grashof, J.; Longo, J.: Mesh generation strategies for CFD on complex configurations. Paper presented on the 8th International Conference on Numerical Methods in Fluid Dynamics, Aachen, West Germany, June 28 - July 2 1982.

[3] Moon, P.; Spencer, D.E.: Field Theory Handbook. Berlin, Heidelberg, New York: Springer-Verlag 1971.

[4] Rizzi, A.W.; Inouye, M.: Time-split finite-volume method for three-dimensional blunt-body flow. AIAA J. 11 (1973) 1478-1485.

[5] Jameson, A.; Schmidt, W.; Turkel, E.: Numerical solutions of the Euler equations using Runge-Kutta time-stepping. AIAA Paper 81-1259, 1981.

[6] Schmidt, W.; Jameson, A.: Recent developments in finite-volume time-dependent techniques for two- and three-dimensional transonic flows. Von Kármán Institute for Fluid Dynamics, Belgium: Lecture Series 1982-04, March 29-April 2, 1982.

Calculation of a Laminar-Turbulent Two-Dimensional Turbine Blade Boundary Layer Including Surface Curvature Effects

R. GRUNDMANN

Institut für Theoretische Strömungsmechanik
DFVLR-AVA, Göttingen

Summary

An experimental pressure distribution along the suction side of a turbine blade [1] was used to calculate a boundary layer including second order surface curvature effects. The surface curvature is generally large on turbine blades. This gives rise to study this second order effect. Comparisons of first and second order calculations with a finite difference method [2] resulted in small changes of the location of the laminar or turbulent separation points. A comparison with an integral method [3] indicated differences in the location of the separation points but generally good agreement can be seen.

Introduction

The experimental investigation of a plane non-rotating turbine cascade was carried out in a paper by Gieß [1]. A calculation of the boundary layer for these turbine cascades was done by the integral method of reference [3]. In this computation laminar separation is predicted at the location where the ratio of the boundary layer thickness yields $H_{32} \leq 1.515$, further downstream the calculation was carried on for turbulent flow.

In this integral code the surface curvature is not included, except implicitely in the inviscid velocity distribution which is used as the outer boundary condition for the boundary layer equations. In the case of turbine blades the radius of the surface contour is small compared to a wing geometry, it is assumed that a difference in the results should be expected taking curvature into account.

The finite difference scheme applied in [2] was extended to the second order boundary layer theory, which involved some additional curvature terms due to the centrifugal- and coriolis acceleration. In second order theory the momentum equation normal to the surface did not vanish but an additional information for the pressure distribution normal to the surface is given in the form of an integral equation.

In this paper a comparison is shown between the integral and the difference method, firstly when the curvature effects are neglected following the first order boundary layer theory and secondly including these effects, which correspond to the second order theory.

Governing equations and Boundary Conditions

Fig. 1 gives a sketch of the turbine blade geometry used in this paper and the curvilinear, orthogonal coordinate system - s, n - fixed on the blade surface starting with $s = 0$ at the stagnation point. The coordinate n is always perpendicular to the surface. The governing equations are written in the tensorial contravariant notation, but the variables with superscripts are already replaced by the usual notation of u and v.

$$\frac{\partial(\rho u)}{\partial s} + \frac{\partial(\rho v)}{\partial n} + \underline{\frac{\kappa}{\mathcal{H}} \rho v} = 0 \tag{1}$$

$$\rho u \frac{\partial u}{\partial s} + \rho v \frac{\partial u}{\partial n} + \underline{\frac{n}{\mathcal{H}} \frac{\partial \kappa}{\partial s} \rho u^2} + \underline{\frac{2\kappa}{\mathcal{H}} \rho u v} = - \frac{1}{\mathcal{H}} \frac{\partial p}{\partial s} + \underline{3\mu \frac{\kappa}{\mathcal{H}} \frac{\partial u}{\partial n}} + \frac{d\mu}{dT} \frac{\partial T}{\partial n} \frac{\partial u}{\partial n} + \mu \frac{\partial^2 u}{\partial n^2} \tag{2}$$

$$\underline{\frac{\partial p}{\partial n} = \kappa \mathcal{H} \rho u^2} \tag{3}$$

$$c_p \rho u \frac{\partial T}{\partial s} + c_p \rho v \frac{\partial T}{\partial n} = E_c\, u \frac{\partial p}{\partial s} + E_c\, v \frac{\partial p}{\partial n} + \frac{1}{Pr} \frac{d\lambda}{dT} \frac{\partial T}{\partial n} \frac{\partial T}{\partial n} + \frac{1}{Pr} \lambda \frac{\partial^2 T}{\partial n^2} - E_c \mathcal{H}^2 \mu \left(\frac{\partial u}{\partial n}\right)^2 \tag{4}$$

$$\mathcal{H} = 1 + \kappa n \tag{5}$$

This set of differential equations describes the compressible, laminar boundary layer flow over a curved surface developed by the order of magnitude analysis. The underlined terms are small of the order $O(\varepsilon) = O(1/Re)$ and belong to the second order boundary layer theory. They are directely coupled to the surface curvature κ. In case of the curvature being negligible these terms will vanish and the boundary layer equations become first order.

In these equations u and v are the components of the velocity vector in the s and n direction. The temperature, the density and the pressure are denoted by T, ρ and p, respectively. The transport properties are the dynamic viscisity μ, the thermal conductivity λ and the ratio of the specific heats c_p. Pr and Ec are the Prandtl- and the Eckert-numbers.

In the case of turbulent flow the first order closure model of Michel , Quemard and Durant [4] was used.

$$\mu_t = \sqrt{Re}\,\rho\, F^2 l^2 \left|\frac{\partial u}{\partial n}\right| \tag{6}$$

with the mixing length

$$l = \delta \cdot 0.085 \tanh\left(\frac{\bar{\kappa}}{0.085}\frac{n}{\delta}\right) \tag{7}$$

and the van-Driest damping factor

$$F = 1 - \exp\left[-\frac{1}{26\bar{\kappa}\mu_w}\sqrt[4]{Re}\,\sqrt{(\rho\tau)_w}\right] \tag{8}$$

$\bar{\kappa} = 0.41$ is the von-Kármán constant.
The turbulent heat conductivity is

$$\lambda = \frac{Pr}{Pr_t} c_p \mu_t \; . \tag{9}$$

If the turbulent flow is to be calculated the transport properties μ and λ in the equations (2) and (4) have to be replaced by

$$\mu = \mu_1 + \mu_t \tag{10}$$

$$\lambda = \lambda_1 + \lambda_t \tag{11}$$

In the case of laminar flow μ_1 and λ_1 are described by the Sutherland formulation.

The second order boundary layer theory requires special boundary conditions. Usually in the boundary layer theory the pressure gradient normal to the surface is assumed to be zero. In higher order theory there is an integral equation (3) for this pressure. This involves gradients of the inviscid flow properties normal to the surface. The deviation of the streamlines around bodies even in the inviscid flow makes this a reasonable assumption. From the irrationality and constant enthalpy requirements of the outer flow field these boundary conditions can be developed for the velocity u_e and the temperature T_e.

$$\frac{\partial u_e}{\partial n} = -\frac{2\kappa}{\mathcal{H}} u_e \tag{12}$$

$$\frac{T_e}{T_{e1}} \doteq 1 + \frac{1}{2}\frac{u_{e1}^2}{c_p T_{e1}}\left(1 - \frac{u_e^2}{u_{e1}^2}\right) \tag{13}$$

Applying the isentropic conditions yields the density and the inviscid pressure distribution. The suffix e denotes the conditions of the flow at the outer edge of the boundary layer and the suffix e1 the surface conditions.

Pressure Integral

The pressure distribution normal to the surface is calculated by the integration of equation (3) if the velocity and density profiles are known.

$$p = p_w + \kappa \int_0^{\delta} \mathcal{H} \rho u^2 dn \tag{14}$$

where p_w is the pressure at the wall defined by

$$p_w = \{ p_e - \kappa \int_0^{\delta} \mathcal{H} \rho u^2 dn \}_{n=\delta} \tag{15}$$

Computational results

The solution of the governing equations was carried out by the difference method applied in Ref.[3].

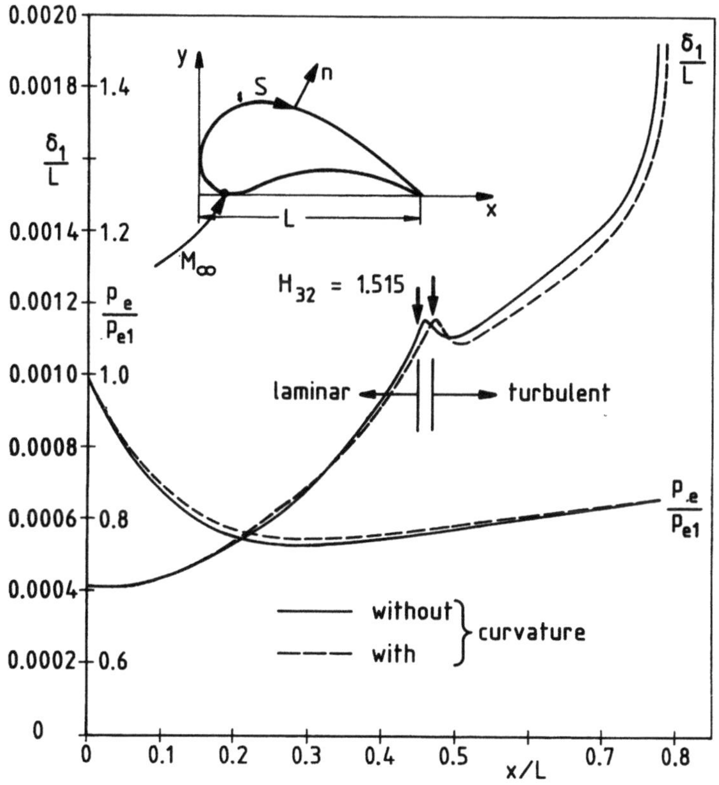

Fig.1. Inviscid pressure distribution and resulting displacement thickness along the chord length

In Fig.1 the contour of a turbine blade given by [1] is sketched with the cartesian and the curvilinear coordinate system fixed at the stagnation point. The non-dimensional, inviscid pressure distributions along the suction side p_e/p_{e1}, where p_{e1} is the measured pressure distribution on the surface in case of no curvature, - i.e. first order boundary layer theory - , and the pressure distribution p_e recalculated by the equations (14) and (15), - i.e. second order boundary layer theory - , are plotted versus the non-dimensionalized x/L coordinate. The displacement thickness δ_1/L shows that the separation point for the case with curvature is down stream of the case without curvature. This is due to the fact, that the recalculated inviscid pressure distribution has a lower adverse pressure gradient than the measured one. The transition point from laminar to turbulent flow was located where the ratio of the boundary layer thicknesses $H_{32} = 1.515$ is reached being a criterion for laminar separation.

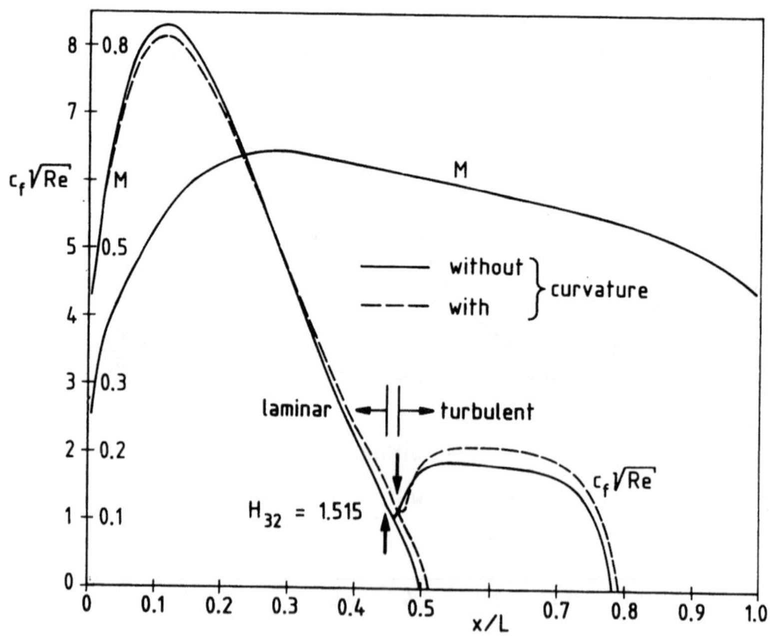

Fig.2. Mach number distribution and resulting wall shear stress along the chord length

Fig.2 shows the inviscid measured Mach number distribution and as a further result the differential wall shear stress $c_f\sqrt{Re}$ along the coordinate x/L. The calculated transition point by the above mentioned criterion as well as the laminar and turbulent separation points differ by less than 2% for the cases with and without curvatures effects.

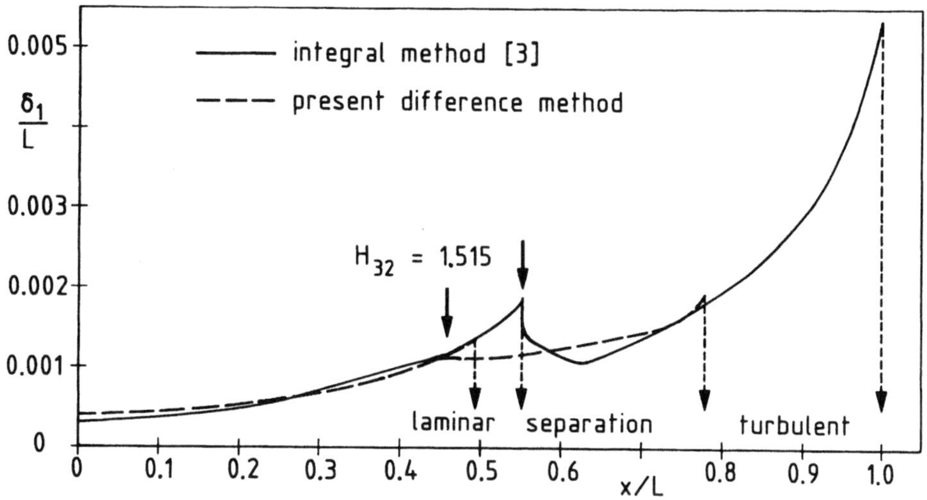

Fig.3. Displacement thicknesses calculated by integral and difference method along the chord length

Fig.3 gives a comparison of the displacement thicknesses calculated by the integral method [3] and the present difference method. The difference in the predicted transition points and the turbulent separation points is significant. They differ about 20%, the laminar separation points about 10%. But in general the quantitative development is in good agreement.

Discussion of results

The effort solving the second order boundary layer equations and its application to highly curved surfaces, keeping in mind the additional iteration step for the pressure integral, has only a slight influence on the results of the boundary layer computations namely a difference of 2% in the location of the separation point. The overall drag coefficient differs in the same order of magnitude. The verification of the findings presented here should lead to further experiments, which focus on transition and separation. A quantitative agreement between the difference and integral method is acceptable until the respective separation points are reached.

References

1. Gieß, P.-A.: Basisprofil T3 für Dampfturbinen- Hoch-und Mitteldruckstufen. Experimentelle Untersuchungen und Vergleich mit den Profilen T2 und P1. DFVLR-IB 222-81 A 42 (1982).

2. Grundmann, R.: Dreidimensionale Grenzschichtberechnungen entlang Symmetrielinien auf Körpern. ZFW, Band 5, Heft 6, (1981) 389-395.

3. Rotta, J.-C.: Fortran IV - Rechenprogramm für Grenzschichten bei kompressiblen ebenen und achsensymmetrischen Strömungen. DLR FB 71-51 (1971).

4. Michel, R.; Quemard, C.; Durant, R.: Application d'un schéma de longueur de mélange a l' étude des couches limites turbulentes d' équlilibre. ONERA, Note Techn. 154 (1969).

Considerations of the Vorticity Fields on Wings

E.H. Hirschel

Messerschmitt-Bölkow-Blohm GmbH
UFE122
Postfach 80 11 60
8000 München 80
Federal Republic of Germany

Summary

Inviscid flow fields of lifting wings usually are calculated with the help of the Kutta condition, which forces the flow smoothly off the wing surface at the trailing edge. The resulting vortex sheet behind the wing in reality carries vorticity which is created at the wing surface due to the non-slip condition. It is discussed qualitatively how the vorticity properties of wing boundary layers fit into the picture of the potential flow model of lifting wings. Especially it is shown how the vorticity vector in the boundary layers, which leave the surface at the trailing edge, corresponds to the (model) potential flow behind the wing. Finally the calculated boundary layer on a swept wing is analysed revealing additional information about the computed flow field.

Introduction

The flow and some of the forces at a lifting wing can be determined by means of potential-flow theory. This is being done with help of the Kutta condition, which forces the flow smoothly off the wing surface at the trailing edge, which is a good simulation of reality (the Kutta condition is employed in all kinds of inviscid flow calculation at lifting wings). In reality the resulting vortex sheet behind the wing, which is being concentrated into a pair of discrete vortices far behind the wing, carries vorticity which is created solely at the wing surface due to the non-slip condition (if the flow is supersonic at least partly, vorticity can be created in resulting shock waves, too).

In the present paper it is discussed how the vorticity properties of the boundary layer at wings fit into the picture of the potential flow model of lifting wings. The aim is to get some insight into the flow especially at the trailing edge which might be useful for considerations of, at least, mildly separated flow in that region.

For the role of vorticity in fluid flow in general the reader is referred to [1, 2]. In the present study Cartesian coordinates are employed for simplicity, that is the wings are considered as flat, with x the downstream (chord) direction, y the spanwise direction and z the direction normal to the surface, while u, v, w are the respective velocity components. The vorticity vector then reads (for the notation in general non-orthogonal, curvilinear, surface-oriented coordinates see [3]):

$$\underline{\omega} = [\omega_x, \omega_y, \omega_z] = \left[\frac{\partial w}{\partial y} - \frac{\partial v}{\partial z}, \frac{\partial u}{\partial z} - \frac{\partial w}{\partial x}, \frac{\partial v}{\partial x} - \frac{\partial u}{\partial y}\right]. \tag{1}$$

If the boundary-layer stretching is introduced, it can be seen that in the three-dimensional boundary layer for high Reynolds-number flow the terms of the components of the vorticity vector are of different magnitude [1, 3]:

$$\underline{\omega} = \left[\frac{1}{\sqrt{Re}}\frac{\partial w}{\partial y} - \sqrt{Re}\,\frac{\partial v}{\partial z}, \sqrt{Re}\,\frac{\partial u}{\partial z} - \frac{1}{\sqrt{Re}}\frac{\partial w}{\partial x}, \frac{\partial v}{\partial x} - \frac{\partial u}{\partial y}\right]. \tag{2}$$

In the following therefore only the two components ω_x and ω_y are considered:

$$\underline{\omega} = [\omega_x, \omega_y, \omega_z] \approx \left[-\frac{\partial v}{\partial z}, \frac{\partial u}{\partial z}, 0\right]. \tag{3}$$

<u>Vorticity Fields on Wings</u>

1. <u>Airfoil</u>

The vorticity development on an airfoil is considered in Fig. 1 . Vorticity is created at the surface which acts as a source of vorticity if $\frac{\partial p}{\partial x} < 0$, and as a sink if $\frac{\partial p}{\partial x} > 0$, [1], Fig. 1A. The vorticity is then transported by diffusion and convection in the boundary layer. Fig. 1B shows qualitatively the profiles of the tangential flow component u and the vorticity component ω_y. In regions where $\frac{\partial p}{\partial x} > 0$ the u(z)-profile exhibits a point of inflection which can be seen from a discussion of the wall-compatibility conditions. To such a profile a ω_y-profile belongs with a maximum at the location of the point of inflection.

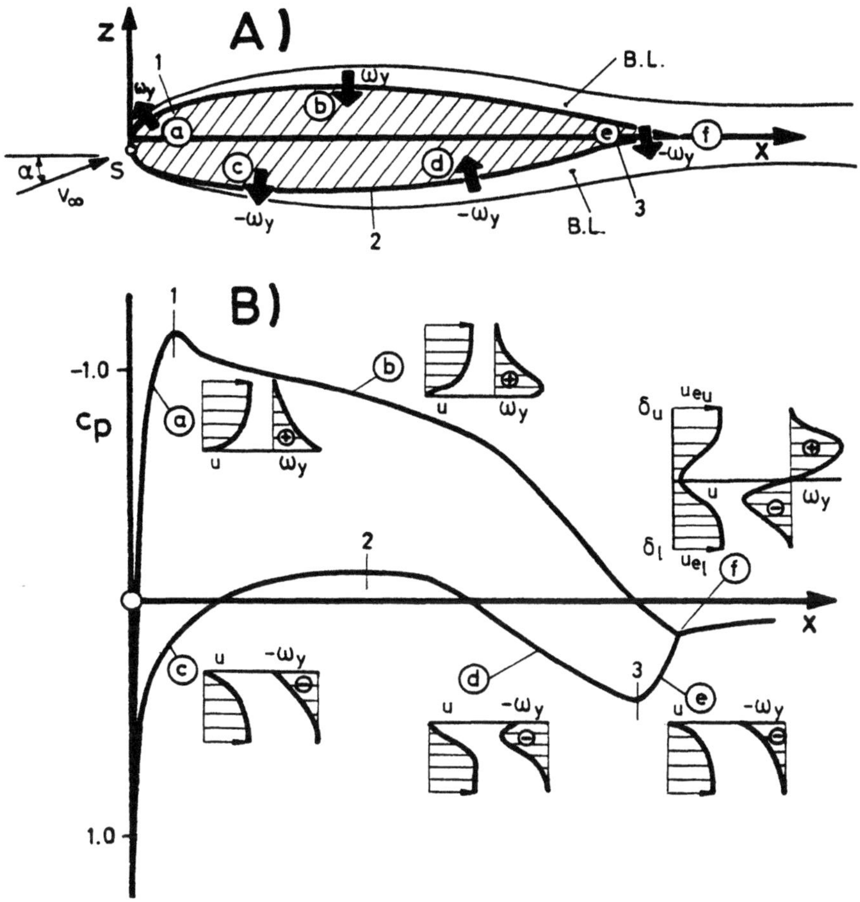

Fig. 1 Schematic of vorticity field at a lifting profile.
A) Vorticity sources and sinks,
B) surface pressure distribution and velocity and vorticity profiles

The vorticity content (per unit area of the boundary layer in high Reynolds number flow at a given point on the surface) is [1] (here written for one component for the three-dimensional case):

$$\bar{\omega}_y(x,y) = \int_0^\delta \omega_y(x,y,z)\,dz = \int_0^\delta \frac{\partial u}{\partial z}(x,y,z)\,dz = u_e(x,y)\,, \tag{4}$$

and the rate of convection in x-direction

$$\int_0^\delta u(x,y,z)\,\omega_y(x,y,z)\,dz = \int_0^\delta u(x,y,z)\frac{\partial u}{\partial z}(x,y,z)\,dz = \frac{1}{2}u_e^2(x,y), \tag{5}$$

with u_e the external inviscid velocity at the edge of the boundary layer.

From this consideration it is seen that the vorticity content of the boundary layer varies along the surface: zero at the stagnation point, high in a pressure minimum (suction peak) and low again, but finite at the trailing edge. This is paralleled by the total pressure at the surface. Because in the boundary layer $\frac{\partial p}{\partial z} = 0$ the total pressure at each point of the surface is equal to the static pressure of the external inviscid flow. In this way the total pressure loss is zero at the stagnation point, high at a pressure minimum and low again, but finite at the trailing edge.

At the trailing edge the boundary layer leaves the surface and carries away vorticity. If the trailing edge is sufficiently sharp pressure and velocity at the upper and the lower side are the same, whereas the vorticity amount is the same but with opposite sign. With eqs. (4) and (5) it follows that no net ω_y-vorticity leaves the airfoil whether it is lifting or not. Only during the starting process or if the lift is being changed net ω_y-vorticity will leave the surface, either positive or negative according to the direction of lift change. This is in accordance with the potential flow model of a lifting airfoil.

2. Infinite swept wing

At the infinite swept wing the inviscid surface flow is composed of the component $\tilde{u}_e$ in $\tilde{x}$-direction (normal to the leading edge, Fig. 2) and the component $\sin\varphi\, v_\infty$ in $\tilde{y}$-direction (along the leading edge), where φ is the sweep angle and v_∞

the flight velocity. At lift the external flow on the upper side changes direction more strongly than that on the lower side, especially the boundary layer gets more skewed on the upper side, Fig. 2A .

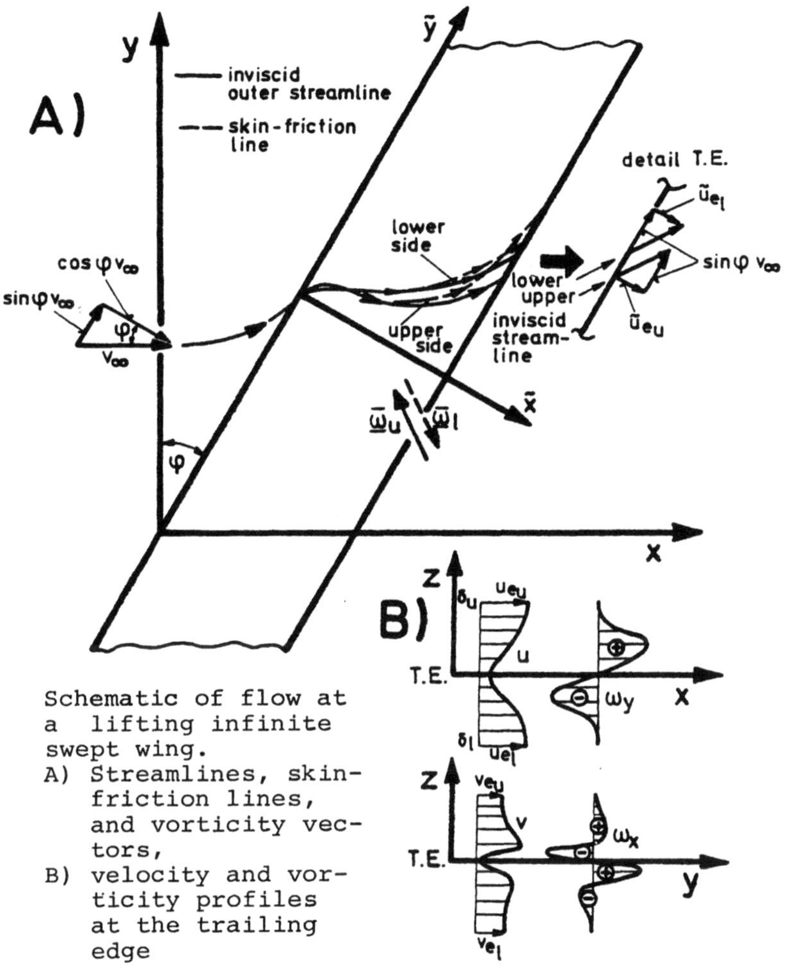

Fig. 2 Schematic of flow at a lifting infinite swept wing.
A) Streamlines, skin-friction lines, and vorticity vectors,
B) velocity and vorticity profiles at the trailing edge

At the trailing edge due to the composition of the inviscid flow field again pressure and velocity vector on the upper and the lower side are the same. Just behind the trailing edge the tangential flow components u,v of the x,y,z-coordinate system schematically have the profiles given in Fig. 2B, together with the pertinent vorticity profiles.

Applying eqs. (4) and (5) one again finds that no net ω_y-vorticity leaves the wing if the lift is not changing. Important is that also no net ω_x-vorticity leaves the surface. This has to be this way because the infinite swept wing has no lift variation in span direction. The global vorticity vectors (eq. (4)) of the boundary-layers $\bar{\underline{\omega}}_u$, $\bar{\underline{\omega}}_l$ at the upper and the lower side at the trailing edge are therefore of equal length but opposite direction, Fig. 2A. A conclusion is that an experimental or theoretical investigation of the wake of a lifting wing ought not to be made with an infinite wing model because no net vorticity, e.g. no vortex sheet, leaves the trailing edge of the infinite swept wing.

3. Finite wing

At a finite lifting wing without sideslip, see e.g. Fig. 3, the boundary layer is quasi-two-dimensional [3] at the root, which is a plane of symmetry. This leads to a vorticity-vector distribution like that for an airfoil, Fig. 1. At the wing tip at the upper side the external inviscid flow is directed toward the root over most of the surface, whereas the contrary is true on the lower side. This is due to the pressure relaxation from the lower to the upper side of the wing around the tip. On the upper side this leads to problems in boundary-layer computations because no exact initial data can be posed along the wing tip. Although here the inviscid flow on the upper side of the wing is directed toward the root its streamlines are curved in the opposite direction at the trailing edge which leads to a boundary-layer flow directed towards the wing tip, Fig. 3. Between root and tip usually a gradual transition between the two extremes occurs.

At the trailing edge of the wing the pressure at the upper and the lower side is the same provided the trailing edge is sufficiently sharp. If that is true, the inviscid velocities are the same in magnitude, but not in direction, $|\underline{v}_{e_u}| = |\underline{v}_{e_l}|$. Assume now a direction s which is the bisector of $\underline{v}_{e_u}$ and $\underline{v}_{e_l}$. The velocity components in this direction on the upper side $\tilde{u}_{e_u}$ and on the lower side $\tilde{u}_{e_l}$ are the same, which means that

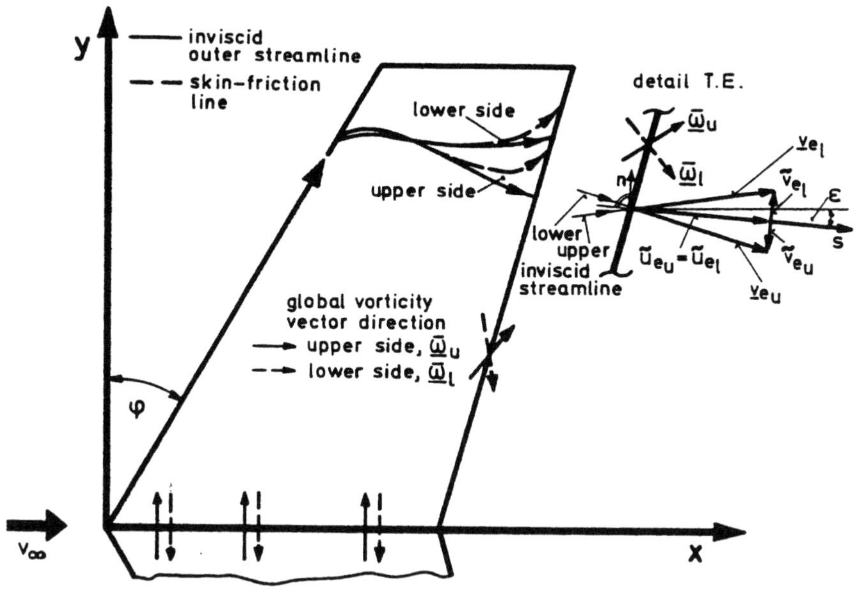

Fig. 3 Schematic of flow at a finite lifting wing (flow turning angles exaggerated)

the global vorticity-vector components at the upper and the lower side in n-direction (n is the direction normal to the s-direction) cancel each other, so that no net ω_n-vorticity leaves the wing. The ω_s-vorticity content of the chord cut is

$$\int_{\delta_l}^{\delta_u} \tilde{\omega}_s \, dz = - \int_{\delta_l}^{\delta_u} \frac{\partial \tilde{v}}{\partial z} dz = - \tilde{v}_{eu} + \tilde{v}_{el} \ . \qquad (6)$$

Because $\tilde{v}_{el} = -\tilde{v}_{eu}$ (due to the choice of the s-direction) the ω_s-vorticity content is $-2\tilde{v}_{eu}$. Because $\tilde{v}_{eu} < 0$ positive net ω_s-vorticity leaves the trailing edge. With the theorem of Stokes [2], which connects the circulation around a vortex tube with its vorticity, it is seen that this net ω_s-vorticity is the vorticity shed due to the change of circulation in spanwise direction of the wing.

This qualitative picture may undergo many changes in reality due to wing planform etc.. The angle ε between the s-direction and the x-direction, which in potential lifting wing theory usually is assumed to be zero, may have positive or negative sign. The upper and lower global vorticity vector at the trailing edge, however, will become increasingly co-directional toward the wing tip, as shown in an exaggerated way in Fig. 3. At a two-axial symmetric wing at zero angle of attack of course flow and vorticity fields on the upper and the lower wing side will be congruent. At a general cambered and twisted wing at zero lift net ω_S-vorticity may leave in some spanwise regions but is canceled globally on one wing half.

Finally the vorticity field on a finite lifting wing resulting from a numerical investigation [4] is considered. In Fig. 4 the direction of the vorticity vector at the z-station where the vorticity has its maximum in direction normal to the wing surface is given in discrete points on the upper and the lower

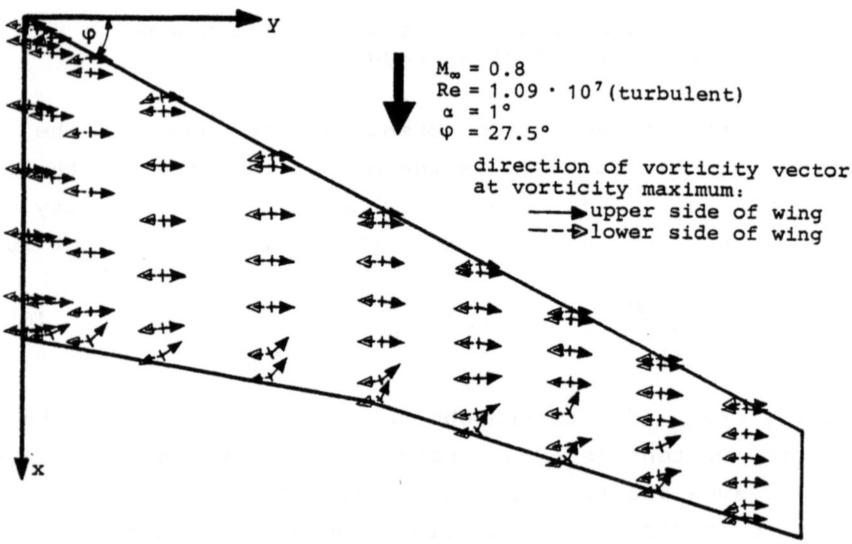

Fig. 4 Vorticity vector direction at vorticity maximum on finite lifting wing (basic data from [4])

side (in all cases these stations are quite near to the surface). At the root the directions are as expected, also at the trailing edge up to approximately half span. Then some irregularities appear insofar as the direction of the upper side vectors don't follow the expected trend. The reason for this lies in the pressure adjustments which were necessary in order to prevent separation during the iteration procedure in [4]. Neither the resulting pressure field nor the boundary-layer profiles did show directly the qualitatively wrong result in that area. At the upper side of the wing in the station nearest to the tip the boundary-layer behaviour due to the aforementioned reasons (initial data) appears to be completely wrong.

Conclusions

The vorticity appearing in the wake of a lifting wing is created at the surface of the wing. In steady flight no net ω_y- (ω_n-) vorticity leaves the wing. The circulation change in spanwise direction gives rise to a net ω_x- (ω_s-) vorticity shedding which can be identified with the vorticity content of the boundary layers at the trailing edge. Infinite swept wing models in experiment and theory do not have the vorticity shedding properties which are found on finite wings.

References

1. Lighthill M.J.: Introduction boundary-layer theory. In: "Laminar boundary layers", L. Rosenhead (ed.), Oxford at the Clarendon Press, 1963, pp. 46 - 113.

2. Batchelor G.K.: An introduction to fluid dynamics. Cambridge University Press, Cambridge, 1967.

3. Hirschel E.H.; Kordulla W.: Shear flow in surface-oriented coordinates. Vieweg Verlag, Braunschweig, 1981. Vol. 4 of Notes on Numerical Fluid Mechanics.

4. Kordulla W.: Viscous transonic flows about 3-D wings. ZFW 2 (1978), Heft 3, pp. 306 - 312.

Slender Wings with Leading Edge Vortex-Separation

Stephan M. Hitzel

Dornier GmbH, Theoretical Aerodynamics Dept.
Postfach 1420
7990 Friedrichshafen, Germany

Summary

The demands of high speed performance and maneuverability on modern transport- and fighter-aircraft require slender wings, whose leading edge vortex flows result in increased lift at high angle-of-attack. Beginning with the experimental and theoretical development and the history of application the article will discuss the application of panel- and vortex-lattice-methods and the latest employment of an Euler-code to predict the behaviour of the wing-vortex-system. Additionally the vortex-breakdown-phenomenon will be discussed.

Performance and maneuverability by slender wings with leading-edge-vortices

When the possibility of supersonic flight was seriously considered in the late forties and early fifties, it seemed inevitable that, on the aerodynamics alone, supersonic long-range flight would be uneconomic if not altogether impossible. It became clear that, for the same physical reason that makes the aerodynamics of conventional wings and bodies deteriorate at supersonic Mach-numbers, the aerodynamics of jet engines improve (1). The end of the war 1945 allowed a fruitful merging of both sides research work, forming a new set of aerodynamics, which became necessary to overcome the problems of the trans- and supersonic flight regime. The results of the new aerodynamics showed that pointed delta wings of small aspect-ratio could have lower wave drag at supersonic speeds than unswept wings. A theoretical approach revealed that a subsonic leading-edge (suction effect), while travelling inside the Mach-wave cone, was a possible mean for the reduction of supersonic drag. Also very important for a supersonic configuration was the use of thin wing sections and the delta planform turned out to be the best structural shape. Consequently most of the first supersonic fighters and bombers became deltas (fig. 1,2) [2,3].

Fig. 1 Convair F-102

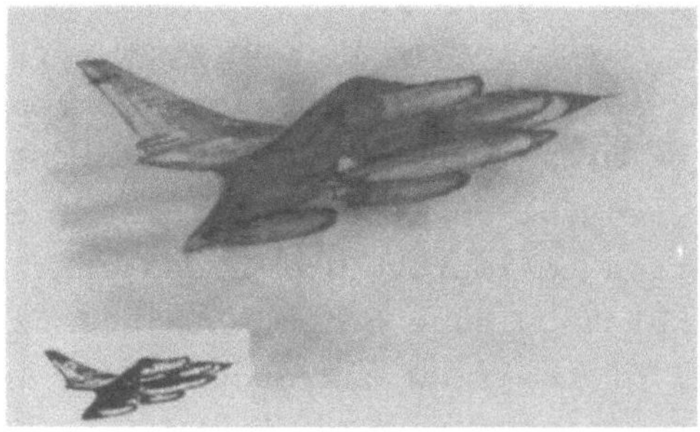

Fig. 2 Convair B-58 Hustler

Low aspect-ratio wings, - by definition having a poor lift gradient, - were expected to present difficult problems in the approach configuration and during maneuvering flight. For a slender wing, designed for high-speed-performance, the wing's leading-edge is quite sharp, allowing flow separation along the leading-edge even at low incidende. This separation rolls up into strong vortices, thus inducing velocities on the wing's upper surface, causing the well-known nonlinear aerodynamic characteristics being exhibited by slender wings (fig. 3). The effect was more or less discovered accidentally

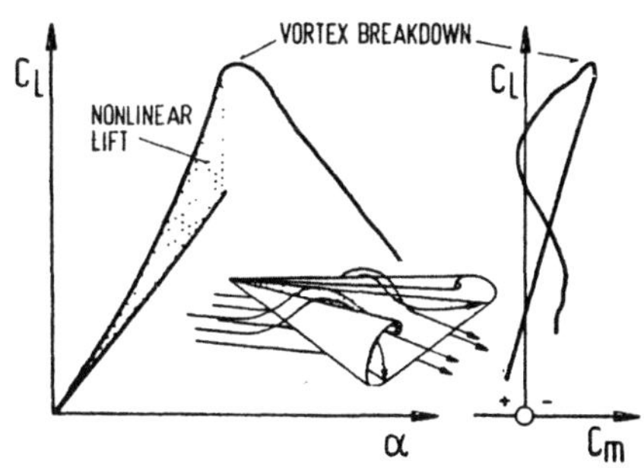

Fig. 3 Aerodynamics of a slender wing

Fig. 4 F-86

by experiment, while at the same time schools of thought had suspected the existance of such a leading-edge flow. Thus in 1951 the discovery of some unexplained vortices peeling off the wing's leading edge of a version of a F-86-fighter (fig. 4) providing extra lift, led to some modifications which doubled the aircraft's combat-maneuverability (4), while at the same time in Europe some very slender wings tested in an ONERA-windtunnel revealed a strong vortex-flow on the surface of an arrow wing (5) and the researchers at RAE (6) discussed the constructive role of controlled leading-edge-separation to improve low-speed-characteristics of slender wings. First knowledge being obtained from experiments, a first model of the flow was proposed by Legendre (7) based on Jone's slender wing theory (8) by superimposing a pair of vortices above the wing. However as Adams (9) discussed the non-unique solution concerning lift and pressure-distribution of Legendre's model, Edwards (10) proved for conical slender wings that a reasonable model must also regard the mechanism of the feeding vortex-sheets emanating from the wing's leading-edge. The position of this vortex-system is determined by the fact that it must remain force-free. Brown & Michael (11) finally applied the feeding-sheet-vortex system in a conformal-mapping-procedure in the cross-flow-plane. Some ten years later military programs (fig. 5) and the supersonic transport program (12) were exeptional opportunities to look at the vortex flow in more detail. Intensive flight tests (fig. 6) and a lot of experimental work (e.g. 13, 14, 15) helped to understand leading-edge-vortices, their structure (fig. 7) and their consequences (fig. 3) [18].

Fig. 5 SR-71

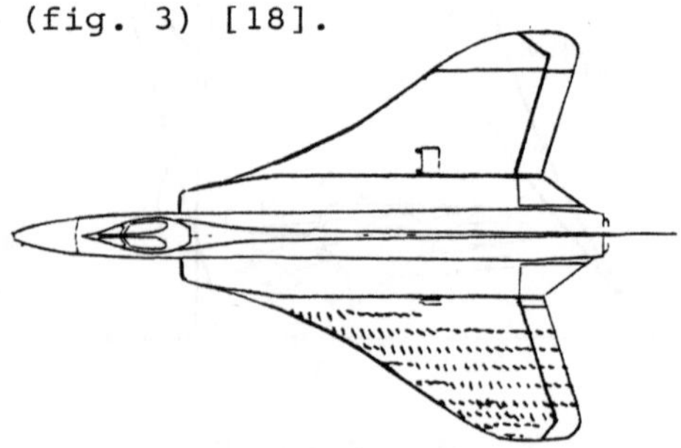

Fig. 6 NASA-F-5D testflight (16)

The analyses pointed out a very severe limitation of vortex-stability called vortex-breakdown. At high angle-of-attack a rapid change of the vortex-structure occurs above the wing (fig. 7), which results in a decrease in lift and a nose-up pitching moment, limiting maneuverability (fig. 3). This

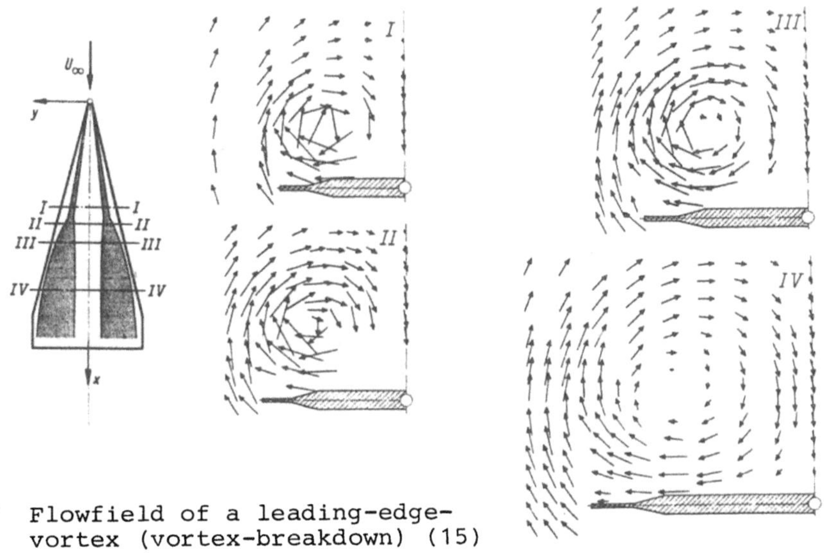

Fig. 7 Flowfield of a leading-edge-vortex (vortex-breakdown) (15)

bursting moves forward from the trailing-edge to the apex with increasing incidence, causing a loss of suction on the upper surface. As Hummel (17) turns out, the pressure gradient along the vortex-path has an important influence on breakdown. An increasing pressure leads towards breakdown, a decreasing one stabilizes the vortex structure, meaning breakdown will be avoided. In recent years a number of different explanations have been proposed of which the following is partial list considered in (18). Ludwieg (19) derived his work from Rayleigh's (20) and Taylor's (21) instability-examinations of a flow between two coaxial rotating cylinders. Benjamin's (22) interpretation considers the analogy with the hydraulic jump or the shock-wave to describe flow conditions ahead and behind the breakdown-position, while Hall (23) tries to explain the problem by developing a numerical method, which fails during the evaluation of corresponding velocity-profiles of a critical vortex, identifying breakdown. However, a lot of questions are left (18) and neither the theories nor a computational method are capable to explain or at least predict vortex-breakdown till today.

Simplified models and their application in panel-methods in cooperation with an Euler-code to describe leading-edge vortex flows and to investigate the vortex-breakdown-phenomenon

A very suitable method for calculating the magnitude of the nonlinear vortex-lift on a rather broad class of slender wing-planforms is provided by the suction-analogy. Polhamus (24) reasoned that the normal force needed for the flow around a leading-edge to reattach to the wing is equivalent to the suction-force necessary to force the flow to be attached to the leading-edge in an unseparated condition. Unfortunately this hypothetical method gives no flow-field details or surface pressure-distributions. Approximately at the same time Smith (25) published the best known model to simulate the vortex-system. By the introduction of a free vortex-sheet (fig. 10) in between the wing's leading-edge and the feeding-sheet, he overcame the deficiencies of the Brown & Michael method, which was unable to satisfy the full Kutta-condition along the leading-edge. Demanding the free sheet to be stream-surface and assuming conical flow, he was able to predict the type of pressure-distribution that had been observed experimentally (fig. 8).

Towards the trailing-edge, Smith's model overpredicts the load distribution by a considerable amount, for the Kutta-condition cannot be satisfied by the conical theory.

The development of higher-order panel-methods (27,28) made it possible to translate the physics of Smith's model into a fully 3-dimensional method (fig. 11) (29, 30). Figure 9 depicts the method's capability to determine not only the overall forces but the pressure-distributions correctly. However, the 3-dimensional problem is now solved in principle (29),

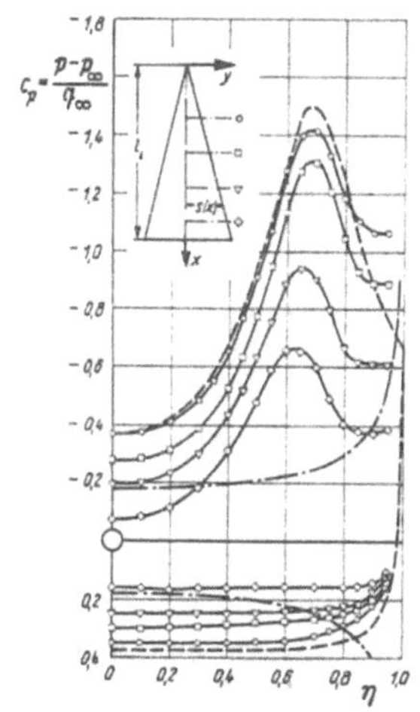

Fig. 8 C_p-distribution
exp. Hummel (26) = 20.5
---- Smith (25)
-.-. Jones (8)

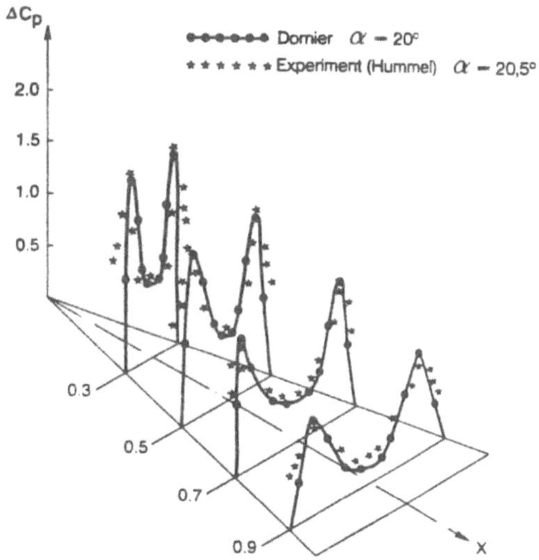

Fig. 9 ΔC_p-distribution for 3-dim. panel method (29)

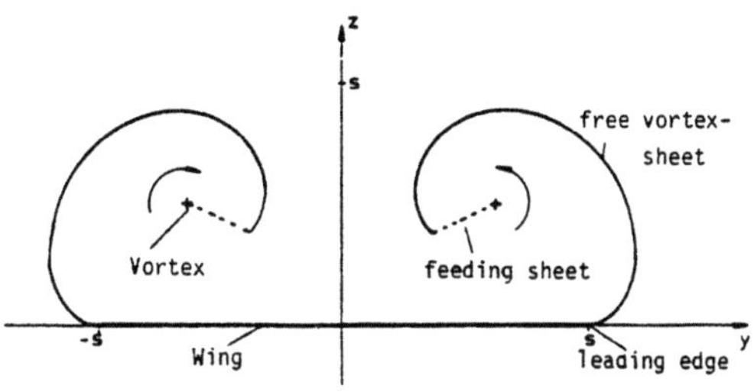

Fig. 10 Smith model

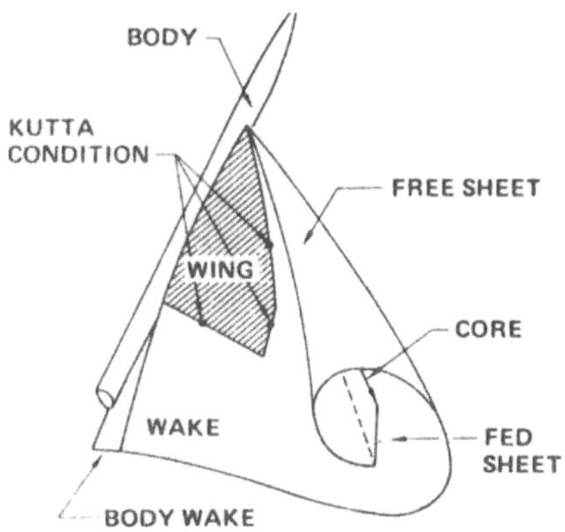

Fig. 11 3-dim Model (30)

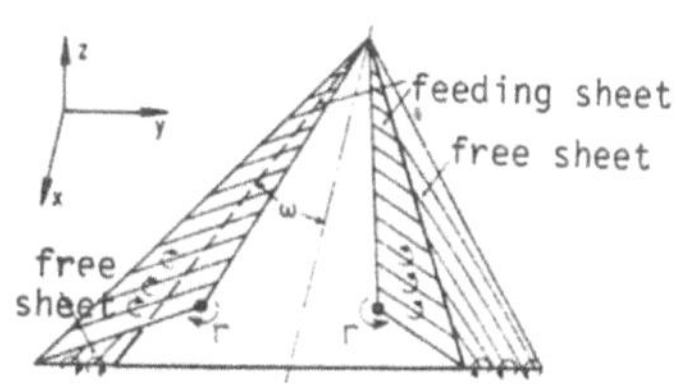

Fig. 12 Simplified model (31)

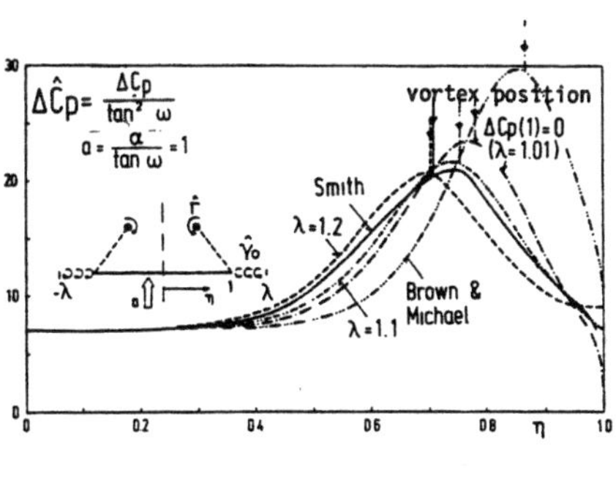

Fig. 13 C_p-distribution by (31)

the method cannot fully satisfy due to the extremely high computational expense which is caused by the necessity of a high number of panels for the simulation of the free vortex-sheet and its iterative procedure to establish the proper position of the vortex-system. This advocates the idea of a simplified model relying on the integral physical effects. In (31) flat free vortex-sheets are introduced near the edges besides the vortex-core above the wing (fig. 12). The flow-field is assumed to be conical and flow is permitted through the free sheet. Instead of applying the usual boundary condition on free vortex-sheets it is attempted to fulfill the suction-analogy by accordingly prescribing normal force coefficients and taking care of the condition that the whole feeding-system should be force-free. One parameter, remaining free, can be used for optimum adaptation of the results to those of Smith's method or for making the vortex-sheet free of forces immediately at the leading-edge. This procedure leads to satisfactory results for the pressure distribution (fig. 13). Further investigations (32) revealed the evident reduction of the Smith's model to the model of Brown & Michael by shortening the free vortex-sheet. Moreover the results demonstrated the importance of this sheet, which can be explained by the near-field influence of the sheet close to the leading-edge to fulfill the Kutta-condition more appropriately, thus regarding the vortex-system really to become streamlined at the edge.

The very simple structure of vortex-lattice-procedures makes them very attractive for the necessary modifications to adopt the simplified models for a 3-dimensional procedure (33). In a first step the Brown & Michael model was employed. The 3-dimensional vortex-lattice-method (34) takes over the geometrical results (conical) - meaning the position of the feeding-sheet - and determines the circulation-distribution of the wing-vortex-sheet-configuration, the Kutta-condition being enforced by vorticity-continuity across the leading-edge. By some slight manipulations - to align the vortex-system to stable (force-free) position - the computational results for

C_N are close to those predicted by the suction-analogy (35) and by the higher-order-panel-method (34) (fig. 14).

The 3-dimensional effect on all transformed cross-flow-models primarily is dominated by the model of the Kutta-condition. By coincidence of vortex- and velocity-vector each free vortex-sheet becomes a real stream-surface as it is for a wake-model. Employing (34) the vector algebra on the formula for the aforementioned condition will disclose two sub-conditions

$$\vec{V} \times \vec{\gamma} = \vec{n} \cdot (\vec{V} \cdot \text{grad}\ \mu) - \text{grad}\ \mu\ (\vec{V} \cdot \vec{n}) = 0$$

while n x grad μ describes the vortex $\vec{\gamma}$.
Proper meeting of the equation demands both parts to become zero independently, because they are not equally directed. The first term corresponds to the condition of no-load, the second one to the cinematic flow-condition. Right behind the wing the assumption of a geometrically "frozen" wake (30) seems to be justified and by meeting the first subcondition $\Delta C_p = 0$ normal to the wake the Kutta-condition can be fulfilled appropriately. Due to the fixed geometry the wing's influence coefficients remain unchanged and only the coefficients of the nonlinear ΔC_p-equations cause iterative computational effort. Unfortunately the convergence of the procedure (34) is configuration-dependent, demanding further work for a satisfying method. By the way, for consideration of a far-field-wake - e.g. wing-tail-interference - the wake-position playes an important role, thus demanding both sub-conditions of the wake-model.

While the significant development of future vector-processors will favour time-dependent Euler-codes over panel-methods, great advances have been achieved with the numerical processing of Euler-equations in recent years. As to (36) these methods can predict separation-phenomena despite neglecting viscosity, if supersonic speeds occure locally somewhere at the considered configuration. The shock-induced losses of total pressure and the vorticity production allow the formation of vortex-sheets flowing off the wing's leading-edge during the unsteady computational cycles, - the converged solution possibly not showing any supersonic regimes left. Figure 15 depicts the result of a very first investigation of an employed code (37). Not being restricted to any speed regimes - as for instance panel-methods or vortex-lattice-procedures - Euler widens the range from subsonic through transonic to supersonic speeds without any modifications. Apart from this no fundamental limitation is posed on thinkable future configurations while panel-methods etc. - normally restricted to one pair of vortices - demand significant manipulations and modifications to check such geometries, because the structure of their vortex-systems is not describable in easy terms (38).

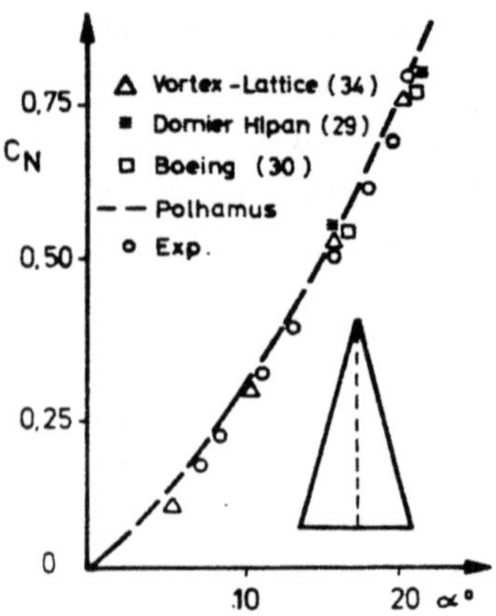

Fig. 14

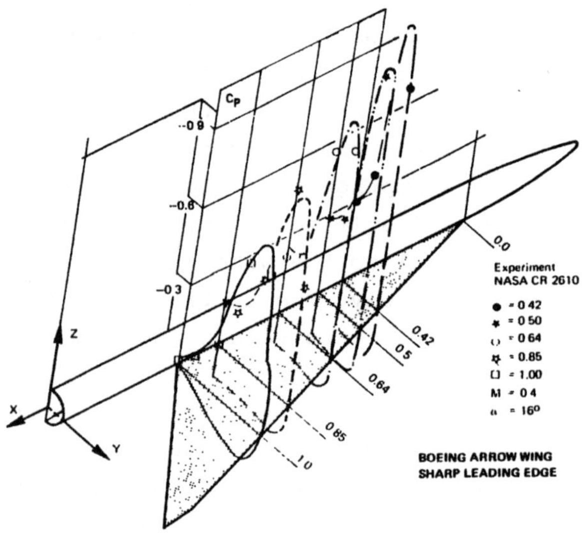

Fig. 15 Euler results

However, the future primarily will charge Euler-codes with the investigation of more complex configurations, the panel- and vortex-lattice-methods will not cede their importance as pilot-codes for further investigations, as for instance the vortex-breakdown-phenomenon. As already mentioned and by (18) the pressure gradient, which is necessarily positive to meet the Kutta-condition at the trailing-edge is probably decisive in triggering breakdown, meaning the causing mechanism is nonviscous in principle. Therefore 3-dimensional nonviscous methods might exhibit such effects. The analysis of some results of a panel-method (29) for a delta-wing exposed to an angle-of-attack regime up to 70 degrees show strong disturbances in the progress of both lift and moment near an angle-of-attack known for vortex-breakdown (fig. 16). Close by, the sum of the residuals, which are obtained from the amount of violation of the cinematic flow condition on the wing and on the free sheet and the no-load condition for the free sheet, tremendously increases. Further investigations revealed the increase of the residuals, mainly to be caused by the cinematic condi-

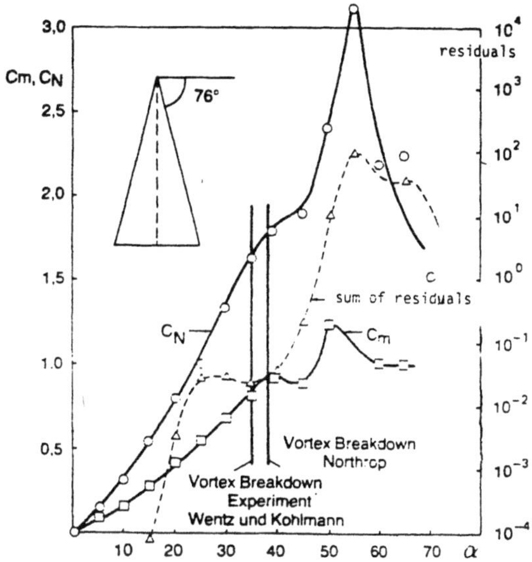

Fig. 16 C_N and C_m versus α by a panel-method (29)

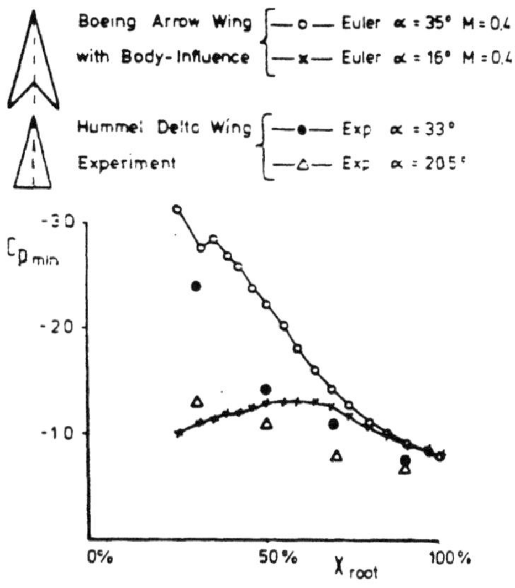

Fig. 17 Minimum C_p underneath the vortex-path by an Euler-code (37)

tion of the vortex-system, while the wing always met its requirements. At very high incidences the vortex-shape and -position even showed severe distortions. A procedure based on the integral physical effects will deliver similar results near the regime of breakdown. The described vortex-lattice-method (34) will provide a model very much suited for corresponding stability-investigations on the vortex-system. It will help to find and verify breakdown and may be a source for simple criterion or a procedure to determine the beginning of vortex-breakdown above the wing. Figure 17 depicts a first trial to fathom the breakdown phenomenon in an Euler-procedure result. Unfortunately no compatible measurements are performed till now, but the comparison of the computational results (34) and Hummel's (15) investigations show similar tendencies. Hummel states vortex-breakdown for his delta wing at $\alpha = 33$ degrees.

Thanks to new computer-generations Euler-codes will probably become the most powerfull tool for detailed investigations on more complex configurations, while more traditional panel- and vortex-lattice-procedures for slender wings with leading-edge-separation will be employed in more simple missile- and preliminary-design efforts in more easy terms. Their principal meaning will be in further detailed research of the physical behaviour of vortex-structures. The mechanism of vortex-breakdown may be revealed by relatively simple stability-investigations.

REFERENCES

1 Schairer, Evolution of Modern Air Transport Wings, AIAA 80-3037

2 Holder, Convair F-106, Aero Series Aero Pub. 1977

3 Robinson, The B-58 Hustler, Aero Pub. Com. Inc. 1967

4 (Mac) Blair, Evolution of the F-86, AIAA 80-3039

5 Poisson-Quinton, Slender Wings for Civil and Military Aircraft, Jour. of Tech. Vol 16, 1978

6 Küchemann, The Aerodynamic Design of Aircraft, Pergamon Press 1978

7 Legendre, Ecoulement au voisinage de la pointe avant d'une aile â forte flèche aux incidences moyennes, Rech. Aero. 1952

8 Jones, Properties of L.A.R. Pointed Wings at Speeds Below and Above the Speed of Sound, NACA-Rept. No. 835 1946

9 Adams, Leading Edge Separation from Delta Wings at Supersonic Speeds, J. Aeron. Si. 20, 1953

10 Edwards, Leading-Edge Separation from Delta Wings, J. Aeron. Si 21 1954

11 Brown & Michael, Effect of Leading-Edge Separation on the Lift of a Delta Wing, NACA TN 3430 1955

12 Rech & Leyman, A Case Study by Aerospatiale and British Aerospace on the Concorde, AIAA Prof. Study

13 Werlé, Flow Visualisation Techniques for the Study of High Incidence Aerodynamics, AGARD Lec. No 121 1982

14 Hummel, Zur Umströmung scharfkantiger schlanker Deltaflügel bei großen Anstellwinkeln, Z. Flugw. 1967

15 Hummel, Untersuchungen über das Aufplatzen der Wirbel an schlanken Deltaflügeln, Z. Flugw. 1965

16 Rolls, Koenig & Drinkwater, Flight Investigation of the Aerodynamic Properties of an Ogee Wing, NASA TN D-3071 1965

17 Hummel, Private Communication

18 Stephan M. Hitzel, Vortex Breakdown of Leading-Edge-Vortices on Slender Sharp-Edged Wings. Some Ideas on the Physical Explanations and Theories, Dornier Aktenvermerk BF 30-1926/81 1981

19 Ludwieg, Vortex Breakdown, DFVLR 70-40

20 Rayleigh, On the Stability or Instability, of Certain Fluid Motions Scientific Papers Proc. Roy. Soc. London 1880

21 Taylor, Stability of a Viscous Liquid contained between Two Rotating Cylinders, Phil. Trans. Roy. Soc. London 1923

22 Benjamin, Theory of the Vortex Breakdown Phenomenon, J. Fluid Mechanics 14. 1962

23 Hall, A new Approach to Vortex Breakdown, Stanford University Press. 1967

24 Polhamus, A Concept of the Vortex Lift of Sharp-Edged Delta Wings, Based on a Leading-Edge Suction Analogy. NASA TN D-3767, 1966

25 Smith, Improved Calculations of Leading-Edge Separation from Slender Delta Wings. RAE Technical Report, 1966

26 Hummel, Redeker, Experimentelle Bestimmung der gebundenen Wirbellinien sowie des Strömungsverlaufs in der Umgebung der Hinterkante eines schlanken Deltaflügels, Sonderdruck 1971

27 Lucchi, Ein Panelverfahren höherer Ordnung für kompressible Unterschallströmungen. Dornier FB 78/13B

28 Johnson, Ehlers, and Rubbert, "A Higher Order Panel Method for General Analysis and Design Applications in Subsonic Flows", Proceedings of the Fifth International Conference on Numerical Methods in Fluid Dynamics, Enschede, 1976

29 Hitzel, Wagner, Berechnung der 3-dimensionalen Strömung am Flügel mit Vorderkantenablösung Teil 1 Panelverfahren höherer Ordnung, Dornier-Bericht 80/18 B

30 Johnson, Lu, Tinoco, Epton, An Improved Panel Method for the Solution of Three-Dimensional Leading-Edge Vortex Flows, NASA-CR-3278, July 1980

31 Wagner, Vereinfachte Wirbelmodelle für schlanke Tragflächen mit Vorderkantenablösung, ZFW Band 4, 1980

32 Lämmlein, Untersuchung der Vorderkantenablösung an schlanken Flügeln für den Fall konischer Strömung mit Hilfe des Strömungsmodells nach Smith. Dornier AKV BF 30-2205/82

33 Lucchi, Dornier Vortex-Lattice-Program VORLAT 1977

34 Hitzel, Berechnung der dreidimensionalen Strömung um Flügel mit Vorderkantenablösung II, Vereinfachte Verfahren, Euler-Code, Wirbelplatzen. Dornier Bericht 82 BF/63

35 Hitzel, Application of the Dornier Vortex-Lattice Method for the Evaluation of the Vortex-Lift-Coefficients according to the Polhamus Suction Analogy, Dornier AK BF 30-1921/81

36 Schmidt, and Jameson, Euler Solutions as Limit of Infinite Re-Number for Separated Flows and Flows with Vortices, Paper given at 8th Inter. Conf. on Num. Meth. in Fl. Dyn., Aachen 1982

37 Euler-code Dornier-FLO57, 1982

38 Brennenstuhl, Hummel, Untersuchungen über die Wirbelbildung an Flügeln mit geknickten Vorderkanten, ZFW Band 5, 1981

Turbulent Heat Transfer

M. Jischa and H.B. Rieke

Institut für Technische Mechanik der
Technischen Universität Clausthal
and
Fachbereich Maschinentechnik der
Universität Essen-GH

Summary

Heat transfer problems in turbulent flows are mainly described by empirical relationships Nu = Nu(Re, Pr). The progress of modeling the turbulent heat transfer is quite poor. For the turbulent momentum transfer on the other hand we have a lot of semiempirical modeling assumptions which lead to good results for engineering purposes. After an introduction into the closure assumptions for the turbulent momentum transfer the turbulent heat transfer, in particular the turbulent Prandtl number are discussed. Finally a model to predict the turbulent Prandtl number is proposed. It can be used to calculate heat transfer correlations Nu(Re, Pr).

Introduction

In incompressible two-dimensional turbulent boundary layer type flow with constant fluid properties the governing equations which have to be solved are:

The continuity equation

$$\frac{\partial \bar{u}}{\partial x} + \frac{\partial \bar{v}}{\partial y} = 0, \qquad (1)$$

the momentum equation

$$\rho\left(\bar{u}\frac{\partial \bar{u}}{\partial x} + \bar{v}\frac{\partial \bar{u}}{\partial y}\right) = -\frac{dp_\delta}{dx} + \frac{\partial}{\partial y}\left(\mu\frac{\partial \bar{u}}{\partial y} - \rho\,\overline{u'v'}\right) \qquad (2)$$

as well as the energy equation

$$\rho\, c_p\left(\bar{u}\frac{\partial \bar{T}}{\partial x} + \bar{v}\frac{\partial \bar{T}}{\partial y}\right) = -\frac{\partial}{\partial y}\left(-\lambda\frac{\partial \bar{T}}{\partial y} + \rho\, c_p\,\overline{T'v'}\right). \qquad (3)$$

The prime (...') denotes fluctuating and the bar $(\overline{...})$ denotes time mean quantities. u and v are the x- and y-components of the velocity vector, T is the temperature, ρ is the density, p_δ is the pressure at the outer edge of the boundary layer and μ and λ are the viscosity and thermal conductivity respectively.

The resulting fluxes on the right hand side of the eqs. (2) and (3), the shear stress

$$\tau = \mu \frac{\partial \bar{u}}{\partial y} - \rho \overline{u'v'} \tag{4}$$

and the heat flux

$$q = - \lambda \frac{\partial \bar{T}}{\partial y} + \rho c_p \overline{T'v'} = \lambda (- \frac{\partial \bar{T}}{\partial y} + \frac{Pr}{\nu} \overline{T'v'}) \tag{5}$$

consist of two parts, a molecular and a turbulent one. Except for regions very near walls, the turbulent or Reynolds fluxes are much larger than the molecular parts. Only for very low molecular Prandtl numbers $Pr = \mu c_p/\lambda << 1$ (liquid metals) the molecular part of the heat flux becomes more and more important and cannot be neglected anymore.

Following Boussinesq's approach a turbulent kinematic viscosity (or eddy diffusivity for momentum) ε_τ as well as a turbulent thermal diffusivity ε_q are defined by

$$\tau_t = - \rho \overline{u'v'} = \rho \varepsilon_\tau \frac{\partial \bar{u}}{\partial y} ,$$

$$q_t = \rho c_p \overline{T'v'} = - \rho c_p \varepsilon_q \frac{\partial \bar{T}}{\partial y} , \tag{6}$$

that means

$$\varepsilon_\tau = - \frac{\overline{u'v'}}{\partial \bar{u}/\partial y} \quad \text{and} \quad \varepsilon_q = - \frac{\overline{T'v'}}{\partial \bar{T}/\partial y} . \tag{7}$$

In analogy to the molecular Prandtl number $Pr = \mu c_p/\lambda$ a turbulent Prandtl number Pr_t is introduced by

$$Pr_t = \frac{\varepsilon_\tau}{\varepsilon_q} = \frac{\overline{u'v'}}{\overline{T'v'}} \frac{\partial \bar{T}/\partial y}{\partial \bar{u}/\partial y} . \qquad (8)$$

Now the question is how to get information about the turbulent diffusivities ε_τ as well as ε_q or Pr_t respectively.

The turbulent momentum transfer

All modeling proposals can be roughly devided into
- first order and
- second order closure assumptions.

The most famous and successful first order closure assumption is Prandtl's mixing length concept [1]:

$$v' \sim u' = \pm \ell \frac{\partial \bar{u}}{\partial y} \wedge - \overline{u'v'} = \ell^2 \left|\frac{\partial \bar{u}}{\partial y}\right| \frac{\partial \bar{u}}{\partial y} . \qquad (9)$$

Here ℓ is the mixing length. The main disadvantage of eq. (9) and other first order methods is the connection between the turbulent shear stress and the velocity profile which is only approximately valid for turbulent flows near equilibrium. Besides more or less complicated empirical approximations have to be made for the mixing length ℓ (or the eddy viscosity ε_τ) dependent on the distance normal to the wall.

The second order closure assumptions start with balance equations, the transport equations, for the unknown Reynolds fluxes. Here the original work of Prandtl from 1945 is discussed briefly [2], who presented the basic ideas together with Kolmogorov in 1942.

The starting-point is the transport equation for the turbulence energy

$$k = \frac{1}{2} (\overline{u'^2} + \overline{v'^2} + \overline{w'^2}) , \qquad (10)$$

which reads

$$\bar{u} \frac{\partial k}{\partial x} + \bar{v} \frac{\partial k}{\partial y} = - \overline{u'v'} \frac{\partial \bar{u}}{\partial y} - \varepsilon - \frac{\partial}{\partial y} [\overline{(k+\frac{p'}{\rho}) v'} \ldots] . \qquad (11)$$

The three terms on the right hand side mean production, dissipation and diffusion of turbulence energy. In detail the turbulent dissipation is given by

$$\varepsilon = \nu \overline{\frac{\partial v_j'}{\partial x_k} \left(\frac{\partial v_j'}{\partial x_k} + \frac{\partial v_k'}{\partial x_j}\right)} \ . \tag{12}$$

With Prandtl's modeling assumptions eq. (11) passes into

$$\bar{u} \frac{\partial k}{\partial x} + \bar{v} \frac{\partial k}{\partial x} = C_\tau \sqrt{k}\, L \left(\frac{\partial \bar{u}}{\partial y}\right)^2 - C_D \frac{k^{3/2}}{L} + C \frac{\partial}{\partial y} \left(\sqrt{k}\, L \frac{\partial k}{\partial y}\right) \tag{13}$$

together with

$$\frac{\tau_t}{\rho} = - \overline{u'v'} = C_\tau \sqrt{k}\, L \frac{\partial \bar{u}}{\partial y} \ , \tag{14}$$

which relates the Reynolds shear stress to the turbulence energy. C_τ, C_D and C are modeling constants and $L = L(y)$ is a turbulent length scale. They have to be adjusted to experiments.

Other and partly more complicated models have been presented after Kolmogorov's and Prandtl's work, most of them use more than one transport equation.

The turbulent Prandtl number

About 100 years ago Reynolds assumed that the same mechanism governs the turbulent flux of momentum and of heat. This leads to the assumption $\varepsilon_\tau = \varepsilon_q$, that means $Pr_t = 1$. Transferring Prandtl's mixing length concept (9) to the Reynolds heat flux it follows

$$T' = \pm \ell \frac{\partial \bar{T}}{\partial y} \wedge - \overline{T'v'} = \ell^2 \left|\frac{\partial \bar{u}}{\partial y}\right| \frac{\partial \bar{T}}{\partial y} \tag{15}$$

together with eq. (9). Introducing into eq. (8) this leads to $Pr_t = 1$ too. But why should the mixing length $\ell = \ell_m$ for momentum be equal to the mixing length $\ell = \ell_h$ for heat? This will be discussed later.

Experiments have shown that the value $Pr_t = 1$ is not too far from the experimental one which is about 0.9 for boundary layer, pipe and channel flow (in free shear layers Pr_t is about 0.5) of air, other gases, water, and other liquids like oil; that means for fluids with molecular Prandtl numbers $Pr \gtrsim 1$. On the other hand experiments have shown $Pr_t > 1$ for liquid metals, that means for fluids with $Pr << 1$.

Generally speaking experiments in boundary layer and pipe flow indicate

$$Pr_t = f(Pr, Re, \frac{y}{\delta} \text{ or } \frac{y}{R}) . \qquad (16)$$

The influence of the wall distance (y/δ in boundary layer and y/R in pipe flows) seems not to be great. Beyond it it is doubtful whether Pr_t increases or decreases with increasing wall distance, although most experiments show a slightly higher value near the wall compared with the outer edge of the boundary layer.

The influence of the Prandtl and Reynolds number can be summarized in the following way:

- For $Pr \gtrsim 1$ (air, water, and other liquids) Pr_t seems to be constant ($Pr_t \approx 0.9$).
- The influence of the Prandtl and Reynolds number increases with decreasing Pr and Re.
- Pr_t increases with decreasing Prandtl and Reynolds number.

Among others this has been discussed in detail by Rieke and Jischa [3, 4, 5]. In order to understand this behaviour we write the turbulent Prandtl number (8) as

$$Pr_t = \frac{\ell_m}{\ell_h} . \qquad (17)$$

For $Pr << 1$ a turbulent eddy has a very high thermal conductivity compared with its viscosity. So, the eddy exchanges much quicker heat than momentum with its surroundings and ℓ_m/ℓ_h decreases with increasing Prandtl number. But this dependence is influenced by the Reynolds number, too. The intensity of the velocity fluctuations increases with increasing Reynolds number, so the growing turbulent mixing hinders the influence of the rapid molecular heat conduction. For Pr $\gtrsim$ 1 on the other hand the molecular heat conduction of a turbulent eddy becomes negligible. The turbulent exchange of heat and momentum is then only governed by the turbulent fluctuations which influence both transport processes in a similar way, and Pr_t becomes independent of the Prandtl and Reynolds number.

Now the question is how to formulate a relationship for Pr_t quantitatively. Reynolds [6], Launder [7] and Sideman and Pinczewski [8] have reported on different closure assumptions which can be roughly divided into the following groups:
- Modified mixing length concepts;
- Formal analysis based on exact equations for isotropic turbulence;
- Diffusivity models;
- Wall-layer models;
- Empirical formulae.

All of them agree more or less qualitatively with known experiments. This has been discussed by Rieke and Jischa [3, 4, 5], too, who started to develop a second order closure assumption for the turbulent heat transfer. In the following, a summary of these ideas is given.

Modeling assumptions for the turbulent heat flux

Following Prandtl's ideas we start with the transport equation for the Reynolds heat flux $\overline{T'v'}$. Here it is convenient to write this equation in the following form [4, 5]:

$$\underbrace{\bar{u}\frac{\partial}{\partial x}(\overline{T'v'}) + \bar{v}\frac{\partial}{\partial y}(\overline{T'v'})}_{\text{Convection}} = \underbrace{-\overline{v'^2}\frac{\partial \bar{T}}{\partial y}}_{\text{Production}}$$

$$\underbrace{+\frac{1}{\rho c_p}\overline{v'\Phi'} - \nu\frac{Pr+1}{Pr}\overline{\frac{\partial v'}{\partial x_j}\frac{\partial T'}{\partial x_j}}}_{\text{Dissipative terms}}$$

$$\underbrace{+\overline{\frac{p'}{\rho}\frac{\partial T'}{\partial y}} + \ldots}_{\text{Redistribution}} \underbrace{-\frac{\partial}{\partial y}\left[\overline{T'v'^2} - \ldots\right]}_{\text{Diffusion}} = 0 \,. \qquad (18)$$

The essential terms for the following considerations are the dissipative terms. The first one is due to the fluctuating part of the dissipation $\Phi = \tau_{jk}\ \partial v_j/\partial x_k$ in the energy equation:

$$\Phi' = \mu\Big[\frac{\partial v_j'}{\partial x_k}\Big(\frac{\partial v_j'}{\partial x_k} + \frac{\partial v_k'}{\partial x_j}\Big) + \frac{\partial v_j'}{\partial x_k}\Big(\frac{\partial \bar{v}_j}{\partial x_k} + \frac{\partial \bar{v}_k}{\partial x_j}\Big) + \frac{\partial \bar{v}_j}{\partial x_k}\Big(\frac{\partial v_j'}{\partial x_k} + \frac{\partial v_k'}{\partial x_j}\Big)\Big] . \tag{19}$$

Whereas the molecular part of the dissipation is negligible in incompressible flows the turbulent or indirect dissipation ε, eq. (12), which is the time mean value of eq. (19), multiplied with the density, plays a dominant role in turbulent flows and cannot be neglected anymore. The second dissipative term is due to the fluctuating part of the viscous term in the momentum equation and the heat transfer term in the energy equation.

The redistribution terms are due to correlations of pressure fluctuations with other fluctuating quantities, they are like the diffusion terms not given here completely; for details see [5].

It is known from experiments that the convective and the diffusive terms are normally much smaller than the remaining terms. This is approximately valid for boundary layer flows near equilibrium, as well as for fully developed pipe and channel flow (where the convective terms are exactly zero).

For the remaining terms we suggest the modeling assumptions:

$$\overline{v'^2}\ \frac{\partial \bar{T}}{\partial y} = C_P\ k\ \frac{\partial \bar{T}}{\partial y} \tag{20}$$

$$-\frac{1}{\rho c_p}\ \overline{v'\Phi'} - \overline{\frac{p'}{\rho}\ \frac{\partial T'}{\partial y}} + \ldots = C_U\ \frac{\sqrt{k}}{L}\ \overline{T'v'} \tag{21}$$

$$\frac{\nu}{Pr}\ \overline{\frac{\partial v'}{\partial x_j}\ \frac{\partial T'}{\partial x_j}} = C_D\ \frac{\nu}{Pr}\ \frac{\overline{T'v'}}{L^2} . \tag{22}$$

Eq. (20) makes the assumption $\overline{v'^2} \sim k$. The first dissipative term is modeled together with the redistribution terms similar to Prandtl's modeling $\varepsilon \sim k^{3/2}/L$, see eq. (13). The dissipation of the Reynolds heat flux $\overline{T'v'}$ must be proportional to $\overline{T'v'}$ itself, so we put $\sqrt{k}\,\overline{T'v'}$ instead of $k^{3/2}$. The second dissipative term consists of a ν-proportional and a $\nu/Pr = \lambda/(\rho c_p)$-proportional part. According to models which use the transport equation for the Reynolds shear stress $\overline{u'v'}$ the ν-proportional part is negligible compared with the term (21). The ν/Pr-proportional part cannot be neglected specially for liquid metals because of $Pr \ll 1$. The expression $\overline{T'v'}/L^2$ in eq. (22) is taken from dimensional reasons. Finally we introduce a turbulent Reynolds number

$$Re_t = \frac{\sqrt{k}\,L}{\nu} = \frac{\varepsilon_\tau}{C_\tau \nu} . \tag{23}$$

Using this and introducing the modeling assumptions into the transport equation (18) we get

$$C_P\, k\, \frac{\partial \bar{T}}{\partial y} + \left(C_D \frac{1}{Pr\, Re_t} + C_U\right) \frac{\sqrt{k}}{L}\, \overline{T'v'} = 0 , \tag{24}$$

where the convective and diffusive terms have been neglected. It follows

$$-\overline{T'v'} = \frac{C_P}{C_D \dfrac{1}{Pr\, Re_t} + C_U} \sqrt{k}\, L\, \frac{\partial \bar{T}}{\partial y} = \sqrt{k}\, L_h\, \frac{\partial \bar{T}}{\partial y} \tag{25}$$

in analogy to eq. (14):

$$-\overline{u'v'} = C_\tau \sqrt{k}\, L\, \frac{\partial \bar{u}}{\partial y} \quad \text{with} \quad L = L_m . \tag{26}$$

So we have

$$L_h = \frac{C_P}{C_D \dfrac{1}{Pr\, Re_t} + C_U} L_m \quad \text{with} \quad L = L_m .$$

This is the answer to the question formulated at the beginning, how the Prandtl and Reynolds number influence the ration ℓ_h/ℓ_m or L_h/L_m. Introducing eq. (25) together with eq. (26) into the definition (8) of Pr_t we find

$$Pr_t = \frac{C_\tau}{C_P} \left(C_D \frac{1}{Pr\ Re_t} + C_U \right) . \tag{27}$$

Rieke has shown [4], that specially for pipe flows the approximation $Re_t = a\ Re^m$, where $Re = u_m\ d/\nu$ with d as a pipe diameter, can be made. This approximation neglects the influence of the wall distance on ε_τ/ν. It follows

$$Pr_t = K_1 + K_2 \frac{1}{Pr\ Re^m} \tag{28}$$

where $K_1 = \frac{C_\tau\ C_U}{C_P}$; $K_2 = \frac{C_\tau\ C_D}{aC_P}$.

The exponent m = 0.888 results from the approximation $Re_t(Re)$. The two constants K_1 and K_2 have to be fitted on experimental data. In the literature we find measurements of

- the turbulent Prandtl number $Pr_t(Pr, Re)$;
- $\overline{T'v'}$-, ε_q- and $\bar{T}$-profiles;
- the Nusselt number Nu(Re, Pr).

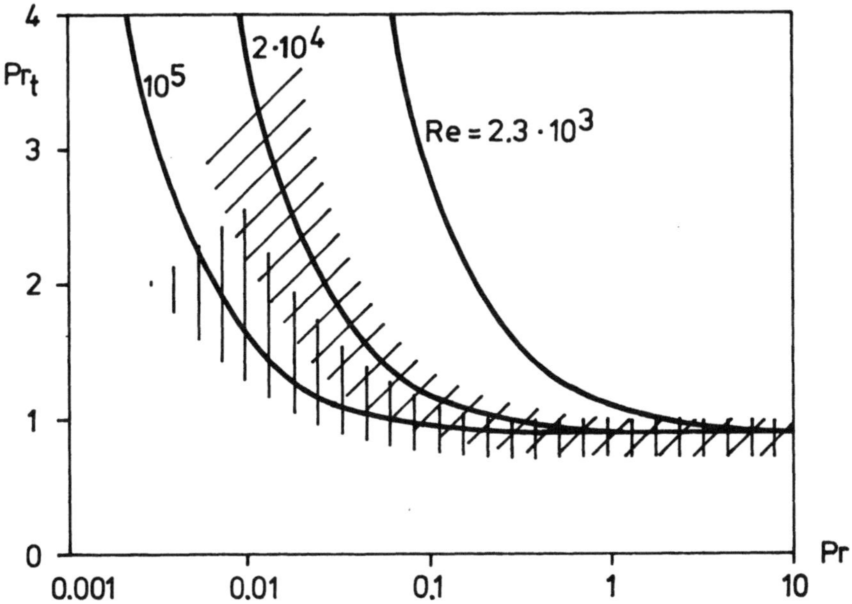

Fig. 1 Turbulent Prandtl number, eq. (28)

Rieke [4] has chosen K_1 = 0.9 and K_2 = 182.4 for a fully developed turbulent pipe flow, where the set of equations to be solved, eqs. (1), (2), (3), (13), (14) and (28), can be reduced to three ordinary differential equations plus equation (28) for the unknown profiles $\bar{u}$, k = f(r/R, Re) and $\bar{T}$, $\overline{T'v'}$ = f(r/R, Re, Pr). For details and numerical results see Rieke [4] or ref. [5], where these profiles as well as the turbulent Prandtl and the Nusselt number are shown compared with experiments. As an example Fig. 1 shows the turbulent Prandtl number, eq. (28), compared with experiments indicated by the hatched lines.

References

1. Prandtl, L.; Gesammelte Abhandlungen; Eds. W. Tollmien, H. Schlichting, H. Görtler, Band 2. Berlin: Springer-Verlag 1961, 736-751.

2. Prandtl, L.; Gesammelte Abhandlungen; Eds. W. Tollmien, H. Schlichting, H. Görtler, Band 2. Berlin: Springer-Verlag 1961, 874-888.

3. Jischa, M.; Rieke, H.B.; About the Prediction of Turbulent Prandtl and Schmidt Numbers from Modeled Transport Equations. Int. J. Heat Mass Transfer 22 (1979) 1547-1555.

4. Rieke, H.B.; Bestimmung des Wärmeübergangs bei turbulenter Rohrströmung mit Hilfe von Transportgleichungen. Thesis Univ. Essen-GH 1981.

5. Jischa, M.; Konvektiver Impuls-, Wärme- und Stoffaustausch. Braunschweig/Wiesbaden: Vieweg 1982.

6. Reynolds, A.J.; The Prediction of Turbulent Prandtl and Schmidt Numbers. Int. J. Heat Mass Transfer 18 (1975) 1055-1069.

7. Launder, B.E.; Heat and Mass Transport, in: Turbulence, Ed. P. Bradshaw. Berlin: Springer-Verlag 1976.

8. Sideman, S.; Pinczewski, W.; Turbulent Heat and Mass Transfer, in: Topics in Transport Phenomena; Ed. Ch. Gutfinger, Washington: Hemisphere Publ. Corp. 1975.

Optical Flow-Velocity Measurement in Irregularly Shaped Cavities

J. LARSEN, and K.G. ROESNER

Math-Tech Institute, Sommervej 7, Charlottenlund, Denmark, and
Institut für Mechanik der Technischen Hochschule Darmstadt, West Germany

Summary

A nearly disturbance-free technique for the measurement of the velocity field of viscous liquids in closed and irregularly shaped cavities is described. This method is based on the photochromic effect. Up to now photochromism was used in experimental fluid dynamics to produce dye lines in a fluid with a photoactive additive of very low concentration. The present method uses a totally saturated solution of indolinespiropyran in a mixture of dekahydronaphthalin and tetrahydronaphthalin. By means of a glass-fiber for the transmission of UV-light of 356 nm it is possible to color small fluid volumes inside the fluid at rest. The motion of the colored spots can be photographed by a movie camera. Thus path-lines of individual particles in the liquid can be determined when the whole body of the liquid is set into motion.

Introduction

Photochromism as a tool for the measurement of velocity fields in viscous liquids is known as flash photolysis. Popovich and Hummel [1] have been the first who applied this technique to analyse the viscous sublayer in turbulent pipe flow. Later Frantisak, Palade de Iribarne, Smith, and Hummel [2] have shown how to use this method for quantitative measurements in turbulent flow. An application to flow properties of a laminar pipe jet was given by Iribarne, Frantisak, Hummel, and Smith [3] for moderate Reynolds-numbers. All these techniques are based on the idea to produce a colored line in the liquid by a laser beam which hits the transparent wall perpendicularly. Using this method a very low concentration of the photochromic compound is necessary to ensure that the UV-light is absorbed on a long distance inside of the liquid. By this method time-lines can be produced and the long back-reaction time of the colored photoproduct allows to take pictures of the distorted colored lines to measure the velocity parallel to the wall. Thus the velocity field near the wall can be determined quantitatively when the path-lines of the individual fluid particles are known. Only for laminar flows along flat walls the path-lines are known a priori. If the flow is analysed along curved walls time lines give only a qualitative picture of the fluid flow. To know the motion of individual small fluid volumes of a viscous liquid another technique must be applied. Studying e.g. the spin-up flow of a liquid in a closed cavity of arbitrary shape it would be of interest to color the liquid pointwise at any position in the fluid volume.

In the following we will describe such a method which is applicable to any finite fluid volume, especially when the volume is very small and of geometrically irregular shape. Such situation is met in biomechanical problems when the flow field e.g. in the human eye is to be analysed. In this case it is impossible to introduce any probe or wire to measure or visualize the velocity field by any of the well-known methods with high accuracy. Therefore using a highly saturated solution of photochromic compounds in organic solvents gives an ideal tool to color individual fluid particles. If UV-light is transmitted into such a liquid the light is absorbed along a very short distance which leads to a small colored spot. This effect makes it possible to create small colored fluid spots by introducing the UV-light into the fluid by means of a glass-fiber which is positioned at the place where an individual fluid particle should be marked. After having prepared the liquid by a pattern of points the total volume can perform any motion and the pathlines of the spots can be registered by a camera. In Fig. 1 the process of coloring small spots in a liquid is shown.

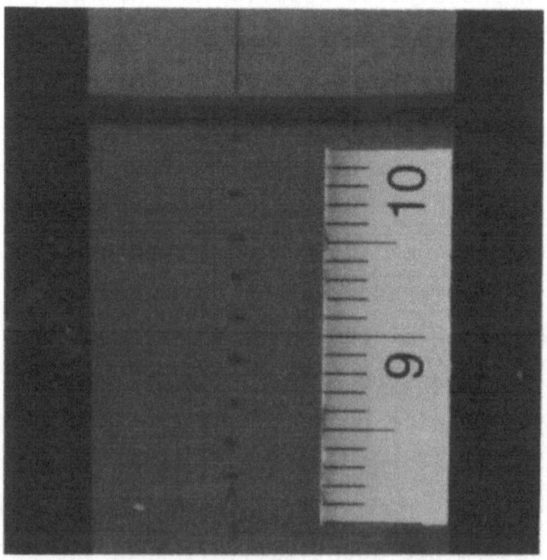

Fig. 1. Small colored spots in a liquid

The chemical compound and the light source

In the experiments which will be described later the well-known photochromic compound 1,3,3-trimethyl-6'-nitroindoline-2-spiro-2'-benzopyran was used. As the walls of the cavity are performed by excavating a plexiglass-cube similarly to the shape of a human eye a mixture of dekahydronaphthalin and tetrahydronaphthalin must be used in order to adjust the index of refraction of the wall material to that of the liquid. In Fig. 2 half of the volume is filled with the solvent. It can be seen that there exists no distortion of

Fig. 2. Partly filled spherical cavity

the line-pattern in the lower part. Therefore if an optical signal from inside of the fluid volume is registered by a camera no influence of refraction of light across the curved walls can lead to a misinterpretation of the position where the signal is coming from. When this mixture of the organic solvents is saturated by the chemical compound indolinespiropyran the liquid becomes yellow, but remains still transparent for visible light. The filled volume of the human eye model is shown in Fig. 3. Photographs are taken by illuminating the plexiglass-cube in front of a yellow screen to have a good contrast for the colored spots during the motion of the model.

Measuring the absorption behavior of the chemical compound one finds a relative maximum around 356 nm for the colorless substance. The light source has

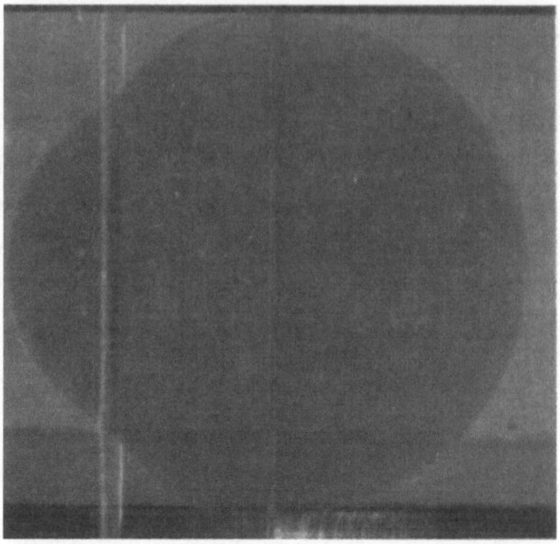

Fig. 3. Model of a human eye

to be chosen such that a maximum of the emitted radiation is found in the region of 356 nm. A mercury lamp e.g. has a spectrum of the desired character. In the experiments concerning the liquid motion in an eye model an UV-laser was used. The laser beam of about 2 Watt power was focused on a glass-fiber of 100 µm diameter. The output cross section of the glass-fiber was mounted on a motor drive which could move up and down to position the end of the fiber in a prescribed height above the bottom of the cavity. In Fig. 1 the glass-fiber was pulled slowly out of the liquid while a number of laser flashes produced the small red spots. When the whole liquid came to rest again the liquid could be set into motion according to a prescribed law. The time for the back reaction of the colored substance was long enough to follow the path of the volume elements for a while. Temperature influences significantly the time until the color disappears completely. Cooling down the photochromic solution the back reaction takes a time of approximately 4 minutes. The fact that the photoreaction is completely reversible makes this method very useful for optical registration. Having filled the model once hundreds of times the reaction can be induced by laser flashes before the molecules of the photochromic compound are destroyed. The color of the spots can be changed by solving the substance in other organic solvents, like kerosene, toluene, ether, cyclohexan and other nonpolar liquids. Choosing the solvent one has to take care of the possible reaction with the wall material.

The registration of path-lines

As an example for the motion of colored spots in a small fluid volume the convective flow motion in a rectangular box of 2x3x4 cm^3 with a free surface on top is shown in Fig. 4 and Fig. 5. The glass-fiber was moved upwards with a velocity of 2 cm/s while seven UV-laser flashes were shot into the liquid to mark the fluid at seven positions by line-shaped volume elements. The result of this marking process is shown in Fig. 4. After a time of about 5 seconds the positions of the individual seven fluid particles were registered again and they were found at the places which are given by Fig. 5. If one measures the displacement of each of the volume elements mean values of the flow velocity at different points can be calculated. Thus for the lowest points one gets a velocity of about 0.4 mm/s and for the highest point a velocity of about 1.25 mm/s. The upper part of the liquid is moving to the right the lower part to the left side of the box. In the same way any type of motion of a liquid-filled cavity can be analysed by using high speed cameras for a better resolution of the flow field.

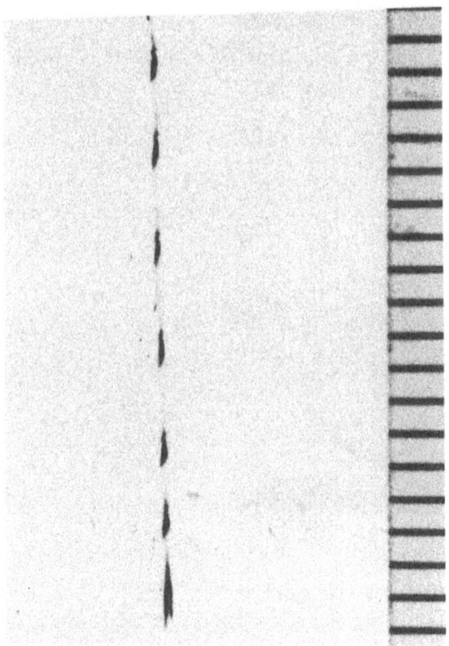

Fig. 4. Line-shaped markers

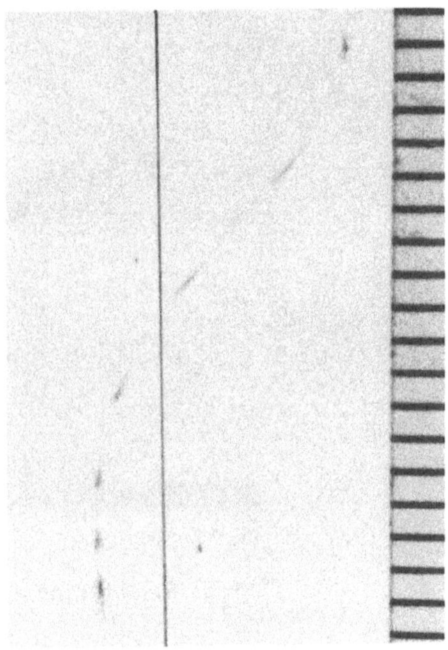

Fig. 5. Displacement after 5 s

Motion in the model of a human eye

According to Larsen's model [4] of a human eye the anterior chamber of a human eye is given by a spherical part which is separated from the vitreous body by a thin membrane of about 10 µm thickness. The whole volume of liquid is shown in Fig. 3. The plexiglass cube from which the cavity is excavated is put on a motor-drive which can perform an impulsive rotation around a vertical axis. The whole body is accelerated and brought to rest within 50 ms after having performed an angular displacement of nearly 50 degrees. As one can see from Fig. 6 a line of arrow-shaped markers is created in the liquid. The rotational motion is registered by a camera taking 4 pictures within one second. The maximal deviation of the arrow-line from the original position gives a measure for the maximal velocity which is induced by the aperiodic motion of the model of the eye-ball.

Fig. 6 gives a front view of the eye model. The inner circle shows the area of the membrane dividing the anterior chamber from the vitreous body.

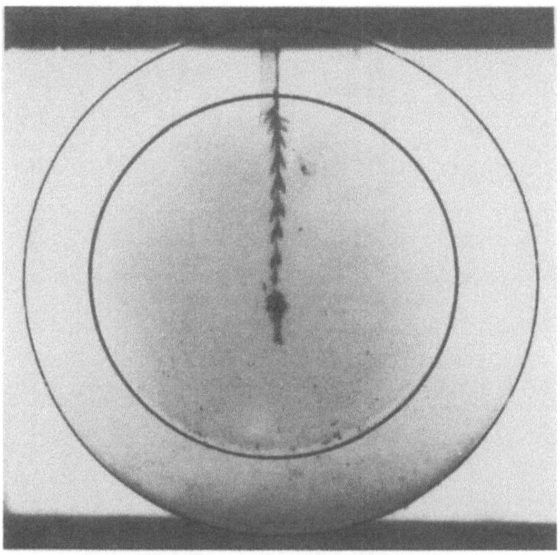

Fig. 6. Anterior eye-chamber with markers

In the liquid used velocities of maximal 2 mm/s were measured near the membrane were the motion is almost parallel to the plane membrane. The described method can be used easily also in the case when an artificial lens is implanted into the anterior eye chamber.

Jacobi and Jagger 5 have investigated the oscillations of the aqueous liquid in the eyes of patients after extracapsular and intracapsular cataract extractions. Measurements of the motion of the artificial lens which was implanted show the oscillating character of the rigid body in the liquid of the anterior chamber. These oscillations are damped by the surrounding liquid. By the method of laser-induced coloring the structure of the flow field in the vicinity of the oscillating lens can be determined pointwise and quantitatively the velocity field can be measured.

Photochromic compounds in aqueous solutions

The use of photochromic compounds in organic solutions of different type have the drawback that sometimes the index of refraction is different from the index of refraction of the material which is surrounding the cavity in which the motion of the liquid is to be studied. If the index of refraction of the wall-material is close to the one of water it would be very convenient to do the experiments with water in which a photochromic compound is solved. There are many possibilities to synthesize such compounds which are called photospirans of the water-solubilizing group [6] . An example of such a compound is given in the pictures below. The water-soluble photospiran in its open form is red in color when solved in water. When this solution is exposed to an intensive light source of visible light the color can be bleached completely. As the maximum of the absorption is still in the region of about 356 nm the solution can be colored again by an intensive UV-laser flash. Thus also in water the process of laser-induced coloring is possible. In Fig. 7 an example is given with a partly bleached solution of a photospiran in water. Visible light was used to create the closed form of the compound in water. After some minutes the whole solution is bleached as shown in Fig. 8. The backreaction is very slow and takes a long time. During this time interval the coloring by UV-light can be done and red spots can be generated in the aqueous solution.

Concluding remarks

It seems that by the use of photochromic compounds together with an intensive UV-light source together with the glass-fiber technique a powerful tool is available for the experimental fluid mechanics to analyse velocity fields in cavities of any irregular shape which may occur in the field of applications e.g. biomechanics, medical science, and the fluid mechanics in its classical sense.

Fig. 7. Water-soluble photo-
spiran (partly bleached)

Fig. 8. Water-soluble photo-
spiran (totally bleached)

Acknowledgement

The authors owe their thank especially to their colleague, Peter Eilbracht, from the Department of Organic Chemistry of the Technische Hochschule Darmstadt who provided them with the photochromic compounds which had to be synthesized in a time-consuming way. A great help came from the Institut für Reaktorbauelemente of the Nuclear Research Center of Karlsruhe. U. Müller, the director of this institute, was kind enough to give the authors the opportunity to carry out the experiments using his Krypton laser.

References

1. Popovich, A.T.; Hummel, R.L.: A new method for non-disturbing turbulent flow measurements very close to the wall. Chem. Eng. Sc., Vol. 22 (1967) 21-25.

2. Frantisak, F.; Palade de Iribarne, A.; Smith, J.W., and Hummel, R.L.: Nondisturbing tracer technique for quantitative measurements in turbulent flow. Ind. Eng. Chem. Fundamentals, Vol. 8 (1969) 160-167.

3. Iribarne, A.; Frantisak, F.; Hummel, R.L.; Smith, J.W.: An Experimental Study of Instabilities and Other Flow Properties of a Laminar Pipe Jet. A.I. ChE.J. Vol. 18 (1972) 689-698.

4. Larsen, J.: A mathematical model of the oscillations in the aphakic eye. Bull. Math. Biology, Vol. 42 (1980) 837-845.

5. Jacobi, K.W.; Jagger, W.S.: Physical Forces Involved in Pseudophacodonesis and Iridodonesis. Albrecht von Graefes Arch. Klin. Ophthalmol., Vol 216 (1981) 49-53.

6. Taylor, L.D.: Composition and process utilizing photospirans. United States Patent Office, 3,320,067, Patented May 16 (1967).

Measured Velocity Fluctuations Inside the Mixing Layer of a Supersonic Jet

H. OERTEL sen.

Deutsch-Französisches Forschungsinstitut Saint-Louis (ISL)
Saint-Louis, France

Summary

The supersonic jet emits Machwaves which are rather regular ones. Previous experiments have shown that they are shock waves accompanying more or less circumferential mixing layer structures which are travelling at three preferred speeds. In order to find out the nature of these structures two special Doppler velocimeters have been used for continuously recording the flow velocity fluctuations inside the mixing layer of a Mach-2-jet. In general the results agree with those obtained by many authors using hot wires or sampling Doppler velocimeters at lower Machnumbers. No indication at all has been found however on the existence of a regular vortex pattern producing the shock waves. But in the present case the three preferred speeds were found to be the convection velocities in the middle and at the two edges of the mixing layer. The difference between those at the edges as well as the highest flow velocity variation within a certain layer equalled the sound speed.

1. Introduction

The supersonic jet emits Machwaves like those visualized in Fig. 1. In the case of almost equal pressures at the nozzle exit and outside the jet these Machwaves are rather regular ones [1, 2,3]. Immobilisations on moving film like that presented in Fig. 2 have shown that such Machwaves do also exist inside the jet. Series of interferograms as well as streak records revealed that they do accompany long living mixing layer structures which are travelling downstream at three preferred speeds. These speeds w, w', w'' measured with the characteristics of both the jet and the ambient gas widely varied fulfill the following simple Machnumber relations

$$\frac{w}{a_\infty} = \frac{u_i - w}{a_i} \; ; \; \frac{w'}{a_\infty} = \frac{u_i - w'}{a_i} + 1 \; ; \; \frac{w''}{a_\infty} = \frac{u_i - w''}{a_i} - 1 \qquad (1)(2)(3)$$

Fig.1. Machwaves

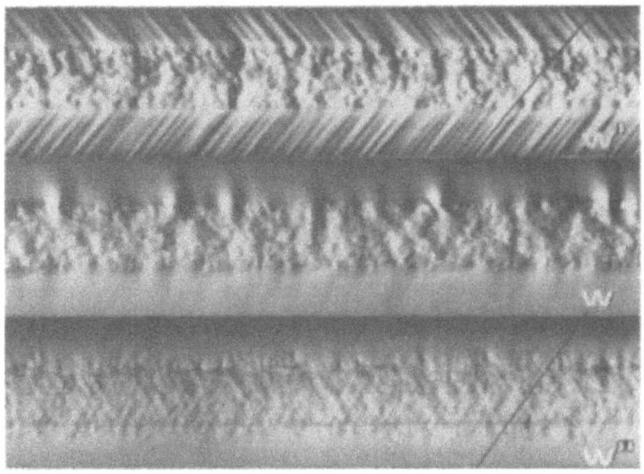

Fig.2. Immobilisations

where u_i is the jet flow velocity and a_i, a_∞ are the sound speeds inside and outside the jet. Local density records taken by means of a laser microphone revealed the Machwaves to be conically spreading shockwaves. The local mean shock distance λ is linearily increasing with the production coordinate ξ and is independent of the propagation coordinate η as defined in Fig. 3:

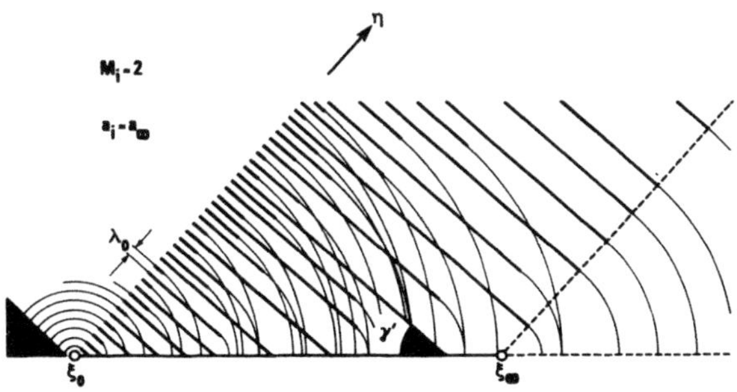

Fig.3. Coordinates ξ and η

$$\lambda - \lambda_0 = C(\xi - \xi_0) \qquad (4)$$

C, ξ_0, λ_0 could be approximated by the relations:

$$C = \frac{1}{12}\left(\frac{a_\infty}{w''} - \frac{a_\infty}{w'}\right) \; ; \quad \frac{\xi_0}{w} = \frac{\lambda_0}{w + a_\infty} = 3.5\ \mu s \qquad (5)(6)(7)$$

The local mean shock compressions π turned out to be determined by the ratio η/λ and an initial mean shock compression $\pi_0 = \pi$ $(\eta = 0)$ which was independent of ξ:

$$\pi = \frac{\frac{2+f}{1+f}}{2\frac{\eta}{\lambda} + 1 + \left(\frac{2+f}{1+f} \cdot \frac{1}{\pi_0} - 1\right)\sqrt{2\frac{\eta}{\lambda} + 1}} \tag{8}$$

The number $f = 2f_i f_\infty/(f_i+f_\infty)$ describes the influence of the caloric equations of state being $f = 3$ in the case of monatomic and $f = 5$ in the case of diatomic perfect gas. π_0 could be approximated by:

$$\pi_0 = \frac{2+f}{4}\left(\frac{w' - w}{a}\right)^2 \quad \text{with} \quad a = \frac{2a_i a_\infty}{a_i + a_\infty} \tag{9}$$

No dependence on the jet exit diameter D_e has been detected. Some of the structures accompanied by the Machwaves appeared to be more or less circumferential ones. But also spiral as well as longitudinal structures have been observed which can more clearly be seen on pictures of subsonic jets like that presented in Fig. 4. For some of the Machwaves no accompanied structures at

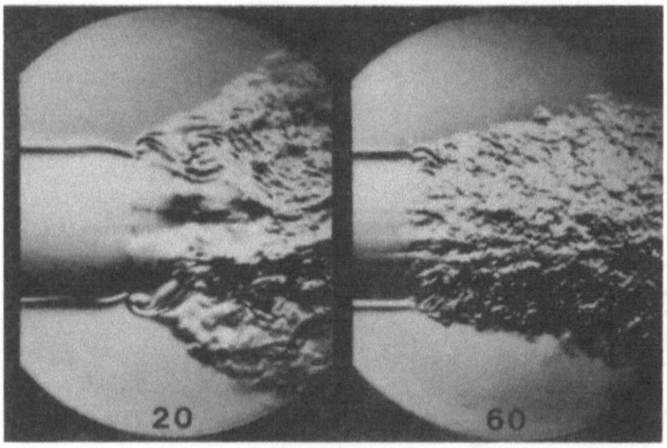

Fig.4. Subsonic jet

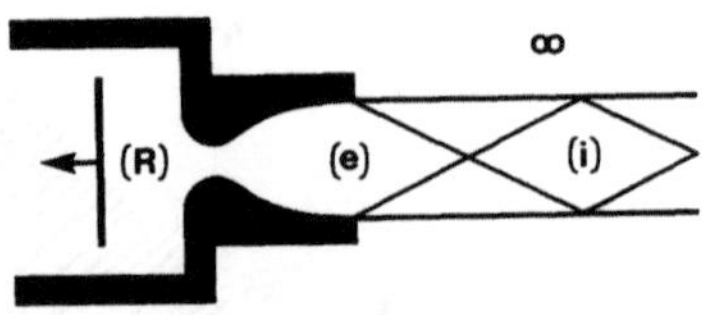

Fig.5. Jet production

all could be detected. The experimental results are compatible with different hypotheses [1] about the reason for the speeds w, w', w''. Local measurements of the velocity fluctuations inside the mixing layer are needed for deciding whether the one or the other or none of the hypotheses holds. A special Doppler velocimeter has been developed for doing such measurements. In the following some first results obtained with this new instrument will be presented [4].

2. Jet Production and Seeding

As for the investigations reported in [1,2,3] the supersonic jet was produced by means of a shock tube described in [5]. Fig. 5 explains the method. The exit pressure equalled the ambient pressure. The nozzle having the exit diameter D_e = 10 mm or 20mm was contoured to produce parallel outflow. The jet Machnumber was $M_i = u_i/a_i = 2$ and the inside to outside sound speed ratio was $a_i/a_\infty = 1$. Strong seeding on both sides of the mixing layer was needed. The short blowing time of about 2 ms allowed for doing this by simply producing the jet inside a case which, prior to the shot, was filled with the filtered smoke of three cigarettes.

3. Doppler Velocimeter and Signal Processing

The special Doppler velocimeter shown in Fig. 6 and described in [6] has been used. This is not a sampling but continuously recording single beam velocimeter. The light scattered by the tracers is conducted to a Michelson interferometer having arms with different optical lengths. There the frequency shift produces a phase shift which, in turn, becomes compensated by means of a phase restorage system. The electric tension driving the Pockels cell of this system is proportional to the frequency shift. Previous experiments with known shock waves produced by sparks revealed that it is the smoke inertia which determines the time resolution. In Fig. 7 the shock velocity jump was recorded as a 5 µs-rise whereas the density jump was recorded by the laser microphone as a 2 µs-rise.

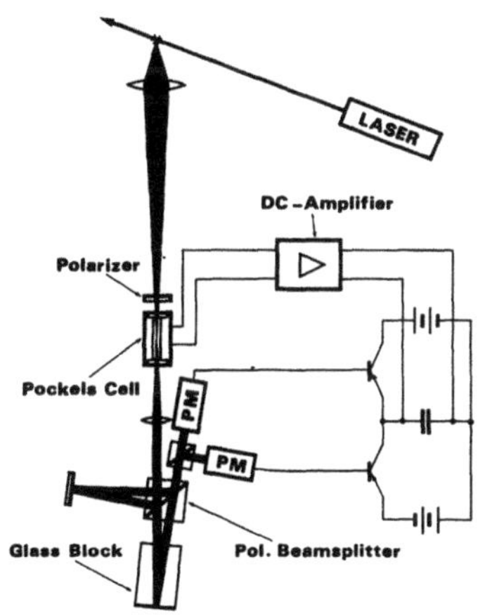

Fig.6. Dopplervelocimeter

By means of two detecting devices two of the three velocity components v_x, v_r, v_φ defined in Fig. 8 at the same point or one of them at two different points were recorded simultaneously. Highly

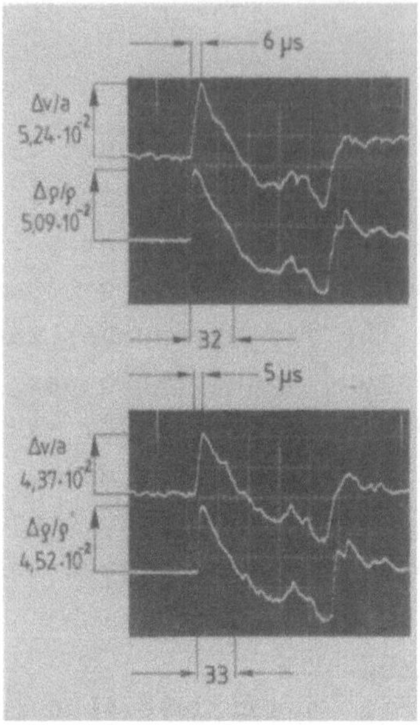

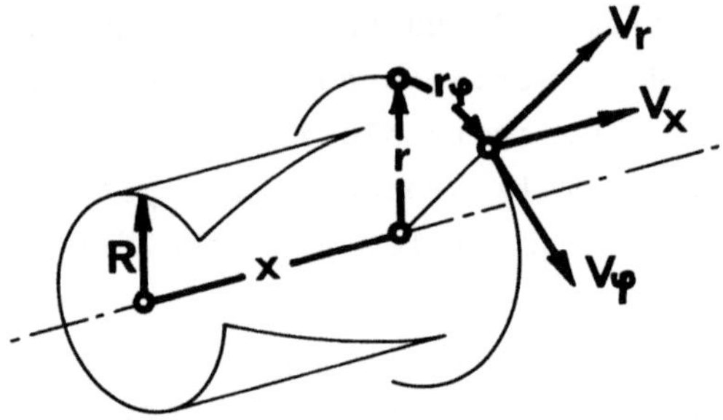

Fig.7. Shockwaves

Fig.8. Velocity components

careful equalization of the transfer functions of the two detecting devices was required for obtaining almost identical signals from the same small volume having diameter 0.3 mm and length 0.5mm. The time interval $t_2 - t_1$ = 1024 µs immediately after the establishment of stationary jet flow was chosen for evaluating the highest value $\hat{e}$ and the lowest value $\check{e}$, the mean value $\bar{e}$, the fluctuations $\Delta e(t) = e(t) = \bar{e}$ and then the rms-value $\widetilde{\Delta e}$, the autocorrelation function $R(\tau)$ and finally, in the case of two signals $e_1(t)$ and $e_2(t)$, the crosscorrelation function $R_{12}(\tau)$. The $R(\tau)$ served for calculating the integrated time scales. The integrated space scales have not yet been precisely determined but estimated from those distances Δx or Δr or $r\Delta\varphi$ at which $R_{12}(0)$ was smaller than 0.1. Finally, in the case of $e_1(t)$ representing $v_x(t)$ at x and $e_2(t)$ representing $v_x(t)$ at $x+\Delta x$ the convection velocity $c = \Delta x/\hat{\tau}$ has been determined with that time shift $\hat{\tau}$, at which the first maximum $\hat{R}_{12}$ did appear. Correlograms of $\Delta e_2(t)$ or $\Delta e_2(t+\hat{\tau})$ over $\Delta e_1(t)$ have been used for visualizing the correlations. This method is explained in Fig.9.

4. Results

Seen as a whole the results were not quite different from those obtained by many authors using hot wires or sampling Doppler velocimeters at lower Machnumbers. But some details have been found which may perhaps be special features of higher Machnumber

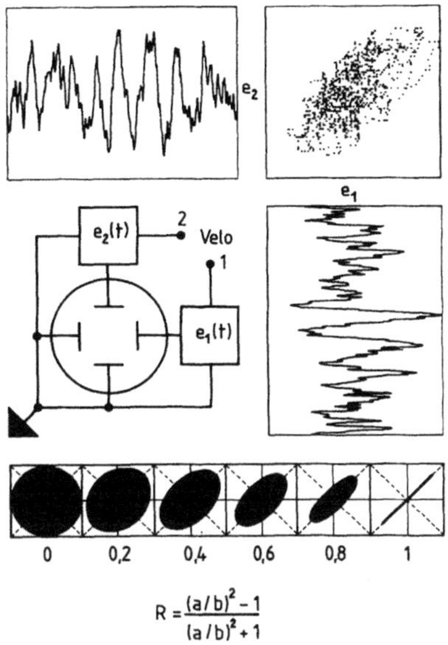

jets and may help to understand the preferred speeds w, w', w''.

As in the case of the subsonic jet the mean velocity profile can be approximated by the Görtler function [7]: $\bar{v}_x/u_i = \frac{1}{2}(1 - \text{erf}\, z)$

with $z = \sigma\,(r - r_{0.5})/x$ (10)

This fact is demonstrated in Fig. 10. But the scaling factor $\sigma = 17$ was considerably smaller than expected. In [8] for instance the experimental results obtained by different authors at lower Machnumbers have been approximated by the formula:

$$\sigma = 10.7(1-0.273\, M_i^2)^{-1} \quad (11)$$

Fig.9. Correlogram

Extrapolation predicts σ going to infinity at $M_i = 1.91$. Our value $\sigma = 17$ suggests the formula

$$\sigma = 10.9(1-0.089 M_i^2)^{-1} \quad (12)$$

for interpolations within the range $0 \leqslant M_i \leqslant 2$. Fig.11 shows the mixing

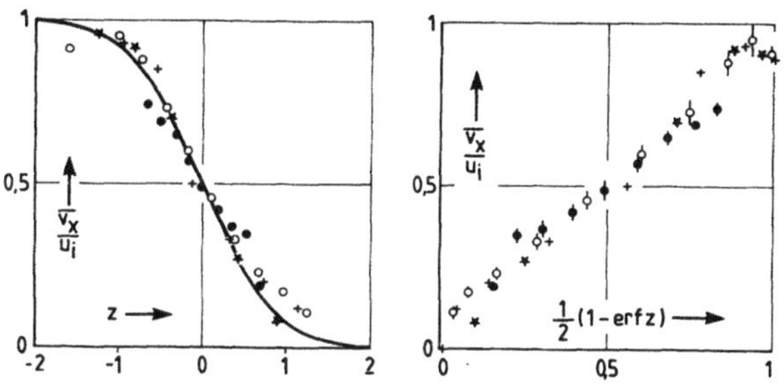

Fig. 10. Mean velocities

layer thicknesses δ defined by

$$\delta = -\left[d\left(\frac{\bar{v}_x}{u_i}\right)\Big/ dr(r_{0.5})\right]^{-1} = \sqrt{\pi}x/\sigma \quad (13)$$

and calculated with these two formulae. Our experimental value was $\delta = 0.104x$. The equation $\delta(x_c) = D_e$ can be used for defining the length x_c of the jet core giving $x_c = 0.96\, D_e$ in our case. This value agrees with that distance from the nozzle

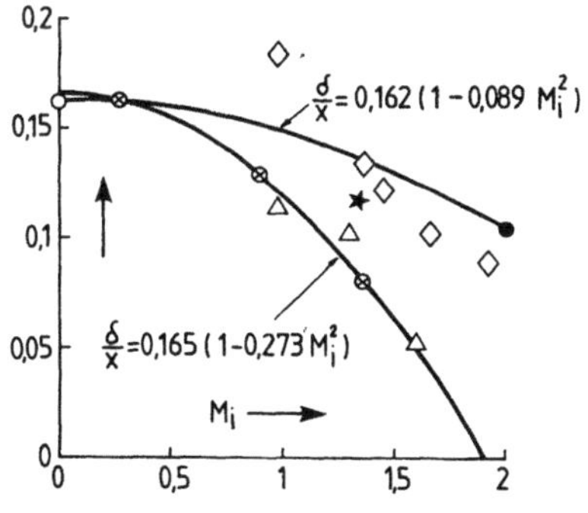

Fig.11. Mixing layer thickness

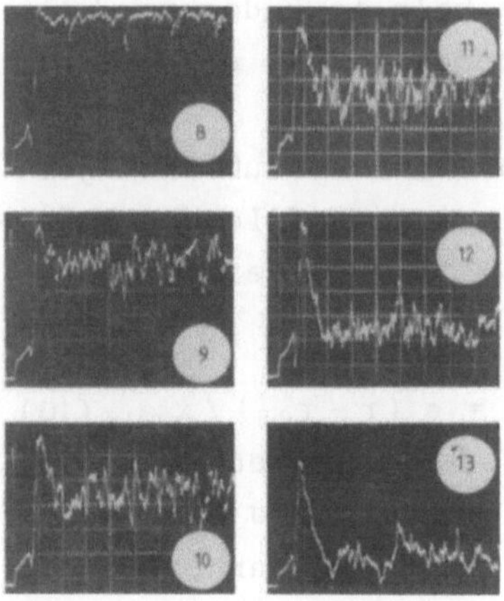

Fig.12. $v_x(t)$ at r/mm
x=60mm; $109\frac{m}{s}$/c; 200μs/c

at which the Machwave emission vanished on interferograms taken from smaller jets [1,11].

As in the subsonic case the $v_x(t)$ recorded in the middle of the mixing layer were quite different from those at its edges, where strong and short signal spikes separated by long intervals of relatively weak fluctuations occurred. This fact can be seen in Fig. 12. In our case these spikes were so short that the physical meaning of rms-values and correlation coefficients was quite questionable. Nevertheless the behaviour of these values was about the same as that found at lower Machnumbers. The rms-values were functions of z as can be seen in Fig. 13. The maxima were somewhat smaller than in the subsonic case [9]. All the correlation functions were oscillating ones like those presented in Fig. 14. The mean time interval $\Delta\hat{\tau}$ between successive maxima was considerably longer than λ/a_∞. Therefore we conclude that the Machwaves are not connected with these oscillations. On the other side the time scales were shorter than λ/a_∞. The very small correlation coefficients measured in the middle of the mixing layer at $\Delta r \geqslant 2$mm

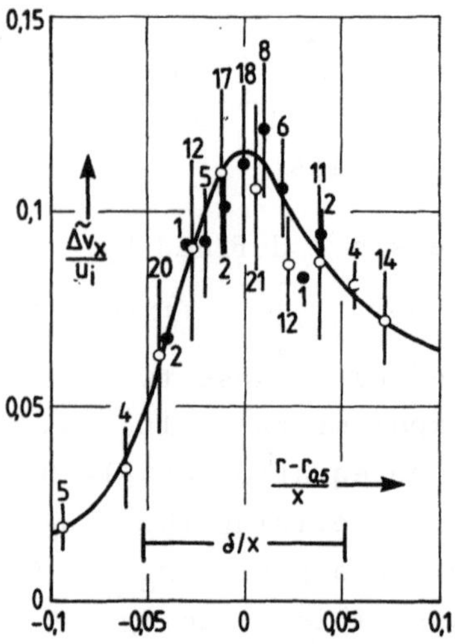

Fig.13. rms-values
x/mm=60(O); 100(O)

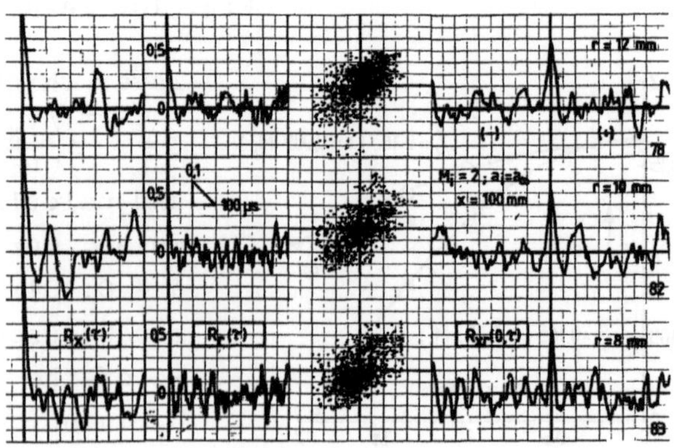

Fig.14. Correlation functions

and $r\Delta\varphi \geq 3$mm agree with the statements in [10] with respect to the space scales inside the mixing layers of subsonic jets. Going to the edges somewhat higher correlation coefficients were found which were not due to the spikes. Fig. 15 shows that even at $r\Delta\varphi$ = 3 mm the spikes did not coincide.

The correlation coefficients $R_{12}(0)$ of the $\Delta v_x(t)$ and $\Delta v_r(t)$ at the same point were found to be between 0.44 and 0.56. These values also agree with those measured at subsonic Machnumbers [9]. When tracing the corresponding correlograms like those presented in Fig. 14 no indication at all of a regular vortex pattern has been detected.

Perhaps, there may be a connection between the convection velocities c and the speeds w, w', w''. The c turned out to equal $\bar{v}_x = u_i/2$ in the middle of the mixing layer and to become about $3\,u_i/4$ when going to the inner edge and about $u_i/4$ when going to the utter one. For our case the formulae (1) to (3) give $w = u_i/2$, $w' = 3u_i/4$, $w'' = u_i/4$. Thus w,w',w'' may perhaps be the mean, maximum and minimum convection velocities c. But about the same values of c/u_i have also been found inside subsonic mixing layers where w' and w'' do not exist.

Finally Fig. 16 revealed that within a certain thick layer inside of the mixing layer the difference $\hat{v}_x - \check{v}_x$ between the highest and lowest instantaneous value of $v_x(t)$ equalled $u_i/2$. In our case the difference

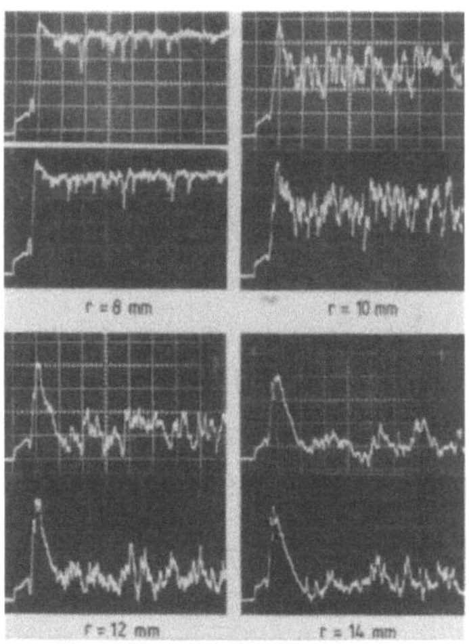

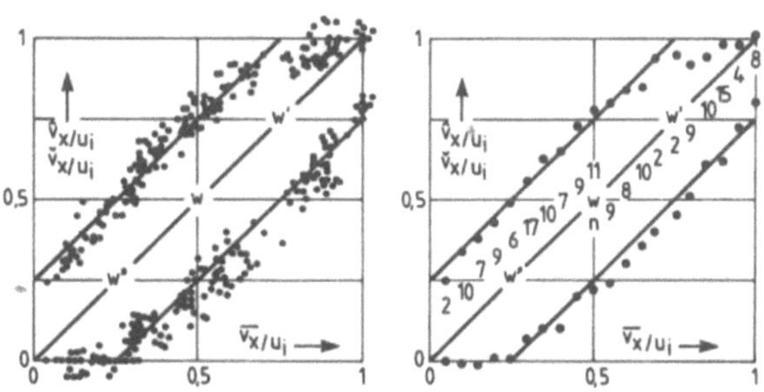

Fig.15. $v_x(t)$ at $r\Delta\varphi$=3.1mm Fig.16. Difference $\hat{v}_x$-$\check{v}_x$ at $\bar{v}_x/u_i$

$$w' - w'' = a = 2a_i a_\infty/(a_i+a_\infty) \qquad (14)$$

had exactly the same value. This fact may mean that $\hat{v}_x-\check{v}_x$ cannot exceed the sound speed.

5. Conclusions

None of these results has given any indication on toroidal vortices neither in a circumferential nor in a spiral direction producing the Machwaves. Taking into account our quite limited knowledge about the entrainment of tracers in supersonic vortical layers, this does not absolutely mean that such vortices do not exist. The smoke has perhaps been completely centrifuged out of long living vortices. But the agreements with results obtained with hot wires in subsonic jets suggest the assumption that the measured tracer velocities were fairely close to the flow velocities. If we adopt this assumption then we come to the conclusion that the more or less circumferential structures visualized on the interferograms are not big vortices, but either rather small ones or pressure waves propagating at the edges of a vortical layer inside of the mixing layer. This is the assumption number one on which the three-layer hypothesis proposed in [1] for explaining w, w', w'' is based. The assumption number two is that the formation of shock waves prevents the difference of the flow velocities on either side of this vortical layer from exceeding the sound speed. It seems that this is also true. But more measurements at varied $M_i > 1$ are needed in order to make sure whether this interpretation of the measured $\hat{v}_x-\check{v}_x$ is allright or wrong.

One of the results only mentioned in passing is directly important for the jet noise prediction. We now know that in the environment of small jets the width of the beam of Machwaves equals the length of the jet core which, in turn, can be calculated with $\sigma = 17$ for the case $M_i = 2$ and $a_i/a_\infty = 1$. This fact has been used for estimating that part of the beam which is not perturbed by diffraction [11]. The equations mentioned in the introduction can only be valid, of course, in this part of the beam.

References

1. Oertel, H.; Kinematik der Machwellen innerhalb und außerhalb von Überschallstrahlen. ISL-R 112/78.

2. Oertel, H.; Machwave radiation of hot supersonic jets investigated by means of the shock tube and new optical techniques pp. 266-275 in "Shock Tubes and Waves"; Proc. 12th Internat. Symp. on Shock Tubes and Waves; Jerusalem 1979, A. Lifshitz and J. Rom Ed., The Magnum Press, Jerusalem 1980.

3. Oertel, H.; Gatau, F.; George, A.; Dynamik der Machwellen in der Umgebung von Überschallstrahlen. ISL-R 124/81.

4. Oertel, H.; Gatau, F.; George, A.; Schwankungsmessungen in der Mischungsschicht eines Überschallstrahls. ISL-R 110/82.

5. Oertel, H.; Stoßrohre - Shock Tubes - Tubes à choc. Springer Verlag Wien - New York, 1966.

6. Smeets, G.; George, A.; Laser-Doppler-Velozimeter mit einem Michelson-Spektrometer. ISL-R 109/80.

7. Görtler, H.; Berechnung von Aufgaben der freien Turbulenz aufgrund eines neuen Näherungsansatzes. ZAMM 22, 5 (1942) 244-254.

8. Lau, J.C.; Morris, P.J.; Fisher, M.J.; Measurements in subsonic and supersonic free jets using a laser velocimeter. J. fluid. mech. 93, 1 (1979) 1-27.

9. Bradshaw, P.; Ferriss, D.H.; Johnson, R.F.; Turbulence in the noise producing region of a circular jet. J.fluid. mech. 19,4 (1964) 591-624.

10. Davies, P.O.A.L.; Turbulence structure in free shear layers. AIAA journ. 4, 11 (1966) 1971-1978.

11. Oertel, H.; Abschätzung der Beugung an den Rändern des Machwellenbündels in der Umgebung des Überschallstrahls. ISL-N 605/81.

Mathematical Model of the Single-File Flow of Red Blood Cells in Capillaries

Heinz-Dieter Papenfuss and Joseph F. Gross

Ruhr-University Bochum, Institute for Thermo- and Fluid-Dynamics, 4630 Bochum, West Germany, and
University of Arizona, Department of Chemical Engineering, Tucson, Arizona 85721, U.S.A.

1. Introduction

An important ingredient for the analysis of the blood flow in the microcirculation is that the suspension of red cells in the plasma should not be treated as a continuum. In capillaries whose diameters range from 3 to 8 μm, the cells flow in single file. In larger vessels the flow is characterized by a distribution of red cells over the intraluminal cross-section. Extensive experimental studies for the latter flow regime have been done by Barbee and Cokelet (3). A theoretical description of this flow, however, has not been developed.

Recently, Albrecht et al. (1) studied the blood flow in glass capillaries ranging from 3.3 to 11 μm in diameter. These in-vitro experiments, therefore, covered the range of true capillaries. The studies were mainly concerned with the determination of tube and discharge hematocrits through simultaneous measurements of the blood and red cell velocity. For the same range of diameters, a number of theoretical studies of the blood flow has been carried out by various authors. Reviews of this work have been presented by Gross and Aroesty (7) and, recently, by Tözeren and Skalak (13). Most of these theoretical investigations deal with the flow of the plasma between two adjacent red cells and around the red cells. Often the red cells are assumed to be rigid having simple shapes such as discs or spheres. In their recent work, Tözeren and Skalak (14) abandon the assumption of rigid particles and consider the motion of closely-fitting elastic spheres instead.

The transport of fluid and substances in the plasma gaps between the red cells and across the capillary wall is greatly influenced by the actual flow behavior of the plasma. Therefore, it is important to determine the solution of the hydrodynamical problem of plasma flow which, subsequently, enters the equations for the mass transfer.

In a recent paper, Gaehtgens et al. (6) demonstrated the existence of a secondary motion in the plasma gap between two adjacent red cells. This vortex motion which is superimposed to the translational movement of the plasma had first been postulated by Prothero and Burton (10) who proposed the name 'bolus flow' for this trapped plasma flow. Gaehtgens et al. (6) added microspheres (1 μm diameter) to red cell suspensions which were used to perfuse glass capillaries of 8 μm diameter and less. Using a high-speed cinematographic technique, it was possible to trace the movement of the latex particles which was characterized by a forward motion near the centerline and a backward motion near the tube wall relative to the red cell motion. A similar bolus flow could be detected by observing the motion of platelets.

The present work is concerned with a theoretical study of the bolus flow. In the course of the development of the theoretical model we will take advantage of the fact that the volume concentration of red cells in the microcirculation is rather low; see G.W. Schmid-Schoenbein and B.W. Zweifach (12). This low hematocrit in the capillaries will make it possible to carry out a singular perturbation analysis of the flow as has been suggested by Zien (16).

2. Theory

In the development of our theoretical model, we consider a straight cylindrical capillary with rigid wall and constant radius R; see FIG. 1. The capillary is perfused by a suspension of red cells and plasma. We assume that the cells flow in single file separated by plasma gaps of equal length L. Further, we assume that the red cells fit tightly in the capillary

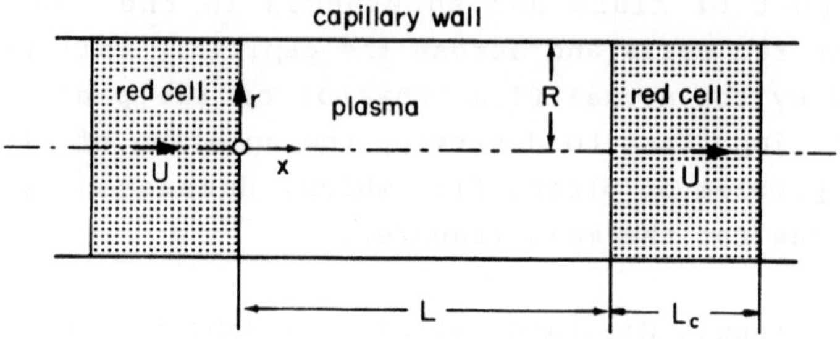

Fig. 1. Geometry of capillary and plasma gap

and that the front and rear surfaces are planar. If the axial length of the red cells is denoted by L_c, then the volume concentration of red cells which is the tube hematocrit is simply

$$H_T = \frac{L_c}{L + L_c} = \frac{L_c/R}{L/R + L_c/R} \quad . \tag{1}$$

The ratio L_c/R is of the order of magnitude $0(1)$ whereas the aspect ratio L/R can attain values varying between 0 and, theoretically, ∞. The walls of the red cells and of the capillary are assumed to be impermeable. The plasma trapped between the red cells is a Newtonian fluid with a viscosity μ which depends on the plasma protein concentration; see Papenfuss and Gross (9). If the red cell velocity is denoted by U , the Reynolds number can be determined from the equation

$$Re = \frac{U \cdot \rho \cdot R}{\mu} \tag{2}$$

to be of the order of 10^{-3} to 10^{-2}. In Eq.(2), ρ is the density of the plasma which is only slightly greater than that of water.

The small values for the Reynolds number suggest that the inertial terms in the Navier-Stokes equations be neglected in comparison with the viscous and pressure terms. This results in the Stokes equations. As is shown in FIG. 1, we use a cylindrical coordinate system (x,r) connected with a red cell. In these coordinates, the Stokes equations are continuity: $$\frac{\partial u}{\partial x} + \frac{\partial v}{\partial r} + \frac{v}{r} = 0 \tag{3}$$

x-momentum: $$\frac{\partial p}{\partial x} = \mu \cdot [\frac{\partial^2 u}{\partial x^2} + \frac{1}{r}\frac{\partial u}{\partial r} + \frac{\partial^2 u}{\partial r^2}] \tag{4}$$

r-momentum: $$\frac{\partial p}{\partial r} = \mu \cdot [\frac{\partial^2 v}{\partial x^2} + \frac{1}{r}\frac{\partial v}{\partial r} - \frac{v}{r^2} + \frac{\partial^2 v}{\partial r^2}] \tag{5}$$

In Eqs.(3-5), u and v are the axial and radial velocity components, respectively, and p is the hydrostatic pressure. The boundary conditions are

$$\left.\begin{matrix} x = 0 \\ x = L \end{matrix}\right\} \quad u = U \ , \quad v = 0 \tag{6}$$

$$r = R: \qquad u = v = 0 \quad . \tag{7}$$

The solution of the above equations yields the flow of plasma between two red cells which move with constant velocity U. The flow relative to the motion of the red cells can simply be determined by superimposing a translational flow in the negative x-direction. The equations for the shear stress τ and axial pressure drop Δp between two adjacent red cells are

$$\tau = \mu \cdot [\frac{\partial u}{\partial r} - \frac{\partial v}{\partial x}] \tag{8}$$

$$\Delta p = \frac{2}{R} \int_0^L \tau_w \, dx \tag{9}$$

where the subscript w refers to the capillary wall.

In the following, we use the nondimensionalized variables

$$x^* = \frac{x}{R}\ , \quad r^* = \frac{r}{R}\ , \quad L^* = \frac{L}{R}$$

$$u^* = \frac{u}{U}\ , \quad v^* = \frac{v}{U}\ , \quad p^* = \frac{p}{\mu U/R}\ , \quad \tau^* = \frac{\tau}{\mu U/R} \tag{10}$$

In addition, we introduce a dimensionless stream function Ψ which is defined by

$$u = \frac{1}{r}\frac{\partial\Psi}{\partial r} \qquad v = -\frac{1}{r}\frac{\partial\Psi}{\partial x}\ . \tag{11}$$

In Eq.(11) and in the following analysis the asterisks are omitted for simplicity. Substituting Eqs.(10) and (11) into Eqs.(3-5) results in the biharmonic equation

$$D^4(\Psi) = 0 \tag{12}$$

where the operator D^4 stands for

$$D^4 = D^2\left[\frac{\partial^2}{\partial x^2} - \frac{1}{r}\frac{\partial}{\partial r} + \frac{\partial^2}{\partial r^2}\right] = \left[\frac{\partial^4}{\partial x^4} + \frac{2\partial^4}{\partial x^2\partial r^2} + \frac{\partial^4}{\partial r^4}\right]$$

$$- \frac{2}{r}\left[\frac{\partial^3}{\partial x^2\partial r} + \frac{\partial^3}{\partial r^3}\right] + \frac{3}{r^2}\frac{\partial^2}{\partial r^2} - \frac{3}{r^3}\frac{\partial}{\partial r}\ . \tag{13}$$

The boundary conditions for Eq.(12) are

$$\left.\begin{array}{l} x = 0 \\ \\ x = L \end{array}\right\} \quad \Psi = \frac{1}{2}r^2\ , \quad \frac{\partial\Psi}{\partial x} = 0 \tag{14}$$

$$r = 1: \quad \Psi = \frac{1}{2}\ , \quad \frac{\partial\Psi}{\partial r} = 0\ . \tag{15}$$

Symmetry at the capillary axis requires

$$\Psi(r=0) = 0 \quad \text{and} \quad \left(\frac{\partial\Psi}{\partial r}\right)_{r=0} = 0\ . \tag{16}$$

In order to obtain the solution for the motion of the plasma relative to the red cells, the stream function of the translational motion, i.e. $-1/2\ r^2$, must be superimposed

$$\hat{\Psi} = \Psi - \frac{1}{2} r^2 \tag{17}$$

where $\hat{\Psi}$ is the stream function of the relative flow. Introducing Eq.(11) into Eqs.(8) and (9) yields

$$\tau_w = \left(\frac{\partial^2 \Psi}{\partial r^2}\right)_{r=1} \tag{18}$$

$$\Delta p = 2 \cdot \int_0^L \tau_w \, dx \quad . \tag{19}$$

Since the flow in the plasma gap is symmetric with respect to the midpoint of the gap at $x = L/2$, the numerical solution of Eq.(12) can be obtained by taking advantage of the symmetry condition

$$\left(\frac{\partial \Psi}{\partial x}\right)_{x=L/2} = 0 \quad . \tag{20}$$

The symmetry makes it possible to solve one quarter of the entire plasma gap. The only parameter in this flow problem is the aspect ratio L/R . FIG. 2 shows the streamlines for different values of L/R which would originate at the red cell surface at $r = 0.6$. The abscissa, ξ, is the reduced axial coordinate given by

$$\xi = x \cdot \frac{R}{L} \quad . \tag{21}$$

The figure shows substantial changes in the curvature with concomitant changes in the direction of the streamline when the aspect ratio L/R is increased. For large values of L/R , the region in which the streamline direction changes is confined to a thin layer next to the red cell wall. This behavior is typical for a singular perturbation problem with

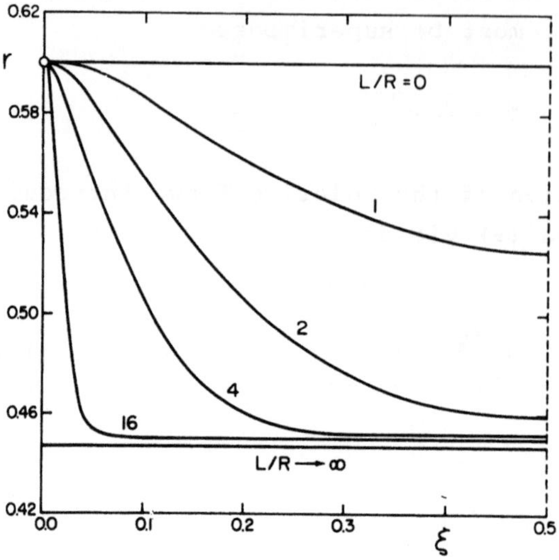

Fig. 2. Streamlines originating at r=0.6 for different values of L/R

$$\varepsilon = \frac{R}{L} \tag{22}$$

as the perturbation parameter.

Intuitively, we expect that in the limiting case $L/R \to \infty$ or $\varepsilon \to 0$, the velocity profile far from the cell surface will be the classical Hagen-Poiseuille profile for fully-developed pipe flow.

Introducing Eqs.(21) and (22) into Eqs.(12-16) yields

$$\varepsilon^4 \cdot \frac{\partial^4\Psi}{\partial\xi^4} + 2\,\varepsilon^2 \frac{\partial^4\Psi}{\partial\xi^2\,\partial r^2} + \frac{\partial^4\Psi}{\partial r^4} - \frac{2}{r}\left[\varepsilon^2 \frac{\partial^3\Psi}{\partial\xi^2\,\partial r} + \frac{\partial^3\Psi}{\partial r^2}\right]$$

$$+ \frac{3}{r^2} \cdot \frac{\partial^2\Psi}{\partial r^2} - \frac{3}{r^3} \cdot \frac{\partial\Psi}{\partial r} = 0 \tag{23}$$

with the boundary conditions:

$$\xi = 0 : \quad \Psi = \frac{1}{2}\,r^2 \; ; \quad \varepsilon\,\frac{\partial\Psi}{\partial\xi} = 0 \tag{24}$$

$$\xi = \frac{1}{2} : \quad \varepsilon \cdot \frac{\partial\Psi}{\partial\xi} = 0 \tag{25}$$

$$r = 0 : \quad \Psi = 0 \; ; \quad \frac{\partial\Psi}{\partial r} = 0 \tag{26}$$

$$r = 1 : \quad \Psi = \frac{1}{2} \; ; \quad \frac{\partial\Psi}{\partial r} = 0 \; . \tag{27}$$

The approach of the present work is to use the method of matched asymptotic expansions to solve the singular perturbation problem for $\varepsilon = R/L \to 0$. According to Eq.(1), this case corresponds to low values of the hematocrit in the tube which is typical for the microcirculation.

3. Matched asymptotic expansions

3.1 Outer expansion

The outer expansion of the stream function has the form

$$\Psi(\xi,r;\varepsilon) = \Psi_1^{(O)}(\xi,r) + \varepsilon^2 \cdot \Psi_2^{(O)}(\xi,r) + \varepsilon^4 \cdot \Psi_3^{(O)}(\xi,r) + \ldots \tag{28}$$

Using the abbreviation

$$A[\Psi] = \frac{\partial^4\Psi}{\partial r^4} - \frac{2}{r} \cdot \frac{\partial^3\Psi}{\partial r^3} + \frac{3}{r^2} \cdot \left[\frac{\partial^2\Psi}{\partial r^2} - \frac{1}{r} \cdot \frac{\partial\Psi}{\partial r}\right] \tag{29}$$

and introducing Eq.(28) into Eq.(12), we obtain the following set of differential equations after collecting terms of equal order in ε:

$$A[\Psi_1^{(O)}] = 0 \tag{30}$$

$$A[\Psi_2^{(O)}] = -2 \cdot \frac{\partial^4\Psi_1^{(O)}}{\partial\xi^2\partial r^2} + \frac{2}{r} \cdot \frac{\partial^3\Psi_1^{(O)}}{\partial\xi^2\partial r} \tag{31}$$

$$A[\Psi_n^{(O)}] = -\frac{\partial^4\Psi_{n-2}^{(O)}}{\partial\xi^4} - 2 \cdot \frac{\partial^4\Psi_{n-1}^{(O)}}{\partial\xi^2\partial r^2} + \frac{2}{r} \cdot \frac{\partial^3\Psi_{n-1}^{(O)}}{\partial\xi^2\,\partial r} \tag{32}$$

with $n = 3, 4, 5 \ldots$. The boundary conditions are:

$$\left.\begin{array}{l} \xi = 0 \\ \xi = \frac{1}{2} \end{array}\right\} \quad \frac{\partial\Psi_n^{(O)}}{\partial\xi} = 0 \qquad n = 1,2,3 \ldots \tag{33}$$

$$r = 0 : \qquad \Psi_n^{(0)} = \frac{\partial \Psi_n^{(0)}}{\partial r} = 0 \qquad n = 1,2,3.. \quad (34)$$

$$r = 1 : \qquad \Psi_1^{(0)} = \frac{1}{2} \; ; \; \Psi_n^{(0)} = 0 \qquad n = 2,3,4.. \quad (35)$$

$$r = 1 : \qquad \frac{\partial \Psi_n^{(0)}}{\partial r} = 0 \qquad n = 1,2,3.. \quad (36)$$

The equation for the first-order terms, Eq.(30), has the following analytical solution

$$\Psi_1^{(0)} = r^2 \cdot (1 - \frac{1}{2} r^2) \quad . \qquad (37)$$

This is the Hagen-Poiseuille equation for fully-developed laminar flow. However, Eq.(37) does not satisfy the boundary condition at the red cell face, Eq.(24). Since Eq.(37) is independent of ξ , the right-hand side of Eq.(31) for the second-order term vanishes also. The analytical solution for the second-order term is

$$\Psi_2^{(0)} = C_2(\xi) \cdot r^2 \cdot (1 - r^2 + 2 \ln r) \quad . \qquad (38)$$

The function $C_2(\xi)$ remains undetermined. The only information we have for C_2 is that $\partial C_2/\partial \xi = 0$ at $\xi = 0$ and $\xi = 1/2$. However, matching the outer with the inner expansion, which will be presented below, will yield the value for $C_2(0)$. The function $C_2(\xi)$ itself can, in principle, be determined from the equation for the third-order terms, Eq.(32), which is a partial differential equation. The present work, however, deals only with the second-order terms.

3.2 Inner expansion

Characteristic of the first- and second-order outer expansions is the fact that the derivatives with respect to ξ vanish. This is particularly true for the highest derivative in ξ which makes it impossible to satisfy the boundary condition expressed by Eq.(24).

The term with the highest derivative in ξ originates from the viscous part of the normal stress, σ_x' , which is given in the dimensionless form by:

$$\sigma_x' = 2\,\mu \cdot \frac{\partial u}{\partial x} \quad ; \tag{39}$$

see Schlichting (11). In a thin layer close to the red cell surface, the x-derivative of the viscous normal stress is of the same order of magnitude as the rest of the terms in the right-hand side of the Navier-Stokes equations. This holds, e. g., for the term $\partial_2 u/\partial r^2$ which results from the r-derivative of the shear stress

$$\tau_{xr} = \mu \cdot \left(\frac{\partial u}{\partial r} + \frac{\partial v}{\partial x}\right) \quad . \tag{40}$$

By taking the ratio

$$\frac{\sigma_x'}{\tau_{xr}} \sim \frac{R}{L} = \varepsilon \tag{41}$$

or the ratio

$$\frac{\partial \sigma_x' / \partial x}{\partial \tau_{xr} / \partial r} \sim \frac{R^2}{L^2} = \varepsilon^2 \quad , \tag{42}$$

it can be seen that the perturbation parameter ε can be interpreted in physical terms in addition to its geometrical meaning.

In the following, we will return to dimensionless notation. The equal importance of viscous tangential and normal stresses in a layer near the red cell surface makes it necessary to consider a second expansion valid for this layer. This so-called inner expansion will be expressed in a stretched variable

$$\bar{x} = \xi/\varepsilon \tag{43}$$

which, in fact, is identical with the physical coordinate x . The inner expansion for the stream function is

$$\Psi(\bar{x},r;\varepsilon) = \Psi_1^{(i)}(\bar{x},r) + \varepsilon^2 \cdot \Psi_2^{(i)}(\bar{x},r) + \varepsilon^4 \cdot \Psi_3^{(i)}(\bar{x},r) + \ldots \qquad (44)$$

If we introduce Eq.(44) into Eq.(12), we obtain a set of biharmonic equations

$$D^4[\Psi_n^{(i)}] = 0 \ , \qquad n = 1,2,3 \ldots \qquad (45)$$

The boundary conditions are

$$\bar{x} = 0 : \qquad \Psi_1^{(i)} = \frac{1}{2} r^2 \ ; \qquad \Psi_n^{(i)} = 0 \qquad n = 2,3,4 \ldots \qquad (46)$$

$$\frac{\partial \Psi_n^{(i)}}{\partial x} = 0 \ , \qquad n = 1,2,3 \ldots \qquad (47)$$

$$r = 0 : \qquad \Psi_n^{(i)} = \frac{\partial \Psi_n^{(i)}}{\partial r} = 0 \ , \qquad n = 1,2,3 \ldots \qquad (48)$$

$$r = 1 : \qquad \Psi_1^{(i)} = \frac{1}{2} \ ; \qquad \Psi_n^{(i)} = 0 \ , \qquad n = 2,3,4 \ldots \qquad (49)$$

$$\frac{\partial \Psi_n^{(i)}}{\partial r} = 0 \ , \qquad n = 1,2,3 \ldots \qquad (50)$$

The boundary condition for $\bar{x} \to \infty$ results from matching the outer with the inner expansion. The matching procedure has been described in detail by Van Dyke (15). For the first-order term we obtain

$$\Psi_1^{(i)}(\bar{x} \to \infty) = r^2 \cdot (1 - \frac{1}{2} r^2) \quad . \qquad (51)$$

The corresponding matching condition for the second-order term is

$$\Psi_2^{(i)}(\bar{x} \to \infty) = C_2(0) \cdot r^2 \cdot (1 - r^2 + 2 \ln r) \ , \qquad (52)$$

where the constant $C_2(0)$ attains the value 0.2 .

4. Results and Discussion

The comparison between the numerical results of the two-term asymptotic expansion and of the solution to the full Stokes equations shows excellent agreement for values of L/R as low as 2; see Papenfuss (8). FIG. 3 shows streamlines in the plasma gap for L/R = 16 and streamlines of the relative flow for L/R = 2. For values of x greater than 1 the streamlines are almost parallel. This result is true for all values of the aspect ratio L/R > 2. For L/R → 0 , a case which is not of the particular concern in the present work, the streamlines are parallel, too, and follow the equation $\Psi = 1/2\ r^2$.

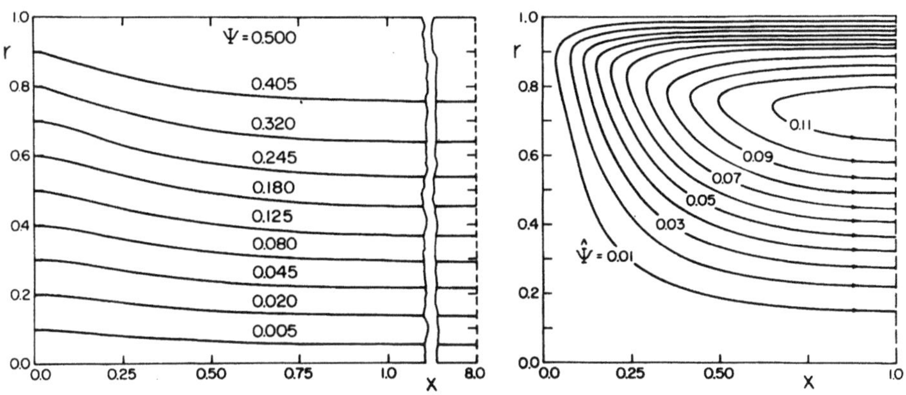

Fig. 3. Left: Streamlines for L/R = 16. Right: Streamlines of relative flow for L/R = 2

The figure on the right-hand side shows that the relative flow is similar to that of a vortex ring. The eddy led to the 'bolus flow' concept first introduced by Prothero and Burton (10) on the basis of experimental observations. FIG. 4 shows the axial development of the profile of the axial velocity component for two values of the aspect ratio in comparison with the parabolic profile for fully-developed flow.

The physiological significance of the bolus flow is that it acts against the build-up of concentration boundary layers as a consequence of induced radial mixing. Aroesty and Gross(2)

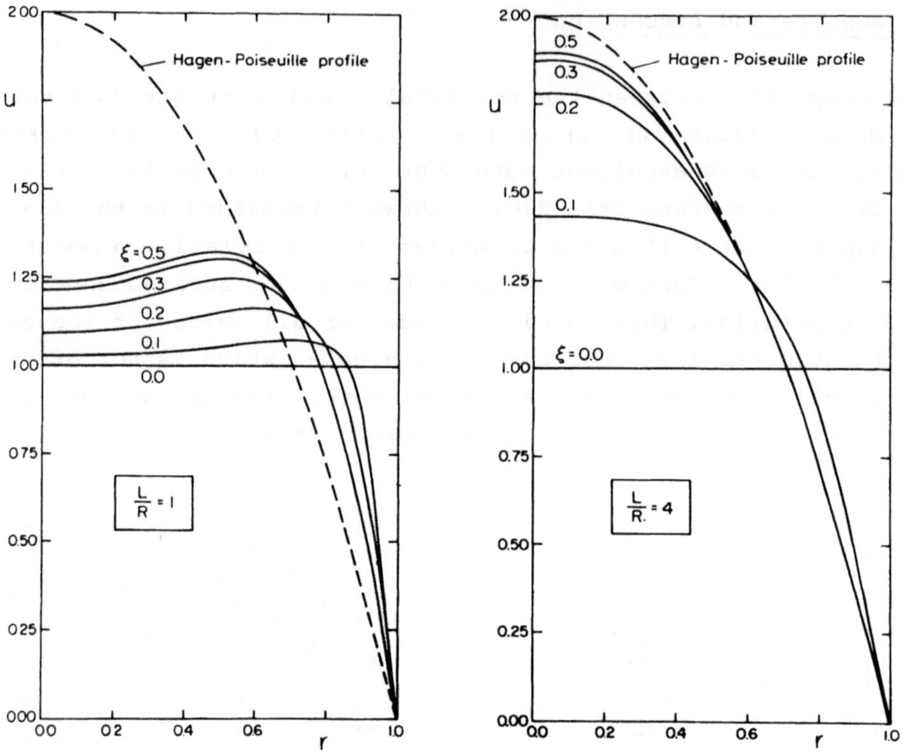

Fig. 4. Axial development of velocity profiles in plasma gap between two adjacent red blood cells

showed, however, that for low molecular weight gases such as oxygen the convective transport is negligibly small in comparison with radial diffusion. On the other hand, for high molecular weight substances, such as plasma proteins, the convective transport plays an important role with respect to transcapillary fluid exchange; see Deen et al. (5).

FIG. 5 shows the wall shear stress τ_w as a function of the reduced coordinate ξ and the dimensionless pressure drop $\Delta p/L$ between two adjacent red cells. The wall shear stress drops from its largest value in the corner at $\xi = 0$ to its lowest value at $\xi = 0.5$. The region in the immediate neighborhood of the corner cannot be determined numerically. A perturbation analysis taylored for the corner problem has recently be carried out by Bruns (4). In the present work, however, we have not persued this particular point and used a quadratic extrapolation as an approximation for the shear stress in the vicinity of the corner.

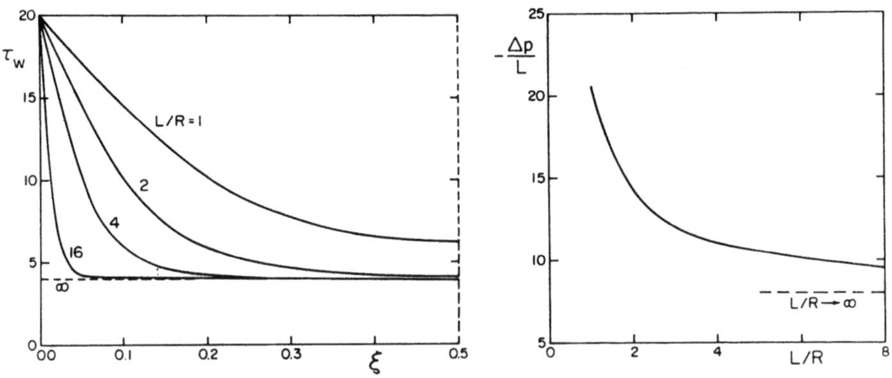

Fig. 5. Left: Wall shear stress as function of the reduced axial coordinate ξ. Right: Mean dimensionless pressure drop between two adjacent red cells

For $L/R \to \infty$, the wall shear stress attains the value -4 which corresponds to the value for fully-developed laminar pipe flow. The figure shows that a value for $L/R = 2$ is already sufficient to approach this asymptotic limit at $\xi = 0.5$.

The results for the pressure drop between two adjacent red cells have been obtained by integration of the wall shear stress with respect to the coordinate ξ. As the red cell spacing decreases, corresponding to an increase of the hematocrit, the pressure drop rises dramatically. As the cell spacing is increased, the pressure drop tends to the asymptotic value of -8 corresponding to Hagen-Poiseuille flow.

5. Summary

The flow of tightly fitting red cells in a capillary has been investigated theoretically. The fluid mechanical problem which describes the flow in the plasma gap between two adjacent red cells is of the Stokes flow type. This requires the solution of the biharmonic equation when the concept of the stream function is used. For large values of the red cell spacing which correspond to low values of the tube hematocrit, the flow leads to a singular perturbation problem. The perturbation parameter is the inverse of the aspect ratio of the plasma gap (axial distance L divided by tube radius R). Numerical solutions to the first two terms of the asymptotic expansions for $L/R \to \infty$ led to results for the stream function, velocity, shear stress and pressure drop in the plasmatic gap. The flow relative to the motion of the red cells is of a vortex-ring type and carries the name 'bolus flow'. The comparison of the results for the two-term expansion with numerical solutions of the full biharmonic equation showed excellent agreement down to aspect ratios as low $L/R = 2$.

References

1. Albrecht, K.H.; Gaehtgens, P.; Pries, A.; Heuser, M.: The Fåhraeus effect in narrow capillaries (i.d. 3.3 to 11.0 μm). Microvasc. Res. 18 (1979), 33-47.

2. Aroesty, J.; Gross, J.F.: Convection and diffusion in the microcirculation. Microvasc. Res. 2 (1970), 247-267.

3. Barbee, J.H.; Cokelet, G.R.: Prediction of blood flow in tubes with diameters as small as 29 μ. Microvasc. Res. 3 (1971), 17-21.

4. Bruns, F.W.: Numerical calculation on creeping flow in a closed cylindrical cavity. Computers and Fluids 7 (1979), 257-265.

5. Deen, W.M.; Robertson, C.R.; Brenner, B.M.: Concentration polarization in an ultrafiltering capillary. Biophysical J. 14 (1974), 412-431.

6. Gaehtgens, P.; Dührssen, C.; Albrecht, K.A.: Motion, deformation, and interaction of blood cells and plasma du-

ring flow through capillary tubes. Blood Cells 6 (1980), 799-812.

7. Gross, J.F.; Aroesty, J.: Mathematical models of capillary flow: A critical review. Biorheology 9 (1972), 225-264.

8. Papenfuss, H.-D.: Die Mechanik der Blutströmung und der Flüssigkeitsaustausch in der Mikrozirkulation. Habilitationsschrift, Ruhr-Universität Bochum (1979).

9. Papenfuss, H.-D.; Gross, J.F.: The interaction between transmural fluid exchange and blood viscosity in narrow blood vessels. Biorheology 14 (1977), 217-228.

10. Prothero, J.; Burton, A.C.: The physics of blood flow in capillaries. I. The nature of the motion. Biophysical J. 1 (1961), 565-575.

11. Schlichting, H.: Grenzschicht-Theorie. 5. Auflage. Verlag G. Braun, Karlsruhe (1965).

12. Schmid-Schoenbein, G.W.; Zweifach, B.W.: RBC velocity profiles in arterioles and venules of the rabbit omentum. Microvasc. Res. 10 (1975), 153-164.

13. Tözeren, H.; Skalak, R.: The steady flow of closely fitting incompressible elastic spheres in a tube. J. Fluid Mech. 87 (1978), 1-16.

14. Tözeren, H.; Skalak, R.: Flow of elastic compressible spheres in tubes. J. Fluid Mech. 95 (1979), 743-760.

15. Van Dyke, M.: Perturbation Methods in Fluid Mechanics. The Parabolic Press, Stanford, Cal. (1975).

16. Zien, T.-F.: Hydrodynamics of bolus flow - an analytical approach to blood flow in capillaries. Bull. Math. Biophys. 31 (1969), 681-694.

Experimental Method for Pressure-Velocity Correlation Measurements in Three-Dimensional Turbulent Boundary Layers

L. PAULSEN and K. O. FELSCH

Institut für Strömungslehre und Strömungsmaschinen
Universität Karlsruhe, D-7500 Karlsruhe, West Germany

Summary

An experimental method is proposed to investigate the lateral correlations of pressure and velocity within a three-dimensional turbulent boundary layer. The measurements are performed by a pressure-velocity correlation probe, consisting of two pairs of closely side by side mounted miniature hot-wire probes and 1/8″-microphones. The distance between the pairs is variable. That allows to analyse as well space correlations perpendicular to the stream line as time correlations. The data processing is done by a digital computer. Thus all correlations and spectra are calculated from one ensemble. By means of digital filter procedures a distinct analysis of micro-scale and macro-scale characteristics is possible.

Introduction

Most boundary layers occuring in practice are turbulent and three-dimensional. Therefore it is not surprising that a large number of investigations has been published concerning with this kind of fluid motion. A recent paper by Fernholz and Vagt [1] presents a list of investigations, in particular of the Reynolds stress tensor. The importance of boundary layer research is intensified by the fact that the behaviour of the outward flow region is determined by the boundary layer, in particular friction, separation, and displacement.

Because of the irregularity of turbulent flows the governing equations are extremely complicated. As actual computers are too slow and in general do not have enough memory address room the calculation is not yet practicable and won't be for a number of years. So different turbulence models have been introduced to reduce the necessary resolution of the computation [2]. These models are made by fitting experimental data by the time-averaged equations. So experimental investigations of different flow configurations have to be done in order to get experience with the application of turbulence models. On the other hand 'subgrid models' of turbulence have been used more widely in the recent time. These models describe the characteristics of small-scale turbulence and therefore are of rather universal nature. Because of the high requirements to the calculation procedure three-dimensional boundary layers are well suited for testing these models in practice.

The behaviour of mean velocities and Reynolds stresses in three-dimensional boundary layers is rather well-known from a number of investigations. Only few papers have been published concerning with the pressure fluctuations. These are, however, an essential characteristic of turbulent flows and correlated to the velocity field via the Navier-Stokes equations. In the case of incompressible flow pressure satisfies the Poisson equation and therefore is not a primary variable but can be expressed in terms of the velocity derivatives. This is utilized in most of the calculation procedures. Uberoi [3] has pointed out that in the case of isotropic turbulence the space-time correlations of pressure and velocity can be expressed by quadruple correlations of velocity fluctuations. Since the turbulence of three-dimensional boundary layers is not isotropic in this case we have to investigate the pressure-velocity correlations experimentally. There have been published several papers concerning with this subject but because of the problems of disturbance the flow by the pressure probe there only have been analysed the pressure fluctuation at the wall [e.g. 4]. So our work will be to measure the pressure-velocity correlation directly at different points within the flow region.

Experimental arrangement

The wind-tunnel provided for the investigation has been described in detail by Dechow [5]. It is a low speed wind-tunnel ($U_\infty < 50$ m/s) with a 3000 × 1500 × 300 mm test section in front of a vertically standing cylinder of 320 mm diameter. The probes are mounted in a traverse gear and can be positioned at any point of the measurement section. Since a number of results of mean velocity, Reynolds stresses,and static pressure at distinct positiones has been published it is suggested to investigate the pressure-velocity correlations at the same points.

The pressure-velocity correlation probe (fig. 1) consists of two pairs of 1/8″-microphones (Brüel & Kjaer 4138-type with nose cone) and miniature hot-wire probes (DISA 55P11-type). The distance between the pairs perpendicular to the mean flow direction is variable, as well horizontal

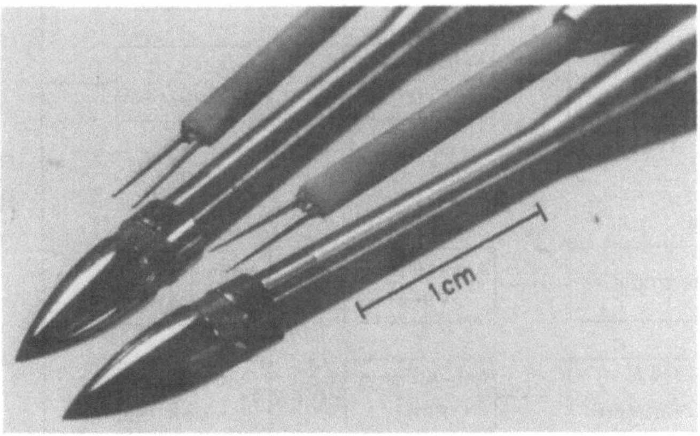

Fig. 1. Pressure-velocity correlation probe

as vertical (by rotating the probe). Thus we are able to calculate lateral space-time correlations along two co-ordinate axes. It will not be successful to put two pressure probes one behind another along the streamline itself to analyse longitudinal correlations because the second one would be interferred by the wake of the first one.

Since the pressure fluctuations are significantly stronger than the magnitude of the dynamic pressure, induced by the interaction of velocity fluctuations on the microphone body, it is assured that the microphones really record the pressure field in the flow. Past experience has shown that the frequency spectrum of velocity fluctuations is almost not influenced by introducing of the pressure probe into the flow and there is only little effect on its magnitude. Since correlation and coherency functions are normalized by the r.m.s. value of the fluctuations, anyhow, the distortion can be neglected in our case.

The schematic data acquisition system is shown in figure 2. The hot-wires are supplied by constant temperature anemometers and the condenser microphones by preamplifiers. In order to avoid aliasing problems low-pass filters with Butterworth characteristic and a decay of 24 dB per octave are employed. The cut-off frequency is chosen to be slightly less than half of the sample rate. Good experience has been made with 4/10th of the sample frequency. The output voltages are digitized by sample & hold converters with a capability of 13 kHz. 8 bit resolution for the microphone channels result in 0.4 % inaccuracy. Since the anemometer output voltage is a nonlinear function of the cooling velocity, in this case we have to use 12 bit converters to obtain the same accuracy of the velocity signal as for the pressure. The relationship between the cooling velocity and the output voltage is approximated by a polynomal shape function as described by

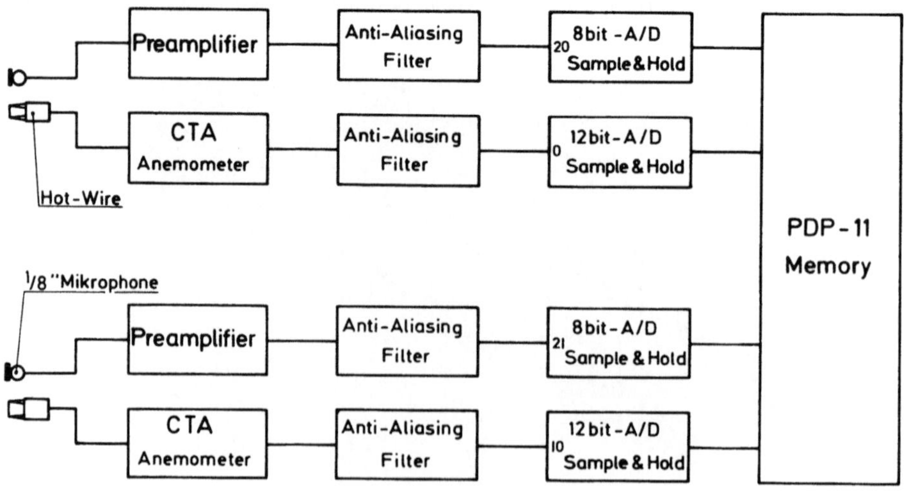

Fig. 2. Data acquisition system

Paulsen [6]. This provides better accuracy than the use of electronic linearizers. Our computer is a PDP 11/20-type with a 64 kB memory. This means in practice that it is not possible to record ensembles with more than 4 k samples. If the statistical behaviour of the flow requires more sample points we have to split up the data into subparts and average over the ensembles.

Spatial resolution of probes

A particular problem of turbulence measurements is the spatial and time resolution of the transducer-sensing element, as well for the pressure field as for the velocity field. Any sensor with finite size integrates over the fine structure of the local turbulent fluctuations and thus neglects physical information. Assuming locally frozen structure ($\partial .../\partial t \approx U_c \cdot \partial .../\partial x$) the geometric expansion of eddies (diameter 2R) is associated with a measured radian frequency ω by the convective velocity U_c. In our case the half sample frequency of 6.5 kHz limits the resolution to structures of approximately 1.5 mm if the velocity is 10 m/s. The effect of pressure transducer size in the case of homogeneous turbulence has been analyzed by Corcos [6]. The ratio of measured to actual spectral density $\Gamma_m(\omega)/\Gamma(\omega)$, i.e. the attenuation of the pressure mean square value at a given frequency, is shown in figure 3.

It must be noted, however, that the diagram is valid only for pressure and not for velocity fluctuations and is not exactly adoptable to our flow configuration because it is based on experimental results of pipe flow boundary layers. As it gives a good impression of the serious effect of finite size probes we shall take it for rough estimations.

A similar consideration is outlined to calculate the correlation between two sensors x and y positioned with a distance of 2 a (fig. 4). It is assumed that inside the 'eddy' pressure or velocity, respectively, are fully correlated and there is no correlation to the outer flow region. The eddies are regarded as coherent structures within a random field and a train of eddies with the diameter 2 R

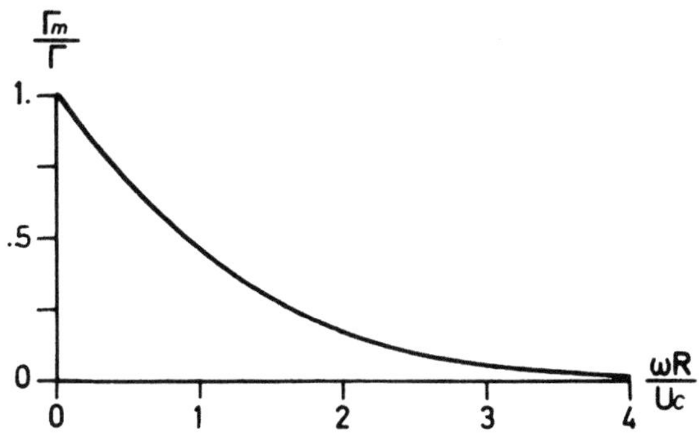

Fig. 3. Attenuation of the spectral density by the probe size

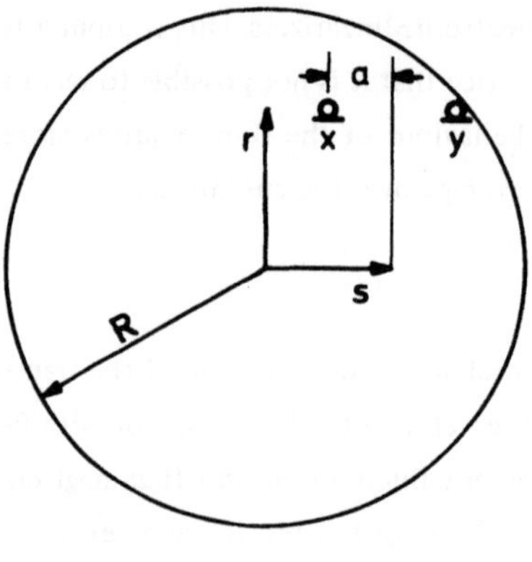

Fig. 4. Probe distance and eddy size

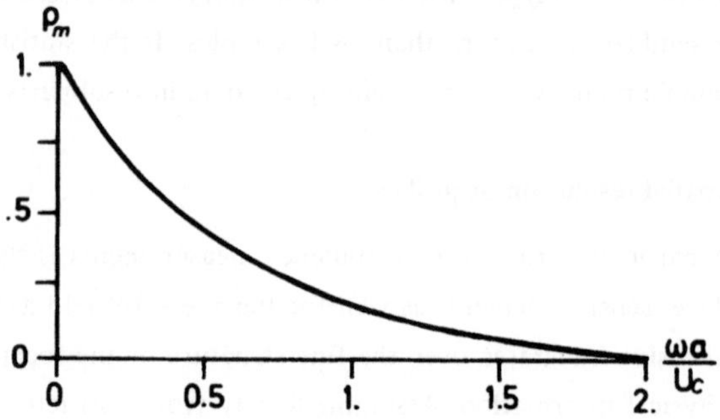

Fig. 5. Attenuation of the cross-correlation by the probe distance

will induce an initial frequency of ω when traveling along the sensors with the convective velocity U_c. Applying this model the attenuation of the measured cross-correlation coefficient ρ_{xy} at a given frequency simply is the fraction of the correlated component to the total signal component. The correlation equals unity if both sensors are placed within the eddy, i.e.

$$A_{corr} : \quad a^2 \leqslant (s+a)^2 + r^2 \leqslant R^2 \tag{1}$$

and a signal is recorded if at least one sensor is affected by an eddy:

$$A_{sig} : \quad a^2 \leqslant (s-a)^2 + r^2 \leqslant R^2 \quad . \tag{2}$$

The quadrature of the two areas yields

$$\frac{A_{corr}}{A_{sig}} = \frac{\frac{\pi}{4} - \frac{a}{2R}\sqrt{1-\frac{a^2}{R^2}} - \frac{1}{2}\sin^{-1}\frac{a}{R}}{\frac{\pi}{4} + \frac{a}{2R}\sqrt{1-\frac{a^2}{R^2}} + \frac{1}{2}\sin^{-1}\frac{a}{R}} \quad . \tag{3}$$

Since the cross-correlation is an energetic form we have to square equation (3) to receive the functional dependence of the frequency (fig. 5). Unlike figure 3 this diagram does not show the actual measured spectral density because we did not introduce experimental data of the distribution of fluctuations.

Computation method

For digital data analysis the time-dependent continuous function $x(t)$ is sampled with a constant sample time Δ according to

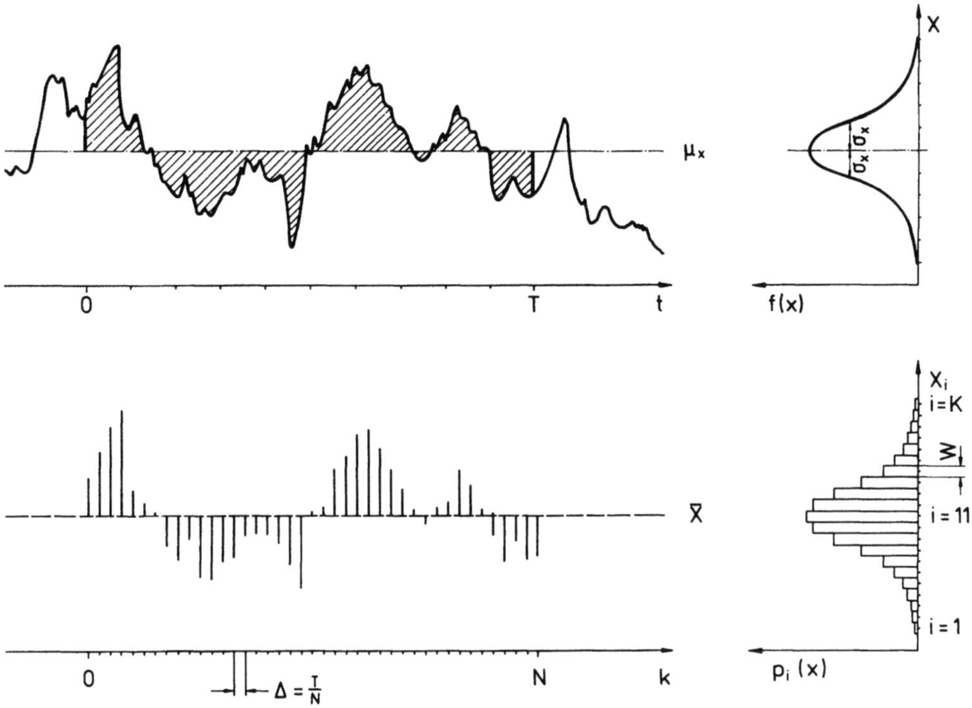

Fig. 6. Continuous and discrete process

$$x(k\Delta) = \sum_{k=-\infty}^{\infty} x(t)\,\Delta\,\delta(t-k\Delta) \tag{4}$$

where δ denotes the delta function. The sampled $x(k\Delta)$ are an infinite sequence of equidistant impulses, each of whose impulse strength is given by $x(t)\cdot\Delta$ for $t = k\Delta$. A finite record length T corresponds to the quantity N of one ensemble (fig. 6). For calculation of the sampled function, time integration is replaced by summation, e.g. the mean value μ_x corresponds to the sample mean $\bar{x}$:

$$\mu_x = \lim_{T\to\infty} \frac{1}{T} \int_0^T x(t)\,dt \quad ; \qquad \bar{x} = \frac{1}{N} \sum_{k=0}^{N-1} x(k\Delta) \ . \tag{5}$$

Because of the finite quantity N the discrete distribution function $p_i(x)$ is a stepped curve which approaches the probability density function $f(x)$ for sufficiently large N, as also the sum (5) approaches the integral.

All spectral and correlation functions are calculated via the well known FFT-procedure which enables a highly efficient computation [8]. The discrete Fourier transform can be written in the form

$$X\left(\frac{n}{N\Delta}\right) = \Delta \sum_{k=0}^{N-1} x(k\Delta)\, e^{-j2\pi\frac{nk}{N}} \tag{6}$$

where $n/N\Delta$ is a discrete frequency. The discrete cross-covariance [1] is

$$c_{xy}(m\Delta) = \frac{1}{N-m} \sum_{k=0}^{N-1-m} x(k\Delta)\, y(k\Delta + m\Delta) - \bar{x}\,\bar{y} \tag{7}$$

with the time-shift $m\Delta$. The same algorithm is valid for space correlations if Δ is replaced by a length scale. The Fourier transformated $C_{xy}(n/N\Delta)$ of the covariance is called the cross-spectrum. The algorithm used in practice is to multiply the complex conjugate transformated of one variable by the transformated of the other one:

$$C_{xy}\left(\frac{n}{N\Delta}\right) = X^*\left(\frac{n}{N\Delta}\right) Y\left(\frac{n}{N\Delta}\right) \tag{8}$$

and the inverse transform yields the cross-covariance:

$$c_{xy}(m\Delta) = \frac{1}{N\Delta} \sum_{n=0}^{N-1} C_{xy}\left(\frac{n}{N\Delta}\right) e^{j2\pi\frac{nm}{N}} \quad . \tag{9}$$

If the covariance function does not damp out rapidly with respect to N, it is necessary to add N zeros to the original functions x and y to avoid overlapping. The reason is that the above described procedure yields a circular function, that means c_{xy} contains as well the results for positive $(m = 0,1,2...,N-1)$ as for negative $(m = N-1, N-2,..., 0)$ shift.

In order to analyse as well the micro-scale as the macro-scale characteristics it is necessary to investigate the correlation as a function of the frequency. This can be done by means of the coherency function κ_{xy}, defined as the ratio of the magnitude of the cross-spectrum Γ_{xy} to the auto-spectra Γ_x and Γ_y of both time functions:

$$\kappa_{xy}^2(f) = \frac{|\Gamma_{xy}(f)|^2}{\Gamma_x(f)\, \Gamma_y(f)} \quad . \tag{10}$$

The analogue expression for the discrete function is

$$k_{xy}^2\left(\frac{n}{N\Delta}\right) = \frac{|C_{xy}(n/N\Delta)|^2}{C_x(n/N\Delta)\, C_y(n/N\Delta)} \quad . \tag{11}$$

Introducing equation (8) for the spectra it is seen that the discrete coherency function is equal unity independent of the actual coherency value:

[1]) The correlation coefficient ρ_{xy} is obtained by normalizing the covariance with the r.m.s. values of x and y.

$$k_{xy}^2 = \frac{[X^* Y]^* \; [X Y^*]}{[X^* X] \; [Y^* Y]} = \frac{X Y^* X^* Y}{X^* X Y^* Y} \equiv 1 \quad . \tag{12}$$

Various methods are known to avoid this peculiar problem. We give preference to the indirect calculation of the coherency function via the spectral-filtered correlation coefficient. The advantage of this method is that we do not have to process several ensembles because the necessary smoothing is processed over frequency components. So the spectral resolution is reduced but this is not important in our flow configuration because no periodic fluctuations occure.

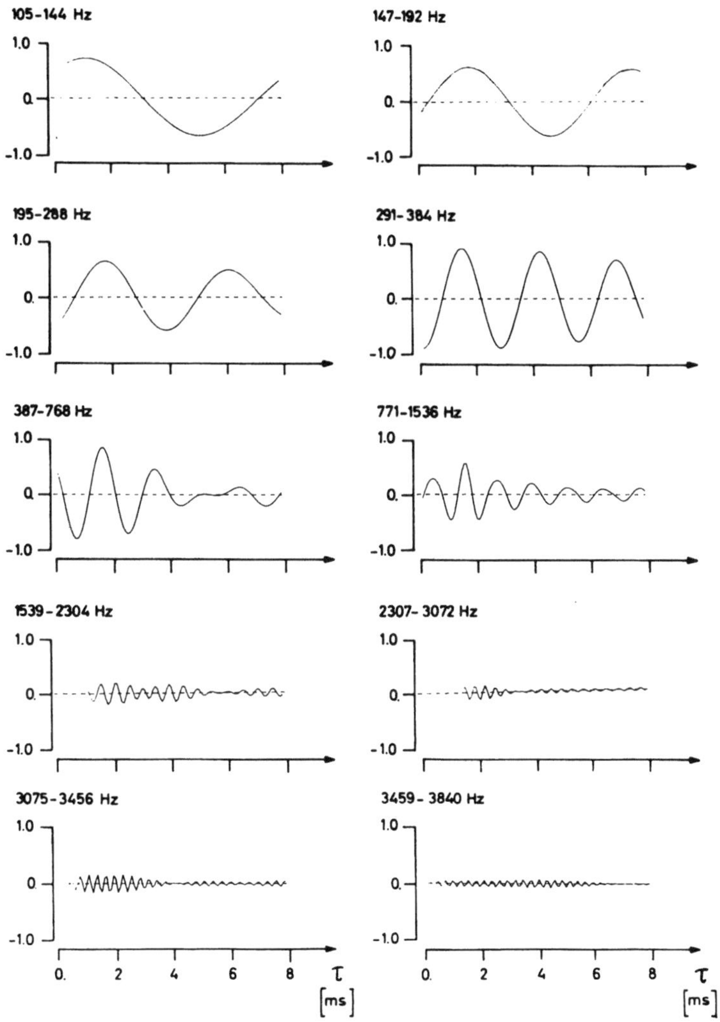

Fig. 7. Band-pass filtered correlation coefficient

An example of band-pass filtered correlation coefficients is shown in figure 7. It is seen that the two signals are best correlated at a time shift $\tau = 1.4$ ms and the coherency function is maximal in the frequency range from 291 to 384 Hz and damps out rapidly at higher frequencies.

It is hoped that the described method enables us to make more detailed analyses of the turbulent pressure field than have been done until now.

References

1. Fernholz, H.H.; Vagt, J.-D.: Turbulence measurements in an adverse-pressure-gradient three-dimensional turbulent boundary layer along a circular cylinder. J. Fluid Mech. 111 (1981) 233–269.

2. Kollmann, W. (Ed.): Prediction methods for turbulent flows. Washington: Hemisphere Publ. Corp. 1980.

3. Uberoi, M.S.: Quadruple velocity correlations and pressure fluctuations in isotropic turbulence. J. Aero. Sci. 20 (1953) 197–204.

4. Rotta, J.C.: Eine theoretische Untersuchung über den Einfluß der Druckscherkorrelationen auf die Entwicklung dreidimensionaler turbulenter Grenzschichten. DLR-FB 79-05 (1979).

5. Dechow, R.: Mittlere Geschwindigkeit und Reynoldsscher Spannungstensor in der dreidimensionalen turbulenten Wandgrenzschicht vor einem stehenden Zylinder. Univ. Karlsruhe, Dr.-Ing.-Diss., Strömungsmechanik u. Strömungsmaschinen 21 (1977).

6. Paulsen, L.: Triple hot-wire technique for simultaneous measurements of instantaneous velocity components in turbulent flows. Subm. to J. Phys. E.: Sci Instrum. (1982).

7. Corcos. G.M.: Resolution of pressure in turbulence. J. Acoust. Soc. Am. 35 (1963) 192–199.

8. Bendat, J.S.; Piersol, A.G.: Random data: analysis and measurement procedures. New York: Wiley & Sons, Inc. (1971).

Random Walk and Diffusion in Two-Dimensional Lagrangian Systems

NORBERT PETERS and HANS-JÜRGEN THIES
Rheinisch-Westfälische Technische Hochschule Aachen

Abstract

The statistical properties of the random vortex method are analysed. This method calculates the Lagrangian motion of vortices and simulates diffusion by a random walk process. It is shown that the probability density function of the spatial distribution of the vortices satisfies a Fokker-Planck equation. This equation can be reduced to the vorticity transport equation for two-dimensional flow. It is concluded that the diffusion coefficient used to simulate two-dimensional turbulent flow fields by the random vortex method should be interpreted as a turbulent diffusion coefficient.

Introduction

Ever since large coherent structures in turbulent shear flows were discovered /1/, /2/, they represented a challenge for theoreticians. There have been attempts to model the turbulent flow by splitting the flow quantities into a mean flow part, a large structure part and a fine-grained turbulence part /3/, /4/. However, the closure problem associated with the interaction of these quantities is far from being resolved.

There exists, however, a numerical method due to Chorin /5/ based on vortex dynamics that seems to be able to simulate the large coherent structures. Ashurst /6/ was the first to show that the method simulates the formation and the merging process of large coherent structures in shear layers. The method uses a Lagrangian formulation to account for the inviscid convection of a large number of vortices. The interaction of the vortices is taken into account by Biot-Savart's law. To simulate diffusion Chorin /5/ has suggested to superimpose a random walk upon the inviscid motion. The step length of the random motion is calculated by analogy between diffusion and random walk according to Einstein's work on Brownian motion /7/.

In using the random vortex method to simulate large coherent structures there has been uncertainty about how to interpret the nature of the diffusion process. If the diffusion of vorticity corresponds to the molecular transport process, as Chorin suggests, then an nonstationary two-dimensional laminar flow field is calculated. The number of vortices would then need to be very large - representing individual molecules of the system. In fact, it has been shown that a very large number of particles is needed to solve even the one-dimensional heat conduction problem /8/. If, on the other hand, the diffusion process is to be considered turbulent, then the choice of the turbulent diffusion coefficient and the turbulent length scale distribution are unknown input parameters. It is the aim of the present paper to discuss these questions from the point of view of a stochastic formulation of the random vortex method.

The Random Vortex Method as a Stochastic Process

The random vortex method is defined by a set of one-dimensional differential equations for the displacement of the n vortices in two dimensions x and y. The motion of the vortex i is given by

$$\begin{aligned} \Delta x_i &= u_i \Delta t + (\Delta x_i)_{random} \\ \Delta y_i &= v_i \Delta t + (\Delta y_i)_{random} \qquad i = 1, 2, \ldots, n \end{aligned} \tag{1}$$

where Δt is the time step. The velocities u_i and v_i are induced according to Biot-Savart's law which is given by

$$\begin{aligned} u_i &= - \sum_{\substack{j=1 \\ j \neq i}}^{n} (y_i - y_j) \frac{\gamma_j}{2\pi r_{ij}^2} \\ v_i &= \sum_{\substack{j=1 \\ j \neq i}}^{n} (x_i - x_j) \frac{\gamma_j}{2\pi r_{ij}^2} \end{aligned} \tag{2}$$

$$r_{ij}^2 = (x_i - x_j)^2 + (y_i - y_j)^2$$

The other vortices j are located at x_j, y_j and have the circulation γ_j. The random part in eq.(1) is calculated by

$$(\Delta x_i)_{random} = (-4\nu\Delta t \ln s_i)^{1/2} \cos(2\pi q_i)$$
$$(\Delta y_i)_{random} = (-4\nu\Delta t \ln s_i)^{1/2} \sin(2\pi q_i) \tag{3}$$

where s_i and q_i are equally distributed random numbers between 0 and 1, such that s_i defines the length and q_i the direction of the random walk of vortex i. The diffusion coefficient ν is the quantity of main interest in this analysis. The distribution function of the random displacements Δx_i, Δy_i can be shown to be a joint normal distribution (cf. /9/, chapter 7.2).

$$p(\Delta x_i,\Delta y_i) = (4\pi\nu\Delta t)^{-1} \exp\{-\frac{\Delta x_i^2 + \Delta y_i^2}{4\nu\Delta t}\} \tag{4}$$

with equal variance

$$\sigma^2 = \sigma_x^2 = \sigma_y^2 = 2\nu\Delta t \tag{5}$$

in the two directions.

In the limit $\Delta t \to 0$ the random walk process defined by eq.(3) turns out to be a two-dimensional Wiener-Lévy process (cf. /9/, p. 292). Replacing Δt by t and Δx, Δy by x and y, respectively, eq.(4) satisfies the diffusion equation

$$\frac{\partial p}{\partial t} = \nu(\frac{\partial^2 p}{\partial x^2} + \frac{\partial^2 p}{\partial y^2}) \tag{6}$$

Statistical Properties of the Random Vortex Method

To analyse the general properties of the method one must go beyond the construction of single realizations and go into statistics. Then it is seen that the quoted properties of the system of equations (1) define a Markov process. In such a process the statistical properties of a system at a time $t > t_0$ are uniquely determined by the properties at time t_0. A quantitative measure of these statistical properties is given by a probability distribution function (pdf). In case of eq.(1) the distribution function

$$p(x_1,y_1,\ldots,x_n,y_n,t\,|\,x_{1,0},y_{1,0},\ldots,x_{n,0},y_{n,0},t_0) \tag{7}$$

describes the probability density that the vortices 1,2,...,n are at the locations $x_1,y_1;x_2,y_2;\ldots;x_n,y_n$ at time t, if they were originally at time t_0 at the locations $x_{1,0},y_{1,0};x_{2,0},y_{2,0};\ldots;x_{n,0},y_{n,0}$. It is a conditioned pdf, but to simplify notation, we will omit in the following the conditioning upon the initial conditions at t_0.

It may be shown (Soong /10/, chapter 7) that the pdf of a process defined by a system of first order differential equations satisfies a Fokker-Planck equation

$$\frac{\partial p}{\partial t} + \sum_{i=1}^{n} \left(\frac{\partial u_i p}{\partial x_i} + \frac{\partial v_i p}{\partial y_i}\right) = \frac{1}{2} \sum_{i=1}^{n} \sum_{j=1}^{n} \left(\frac{\partial a_{ij} p}{\partial x_i \partial x_j} + \frac{\partial b_{ij} p}{\partial y_i \partial y_j}\right) \tag{8}$$

where u_i and v_i are the drift velocities defined by eqs.(2). Since the random walk of vortices is independent of each other one obtains for the diffusion coefficients with eq.(5)

$$a_{ij} = b_{ij} = \lim_{\Delta t \to 0} \left\{\frac{1}{\Delta t} \sigma^2\right\} = 2\nu\delta_{ij} \tag{9}$$

and therefore the equation

$$\frac{\partial p}{\partial t} + \sum_{i=1}^{n} \left(\frac{\partial u_i p}{\partial x_i} + \frac{\partial v_i p}{\partial y_i}\right) = \nu \sum_{i=1}^{n} \left(\frac{\partial^2 p}{\partial x_i^2} + \frac{\partial^2 p}{\partial y_i^2}\right) \tag{10}$$

which, for vanishing drift velocities, reduces to eq.(6). Equation (10) is a 2n-dimensional parabolic equation. Special solution methods have been developed for this equation (cf. /10/). The solution, if it was known, would describe the distribution of the n vortices in a 2n-dimensional space for every time t, when the original distribution at time t_0 is given by the initial conditions to eq.(10).

Reduction of the Fokker-Planck Equation to the Vorticity Transport Equation

Rather than to solve eq.(10) we want to show here that it reduces to the vorticity transport equation of fluid mechanics. To do this we must introduce the following theorems from probability theory:

1.) The integration of the joint probability density p(x,y) over y yields the marginal density p(x)

$$p(x) = \int_{-\infty}^{+\infty} p(x,y)dy \tag{11}$$

2.) Bayes's theorem: Any joint probability density p(x,y) may be represented as the product of the conditional density p(y|x) and the marginal density p(x)

$$p(x,y) = p(x)\, p(y|x) \tag{12}$$

where p(y|x) is the distribution of y under the condition that x is fixed.

3.) The conditional expectation of a function u(x,y) at fixed x is given by

$$\bar{u}(x) = \int_{-\infty}^{+\infty} u(x,y)\, p(y|x)dy \tag{13}$$

One may generalize these theorems for 2n dimensions by writing the marginal density p_i

$$p_i(x_i,y_i,t) = \iint\underset{-\infty}{\overset{+\infty}{\cdots}}\iint p(x_i,y_i,x_1,y_1,\ldots,x_j,y_j,\ldots \\ x_{n-1},y_{n-1},t)\; dx_1dy_1\ldots dx_jdy_j\ldots dx_{n-1}dy_{n-1} \tag{14}$$

The pdf p may be split into the marginal density p_i and the conditional density p_j

$$p = p_i p_j(x_1,y_1,\dots,x_j,y_j,\dots,x_{n-1},y_{n-1},t|x_i,y_i) \qquad (15)$$

where p_i is defined by eq.(14).

The conditional average velocity $\bar{u}_i$ induced at x_i, y_i by all other vortices is then

$$\bar{u}_i(x_i,y_i,t) = \int\int_{-\infty}^{+\infty}\dots\int\int u_i p_j \, dx_1 dy_1 \dots dx_j dy_j \dots dx_{n-1} dy_{n-1} \qquad (16)$$

With these definitions an integration over 2n-2 dimensions x_1, $y_1,\dots,x_j,y_j,\dots,x_{n-1},y_{n-1}$ except x_i and y_i of eq.(10) is performed such that only the marginal density p_i remains. Due to the boundary conditions at $-\infty$ and $+\infty$ all integrals over derivatives in terms of x_j and y_j yield no contribution. One obtains the equation

$$\frac{\partial p_i}{\partial t} + \frac{\partial \bar{u}_i p_i}{\partial x_i} + \frac{\partial \bar{v}_i p_i}{\partial y_i} = \nu\left(\frac{\partial p_i}{\partial x_i^2} + \frac{\partial p_i}{\partial y_i^2}\right) \qquad (17)$$

The quantity p_i is the probability density of finding the vortex i at the position x_i, y_i in the statistical average over all possible positions of the other vortices j. Therefore the integration represents an ensemble average over many realisations by the random method. The velocities $\bar{u}_i$ and $\bar{v}_i$ are average velocities of the motion corresponding to vortex i.

The hydrodynamic vorticity field $\omega_i(x,t)$ originating from vortex i at point x, y is given by (cf. /11/)

$$\omega_i = \gamma_i \delta(x-x_i(t),y-y_i(t)) \qquad (18)$$

where x, y are the spatial coordinates and $x_i(t)$, $y_i(t)$ the Lagrangian position of vortex i at time t. The ensemble average vorticity is given by

$$\overline{\omega}_i = \gamma_i \overline{\delta(x-x_i(t),y-y_i(t))} = \gamma_i p_i(x_i,y_i,t) \qquad (19)$$

Therefore the probability function p_i may itself be interpreted as the average vorticity field originating from vortex i at point x_i, y_i. Then the total vorticity field originating from all vortices is given by

$$\overline{\omega} = \sum_{i=1}^{n} \overline{\omega}_i = \sum_{i=1}^{n} \gamma_i p_i \qquad (20)$$

Likewise the total flow of vorticity is

$$\overline{u}\,\overline{\omega} = \sum_{i=1}^{n} \overline{u}_i\,\overline{\omega}_i = \sum_{i=1}^{n} \overline{u}_i\,\gamma_i\,p_i \qquad (21)$$

where eq.(19) has been used. Multiplying eq.(17) by γ_i, replacing x_i and y_i by x and y, and summing over all i one obtains finally

$$\frac{\partial\overline{\omega}}{\partial t} + \frac{\partial\overline{u\omega}}{\partial x} + \frac{\partial\overline{u\omega}}{\partial y} = \nu\left(\frac{\partial^2\overline{\omega}}{\partial x^2} + \frac{\partial^2\overline{\omega}}{\partial y^2}\right) \qquad (22)$$

This is the vorticity transport equation, which, together with an integral representation of eq.(2) describes the two-dimensional flow field in Eulerian coordinates. In a laminar flow, ν is the molecular viscosity and the mean quantities $\overline{u}$, $\overline{v}$ and $\overline{\omega}$ are averaged over molecular fluctuations. In a turbulent flow, ν is to interpreted as a turbulent viscosity coefficient and the mean quantities are ensemble averages over many turbulent flow fields.

Conclusions

The vorticity transport equation of fluid mechanics is obtained as a special case of the Fokker-Planck equation when a reduction of the dimension and a superposition of the vorticity originating from every vortex is performed. The analogy between diffusion and random walk remains therefore valid even in the presence of convection. As the over-all diffusion process in a turbulent flow field is mainly due to turbulent motion rather than the molecular transport process, the diffusion coefficient appearing in eq.(3) and finally in eq.(17) must be considered as a turbulent viscosity. A turbulent length scale distribution is introduced in an indirect way via eq.(1) and eq.(3) and it is largely dependent on the choice of the time step.

There remains, however, the difficulty of how to determine the turbulent viscosity. A particular interesting application of the random vortex method is of to free shear flows, where the method is able to reproduce the intermittent structure of the flow. In those cases the turbulent viscosity must be that of the turbulent part of flow field and not an average over the turbulent and non-turbulent flow, and is more likely to be constant.

References

/1/ Brown, G.L.; Roshko, A.: On density effects and large structures in turbulent mixing layers. J. Fluid Mech. 64 (1974) 775-816.

/2/ Roshko, A.: Structure of turbulent shear flows: a new look. AIAA J. 14 (1976) 1349-1357.

/3/ Hussain, A.K.M.F.: Coherent structures and studies of perturbed and unperturbed jets. In: The role of coherent structures in modelling turbulence and mixing. Ed. J. Jimenez, Lecture Notes in Physics 136 (1981) 252-291, Springer Verlag Berlin Heidelberg New York.

/4/ Liu, J.T.C.: Internections between large-scale coherent structures and fine-grained turbulence in free shear flows. In: Transition and Turbulence. Ed. R.E. Meyer, Academic Press 1981, 167-214.

/5/ Chorin, A.J.: Numerical study of slightly viscous flow. J. Fluid Mech. 57 (1973) 785-796.

/6/ Ashurst, W.T.: Numerical simulation of turbulent mixing layers via vortex dynamics. In: Turbulent shear flows I. Ed. F. Durst, B.E. Launder, F.W. Schmidt, H.J. Whitelaw, 402-413, Springer Verlag Berlin Heidelberg New York, 1979.

/7/ Einstein, A.: Zur Theorie der Brown'schen Bewegung. Ann. d. Physik 19 (1906) 371-381.

/8/ Peters, N.: Calculation of the instationary turbulent flow past a flat plate using a net-free numerical method. Proceedings of the GAMM-Conference on Numerical Methods in Fluid Mechanics, 134-141, Ed. E.H. Hirschel, Köln 1975.

/9/ Papoulis, A.: Probability, random variables and stochastic processes. Mc Graw-Hill 1965.

/10/ Soong, T.T.: Random differential equations in science and engineering. Academic Press New York - London 1973.

/11/ Leonard, A.: Vortex methods for flow simulation. J. Comp. Phys. 37 (1980) 289-335.

Mesh Influence on Vortex Shedding in Inviscid Flow Computations

ARTHUR RIZZI

FFA The Aeronautical Research Institute of Sweden
Box 11021
S-161 11 Bromma, Sweden

Introduction

There is presently a great amount of activity in the numerical solution of the Euler equations governing aerodynamic flows. Flows computed with this inviscid model have been observed to separate at the trailing edge of airfoils and wings even without recourse to an explicit Kutta condition. In my own work I have seen the flow separate from the leading edge of a sharp delta wing at angle of attack, and perhaps most surprisingly even from the smooth tip of a trapezoidal wing - all without the use of a Kutta condition. Such observations raise the question what is the mechanism in the equations that instigates this separation? There are two most likely possibilities. The first is the argument that if flow were to remain attached around a sharp edge a slipline and shock would appear, a loss in total pressure would ensure, and vorticity would be generated which ultimately would lead to separation. The second explanation is that the discrete equations possess a certain amount of numerical viscosity embodied in their truncation error even though the differential equations are inviscid, and that this viscosity can generate shear which leads to separation. The intent here is to investigate this latter explanation in an indirect way. If numerical viscosity is the responsible mechanism, then it is grid dependent i.e., it goes to zero as the grid is refined. I therefore want to observe the degree to which computed three dimensional solutions are grid dependent and to judge if the sequence of solutions obtained as the grid is refined converges. The numerical solution procedure and grid generation used in this study are ones which we developed previously[1] and the cases which I intend to study under grid refinement are those we computed before with coarser nets[2].

Supersonic Flow Past Delta Wing

Consider supersonic flow, $M_\infty = 1.5$ $\alpha = 15°$, past a delta wing whose chordwise sections are 6% thick circular arcs and whose leading edge is swept 70° and is subsonic. This case was computed before[2] using the mesh depicted in Fig. 1, and contours of constant c_p obtained on the upper surface together with several spanwise distributions of c_p are given by the full lines in Fig. 2. Although the flow separates at the sharp leading edge and rolls up forming the primary vortex over the leeward surface, the detailed comparison with experiment reveals a disagreement in the position and strength of the vortex. The dashed lines in Fig. 2 are the solution computed with the mesh spacing halved in the normal direction near the wing surface. The differences between these two solutions are small, but to judge by the distribution at the 45% section the vortex seems to develop a little further upstream in the contracted-mesh solution, and at the 85% section we see a small re-expansion peak under the core of the vortex and a steepening of the compression slope. This suggests that the vortex is better resolved in the refined discretization, but the computed vortex still is farther outboard and somewhat weaker than the one observed experimentally, for which at the moment I have no explanation. The overall comparison is summed up in the lift and drag coefficients, respectively 0.475 and 0.134 for the original computation and 0.448 and 0.127 for the contracted net. I can well believe this variation to be within the realm of convergence with mesh refinement.

Transonic Flow Past M6 Wing

That the computed solution for the previous case does not vary greatly with a change in the mesh is to be expected since separation from a salient edge is insensitive to Reynolds number. Separation from a smooth surface is a more critical case because the details of the separation line then depend directly on the viscosity in the physical flow. Nevertheless using 0-0 Mesh 1 (Fig. 3) around the ONERA M6 wing with a rounded corner at the leading edge of the tip (Fig. 4) with 64 cells around the chord, 14 cells outward into the field, and 20 along the span we computed previously the solution for transonic flow $M_\infty = 0.84$ $\alpha = 3.06°$ drawn with full lines in Fig. 5, one which compares reasonably well with the experiment even at the tip section. This case then is a good standard to evaluate two grid effects, cell size and tip rounding. The first one is studied by the computation on Mesh 2, identical to Mesh 1 except now the cell density is increased to 96 around the chord and 16 normal to the wing, and plotted with dashed lines in Fig. 5. Here we see an enormous difference in the solutions given on these two nets. It's most startling at the tip (100% section) where the denser-mesh calculation shows a highly supersonic flow that decelerates through a strong shock right at the trailing edge. But even more surprising is the lower pressure on the rearward chord from tip to root that accounts for a 50% increase in drag but very little difference in lift (see Table 1). The only explanation I can offer for this situation is that Mesh 1 induces a level of numerical viscosity which causes the flow fortuitously to separate and roll-up in a manner very near to the measured one, whereas in Mesh 2 the level is too low and a totally unphysical flow is computed at the tip. Unfortunately this unrealistic feature is not confined locally at the tip but has a widespread influence.

Startled by these results I grew suspicious about other details at the tip that might have an influence on the flow separating there, namely the roundness of the leading edge of the tip. In the construction of an O-O mesh we can choose whether to have a rounded or sharp corner at the tip as illustrated in Fig. 4. Our sharp-cornered tip is more similar to the one in a C-H or O-H mesh although that tip has a chisel-like edge, which presumably forces separation, while ours is more rounded away from the corner. Since we wanted to focus on tip flow, the round-cornered mesh always seemed preferable, but now in light of the last results I suspect it delays the onset of inviscid separation. Perhaps the sharp corner would trigger separation farther upstream, so we constructed Mesh 3, identical to Mesh 2 except that the leading edge tip is sharp and the cell density in the span direction is 14 instead of 20. The comparison of the solutions computed on these two grids is presented in Fig. 6. In the two tip sections 95% and 99% the loading is markedly higher for the solution on Mesh 3, a good indication that the flow is separating and rolling up into a vortex nearer to the corner, either because the corner is sharp or the spanwise discretization is more coarse. But look at how this effect is felt in all of the inboard sections too, producing a higher loading on the forward half of the chord and strengthening the shock, reflected in the much higher lift and drag coefficients 0.340 and 0.0218 in Table 1. So once again we see a local detail at the tip causing a non-local effect.

Concluding Remarks

In the case of the sharp-edge delta wing we have observed that the separation at the leading edge and roll-up of the computed flow was independent of the mesh spacings in my two examples. But when the surface is smooth, like the tip of the M6 wing, we have seen that various changes in the grid can produce enormous differences in the computed flow which I attribute to the inviscid separation being dependent upon the numerical viscosity. It would appear then that separating flow can be simulated realistically by the numerical solution of the Euler equations without any Kutta condition only if the obstacle has a sharp edge. Otherwise a Kutta condition is required, in which case it may be simpliest from a practical standpoint to construct a tip that is sharp along part or all of its edge. This may well be the closest we can come to reality with an inviscid flow model, and a more refined simulation will demand the viscous equations.

Acknowledgement

Support for this work, in part by the Office of Naval Research while the author was on sabbatical at the Mecanical and Aerospace Engineering Department of Princeton University and in part by the Swedish Defence Materiel Administration, is gratefully acknowledged.

References

1. Rizzi, A. and Eriksson, L-E.: Transfinite Mesh Generation and Damped Euler Equation Algorithm For Transonic Flow Around Wing-Body Configurations, AIAA Paper No. 81-0999, June 1981.

2. Eriksson, L-E. and Rizzi, A.: Computation of Vortex Flow Around Wings Using the Euler Equations, Proc. IVth GAMM Conf. Num. Meth. Fluid Mech., ed. H. Viviand, Vieweg Verlag, Braunschweig, 1982.

MESH SIZE		LIFT	DRAG
MESH 1	rounded 64x14x20	0.280	0.0115
MESH 2	rounded 96x16x20	0.273	0.0179
MESH 3	sharp 96x16x14	0.340	0.0218

Table 1. Comparison of lift and drag for the ONERA M6 wing computed on three different meshes. $M_\infty = 0.84 \quad \alpha = 3.06$.

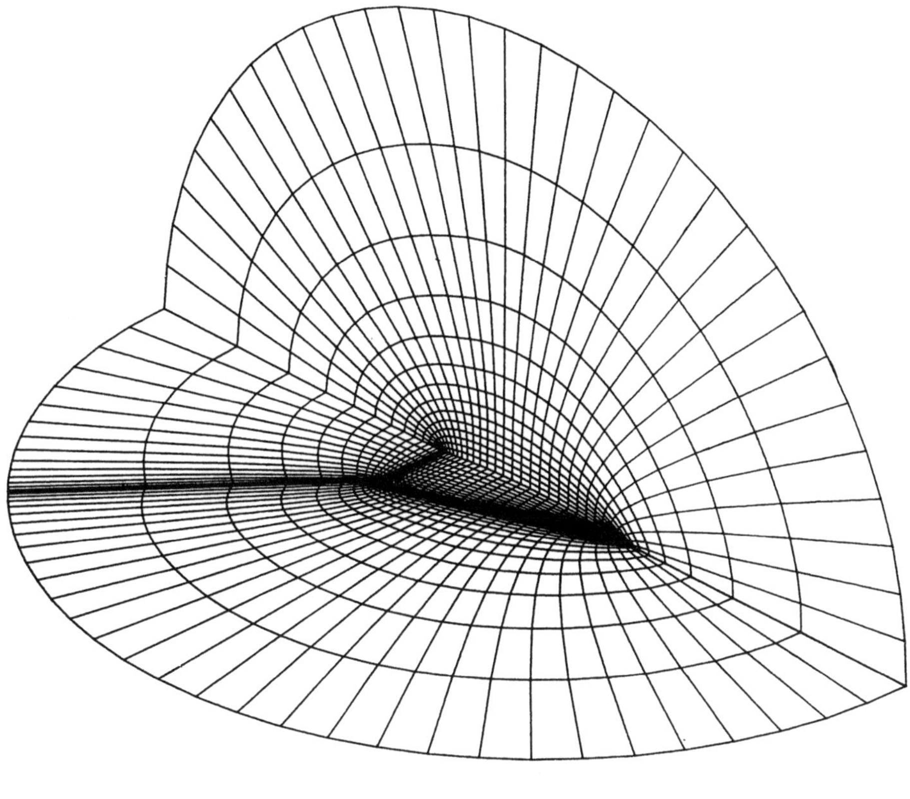

Fig. 1 Topography of the O-O mesh around the delta wing.

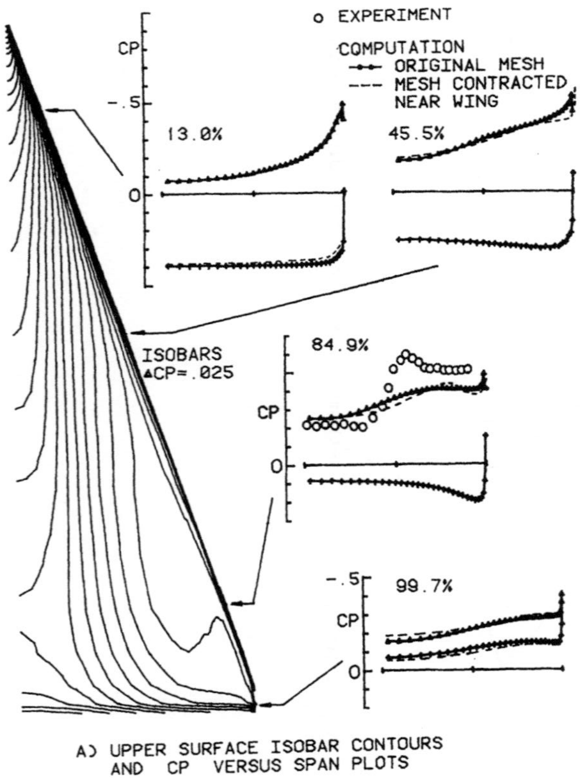

Fig. 2 Comparison of c_p versus span computed on two different meshes. $M_\infty = 1.5$ $\alpha = 15^0$.

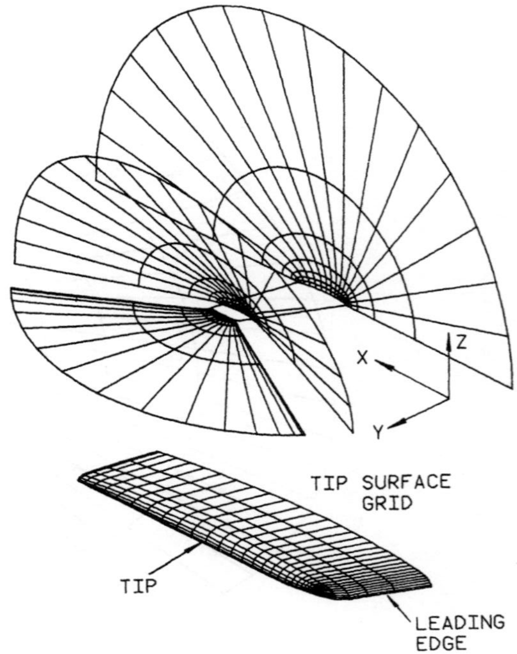

Fig. 3 Topography of the O-O mesh around the ONERA M6 wing.

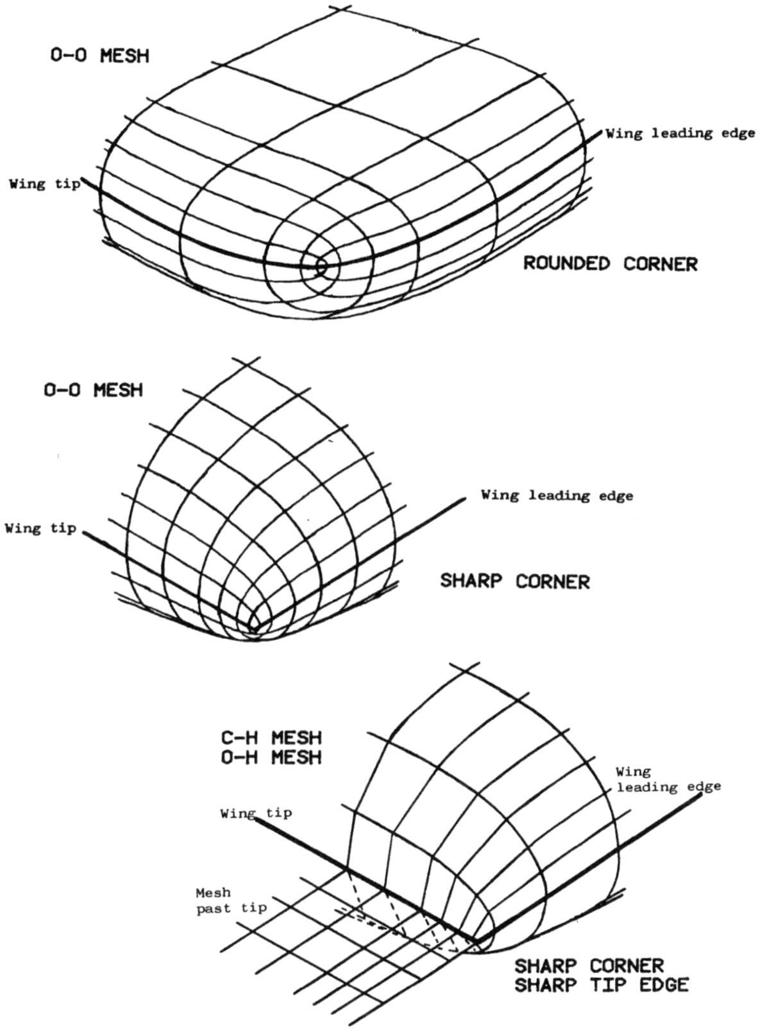

Fig. 4 An O-O- mesh can have either a rounded or sharp corner at the leading edge of the tip. The latter case is more similar to the tip in a C-H or O-H mesh although away from the tip the edge is smooth.

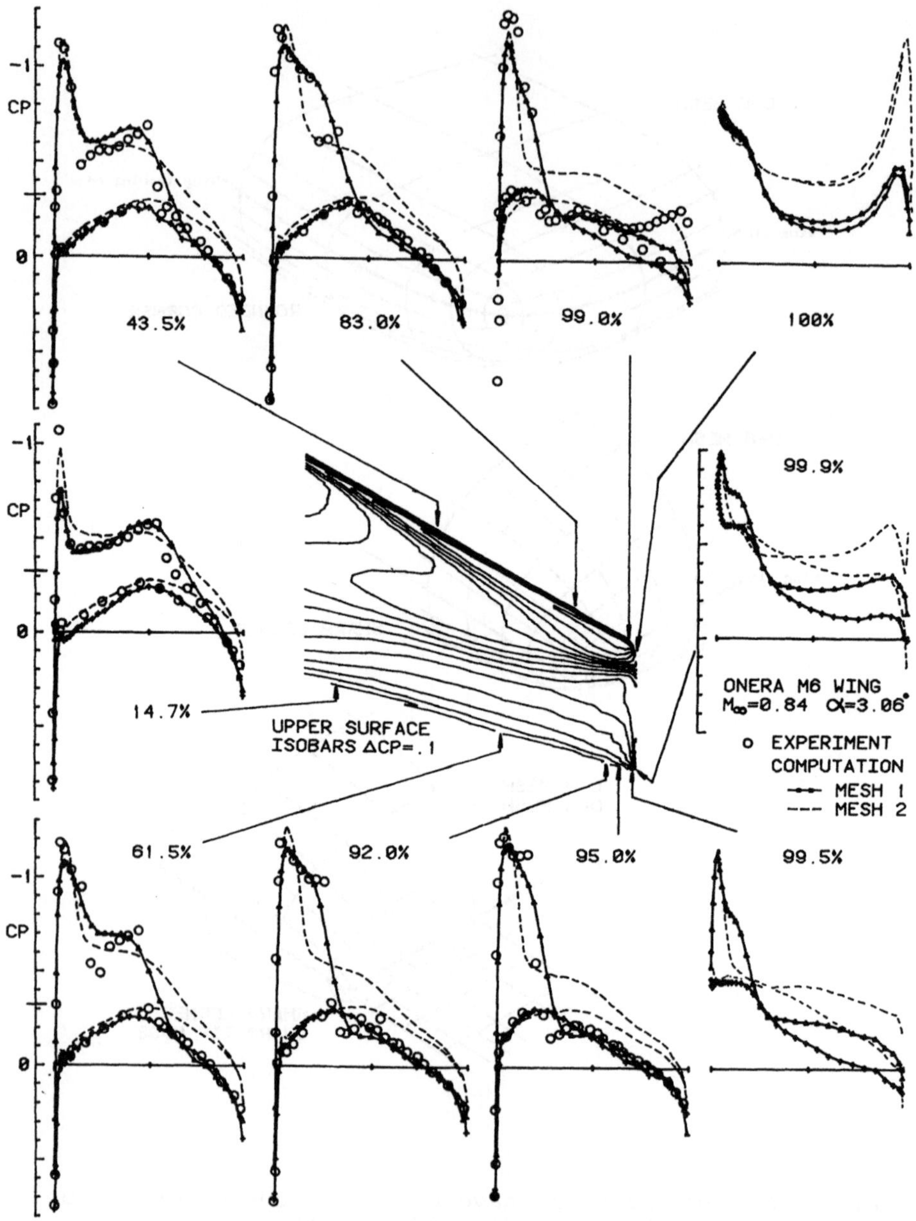

Fig. 5 Pressure distributions on the wing versus chord computed on Mesh 1 and Mesh 2 and compared with experiment. $M_\infty = 0.84$ $\alpha = 3.06$.

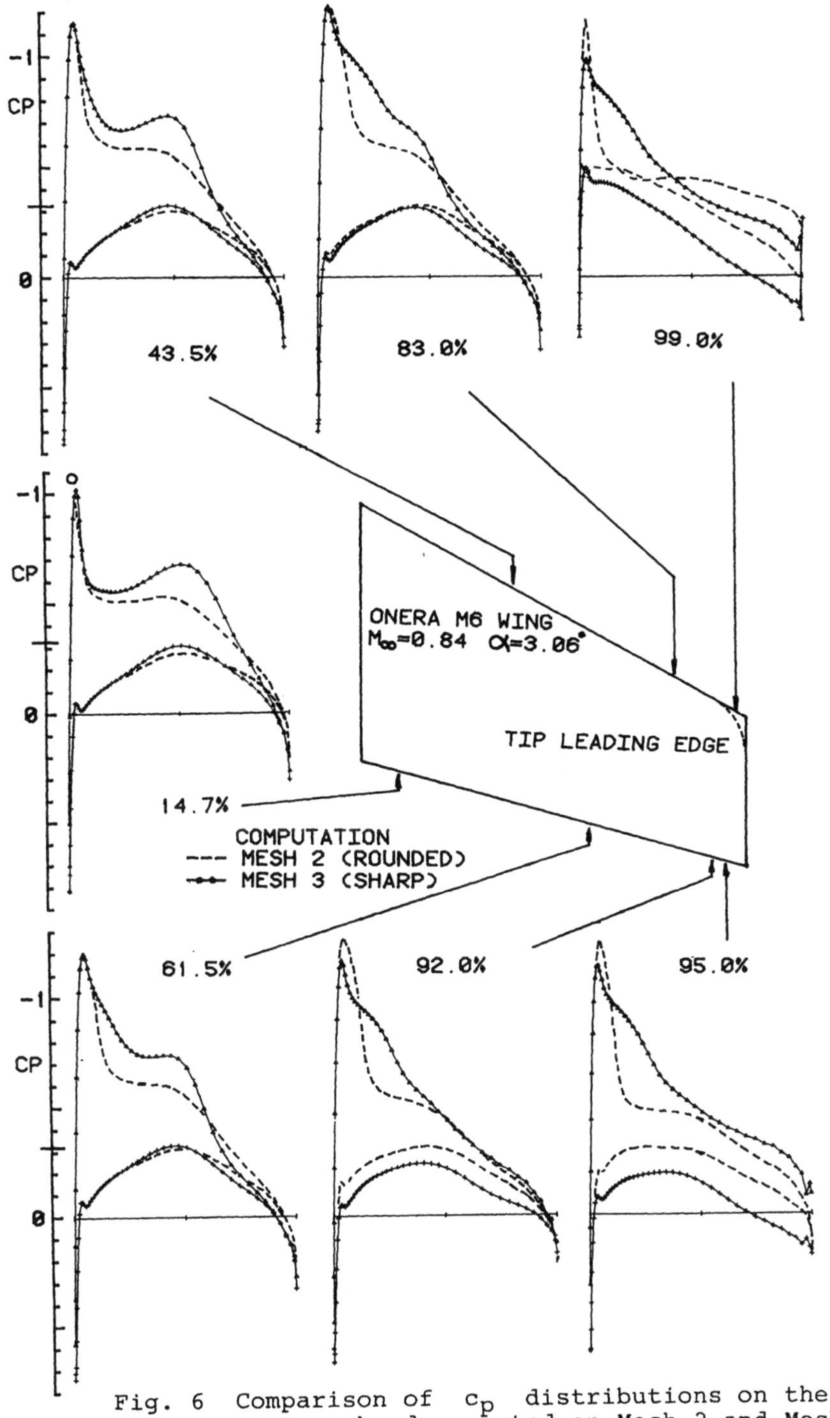

Fig. 6 Comparison of c_p distributions on the wing versus chord computed on Mesh 2 and Mesh 3. $M_\infty = 0.84$ $\alpha = 3.06$.

A Numerical Method to Solve the Steady-State Navier-Stokes Equations for Natural Convection in Enclosures

Y.D. SCHKALLE, F. THIELE and H. WAGNER

Hermann-Föttinger-Institut für Thermo- und Fluiddynamik
Technische Universität Berlin, West-Germany

Summary

A numerical method is presented for thermally driven laminar flow in two-dimensional enclosures. The procedure solves the Navier-Stokes equations in terms of the stream function by using the finite-difference method of Hermitian type. For the solution of the complex system of finite-difference equations a direct solver is developed based on the LU decomposition of the matrix. In order to linearize the nonlinear stream function equation two methods, Biharmonic Driver and Newton-Raphson, are considered. Results with respect to accuracy and rate of convergence demonstrate the reliability of the procedure for low Rayleigh number flows.

Introduction

The problem of steady two-dimensional natural convection arises in many practical applications such as thermopane windows, solar energy collection, etc. For rectangular enclosures with side walls of different uniform temperatures extensive studies have been made and the physical behaviour is reasonably well understood. This flow induced by natural convection is also an excellent singularity free test case for numerical algorithms with respect to viscous flows [1].

The flow in the enclosure is described by the Navier-Stokes equations together with the energy equation. One main difficulty in solving this coupled set of elliptic equations is caused by the nonlinear terms of the Navier-Stokes equations. Most of the numerical procedures developed are based upon differential equations of second order. Recent investigations of Cebeci et al. [2] indicate that the fourth-order formulation in terms of the stream function has certain advantages over the solution of the Navier-Stokes equations written in primitive variables or vorticity and stream function. In order to solve the nonlinear stream function equation Roache and Ellis [3] proposed the Biharmonic Driver method (BID) with $O(h^2)$ finite-difference equations and direct linear solver at each iteration. A similar approximation has been used by Cebeci et al. [2]. Their procedure employs the so called Newton-chord method which converges quite fast and avoids the solution of a new linear system at each iteration.

The calculation procedure proposed in this study applies the finite-difference method of Hermitian type [4] which makes use of the accurate approximation of the derivatives by Hermitian interpolating polynomials. As in addition to the function the first derivatives are the unknowns of the method, no problems arise to incorporate the velocity boundary conditions which are in fact the first derivatives of the stream function. The present investigations are concerned with the improvement of the iteration procedure with respect to the nonlinear terms of the Navier-Stokes equations and the development of a direct solver for the large system of linear finite-difference equations. To solve the energy equation an efficient Poisson solver based upon the Fast Fourier Transform [5] is applied.

Governing Differential Equations

As shown in fig. 1 the natural convection flow in the enclusure is induced by the temperature difference between the hot wall T_h and the cold wall T_c. The horizontal walls of the upright square cavity of length L are adiabatic. For the theoretical treatment of the flow it is convenient to apply dimensionless variables:

$$x = \frac{X}{L} \qquad z = \frac{Z}{L} \tag{1}$$

$$u = \frac{U}{a/L} \qquad w = \frac{W}{a/L} \qquad \theta = \frac{T - T_c}{T_h - T_c} \tag{2}$$

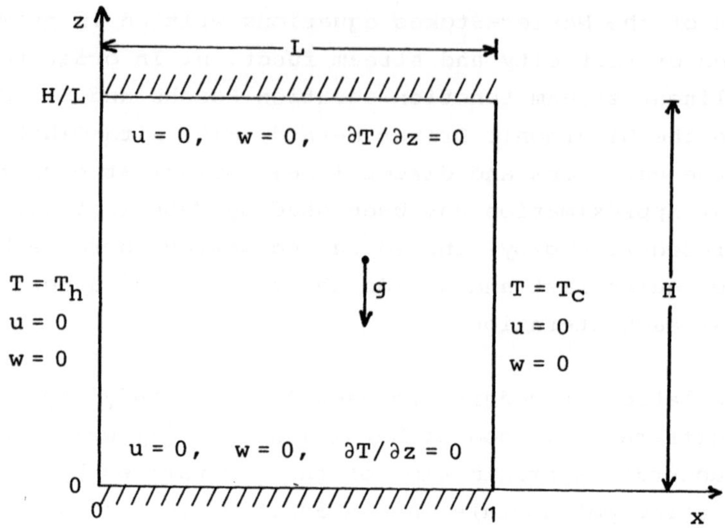

Fig. 1. Geometrical configuration and boundary conditions

The velocity components are u in the horizontal x-direction and w in the vertical z-direction; a is the thermal diffusivity of the fluid.

The governing equations can be considerably simplified by the Boussinesq approximation [6]. That is constant fluid properties for the viscosity ν, thermal conductivity λ and specific heat c_p whereas the density ρ is assumed constant except in buoyancy terms. Introducing the stream function ψ

$$u = \frac{\partial \psi}{\partial z} \qquad w = -\frac{\partial \psi}{\partial x} \tag{3}$$

which satisfies the continuity equation exactly the steady-state laminar flow can be described by the following equations:

$$\frac{\partial^4 \psi}{\partial x^4} + 2\frac{\partial^4 \psi}{\partial x^2 \partial z^2} + \frac{\partial^4 \psi}{\partial z^4} - \frac{1}{Pr}\left[\frac{\partial \psi}{\partial z}\left(\frac{\partial^3 \psi}{\partial x^3} + \frac{\partial^3 \psi}{\partial x \partial z^2}\right) - \frac{\partial \psi}{\partial x}\left(\frac{\partial^3 \psi}{\partial x^2 \partial z} + \frac{\partial^3 \psi}{\partial z^3}\right)\right] = + Ra \frac{\partial \theta}{\partial x} \tag{4}$$

$$\frac{\partial^2 \theta}{\partial x^2} + \frac{\partial \theta}{\partial z^2} = \frac{\partial \psi}{\partial z}\frac{\partial \theta}{\partial x} - \frac{\partial \psi}{\partial x}\frac{\partial \theta}{\partial z} \, . \tag{5}$$

These equations have to be solved with respect to the no-slip condition for the velocity components at all surfaces and the temperature conditions as mentioned below.

$$\left.\begin{array}{lll} x = 0,1 : & \psi = 0 , & \frac{\partial \psi}{\partial x} = 0 \\ x = 0 : & \theta = 1 ; & x = 1 : \quad \theta = 0 \end{array}\right\} 0 \leqq z \leqq 1 \qquad \begin{array}{r}(6)\\(7)\end{array}$$

$$\left.\begin{array}{lll} z = 0,1 : & \psi = 0 , & \frac{\partial \psi}{\partial z} = 0 \\ & \frac{\partial \theta}{\partial z} = 0 & \end{array}\right\} 0 \leqq x \leqq 1 \qquad \begin{array}{r}(8)\\(9)\end{array}$$

The parameters in the stream function equation (4) are the Prandtl number Pr and the Rayleigh number Ra which are defined as

$$Pr = \frac{\nu}{a} , \qquad Ra = \frac{\beta g (T_h - T_c)}{\nu a} \tag{10}$$

where β and g are the coefficient of thermal expansion and the gravitational acceleration, respectively.

<u>Numerical Solution Procedure</u>

The nonlinear stream function equation (4) and the linear energy equation (5) are a coupled system of elliptic equations which can only be solved numerically by an iterative procedure.

In order to linearize the nonlinear stream function equation we investigate the two basic methods: Biharmonic Driver (BID) and Newton-Raphson (NR). The BID method applied to equation (4)

$$\frac{\partial^4 \psi^k}{\partial x^4} + 2\frac{\partial^4 \psi^k}{\partial x^2 \partial z^2} + \frac{\partial^4 \psi^k}{\partial z^4} = \frac{1}{Pr}\left[\frac{\partial \psi}{\partial z}\left(\frac{\partial^3 \psi}{\partial x^3} + \frac{\partial^3 \psi}{\partial x \partial z^2}\right) - \right.$$
$$\left. - \frac{\partial \psi}{\partial x}\left(\frac{\partial^3 \psi}{\partial x^2 \partial z} + \frac{\partial^3 \psi}{\partial z^3}\right)\right]^{k-1} + Ra\,\frac{\partial \theta^{k-1}}{\partial x} \tag{11}$$

solves the biharmonic operator for the new value ψ^k lagging the non-homogenous term. The main advantage of this method is that the linear biharmonic operator has to be generated only once. Thus, the work in solving eq. (11) consists in updating the right-hand side at each iteration.

With the NR method (see e.g. [7]) the nonlinear terms are evaluated with a truncated Taylor's series where second and higher-order terms are neglected. The application of the NR method to the stream function equation (4) results in

$$\begin{aligned}
&\frac{\partial^4\psi^k}{\partial x^4} + 2\frac{\partial^4\psi^k}{\partial x^2\partial z^2} + \frac{\partial^4\psi^k}{\partial z^4} + \frac{1}{Pr}\left[\frac{\partial\psi^k}{\partial x}\left(\frac{\partial^3\psi^{k-1}}{\partial x^2\partial z} + \frac{\partial^3\psi^{k-1}}{\partial z^3}\right) + \right.\\
&\left.\frac{\partial\psi^{k-1}}{\partial x}\left(\frac{\partial^3\psi^k}{\partial x^2\partial z} + \frac{\partial^3\psi^k}{\partial z^3}\right)\right] - \frac{1}{Pr}\left[\frac{\partial\psi^k}{\partial z}\left(\frac{\partial^3\psi^{k-1}}{\partial x^3} + \frac{\partial^3\psi^{k-1}}{\partial x\partial z^2}\right) + \right.\\
&\left.\frac{\partial\psi^{k-1}}{\partial z}\left(\frac{\partial^3\psi^k}{\partial x^2\partial z} + \frac{\partial^3\psi^k}{\partial z^3}\right)\right] = \frac{1}{Pr}\left[\frac{\partial\psi}{\partial x}\left(\frac{\partial^3\psi}{\partial x^2\partial z} + \frac{\partial^3\psi}{\partial z^3}\right) - \right.\\
&\left.\frac{\partial\psi}{\partial z}\left(\frac{\partial^3\psi}{\partial x^3} + \frac{\partial^3\psi}{\partial x\partial z^2}\right)\right]^{k-1} + Ra\,\frac{\partial\theta^{k-1}}{\partial x}\,.
\end{aligned} \tag{12}$$

Although the NR linearization converges quadratically this procedure requires the solution of a new linear fourth-order differential equation at each iteration.

To solve eq. (11) or (12) numerically we apply the finite-difference method of Hermitian type [4] where the rectangular domain is divided into a grid with constant mesh spacing $\Delta x = 1/(N-1)$ and $\Delta z = 1/(M-1)$. The grid point values of the function ψ and the derivatives $\partial\psi/\partial x$, $\partial\psi/\partial z$ and $\partial^2\psi/\partial x\partial z$ are treated as unknowns. This technique gives a nine-point approximation of the differential equation which is second-order accurate. Due to the four unknowns at each grid point the block tridiagonal system of finite-difference equations is of higher complexity than that of conventional procedures. In [4] this large system has been solved by means of a block elimination scheme. Here, we develop a direct solver which makes use of a block LU decomposition of the matrix. The derivation

of the method is presented in the next section.
At each iteration k the solution of the energy equation (5) is obtained by a second-order finite-difference method using the Fast Fourier Transforms [5]. The application of the FFT requires slight modifications so that all boundary conditions (7) and (8) can be directly satisfied. The right-hand side of equation (5) is updated using the latest values for the velocity components and the temperature gradient with central finite-differences.

Linear Direct Solver for Tridiagonal 4 x 4 Block Matrix

The finite-difference method of Hermitian type applied to eq. (10) or (11) gives a linear system of $4\cdot(N-2)\cdot(M-2)$ independent equations. Near the boundaries the system is modified such that the boundary conditions are incorporated at the right-hand side of the finite-difference equations.

With the definitions of the vectors S and R [4] for the unknowns and the right-hand side, respectively

$$S = (S_2, \ldots, S_{N-1})^T \qquad R = (R_2, \ldots, R_{N-1})^T \tag{13}$$

where the S_i and R_i are column vectors of dimension $4(M-2)$:

$$S_i = \left(S_{i,2}, S^x_{i,2}, S^z_{i,2}, S^{xz}_{i,2}, \ldots, S_{i,M-1}, S^x_{i,M-1}, S^z_{i,M-1}, S^{xz}_{i,M-1}\right)^T \tag{14}$$

$$R_i = \left(RS_{i,2}, RN_{i,2}, RW_{i,2}, RE_{i,2}, \ldots, RS_{i,M-1}, RN_{i,M-1}, RW_{i,M-1}, RE_{i,M-1}\right)^T \tag{15}$$

The linear system can be written in the form

$$Q \cdot S = R \,. \tag{16}$$

Here, the $4(N-2)(M-2)$ square matrix Q

$$Q = \begin{bmatrix} q_{2,2} & q_{2,3} & & & & 0 \\ q_{3,1} & q_{3,2} & q_{3,3} & & & \\ & & & q_{N-2,1} & q_{N-2,2} & q_{N-2,3} \\ & 0 & & & q_{N-1,1} & q_{N-1,2} \end{bmatrix} \tag{17}$$

consists of $4(M-2)$ square submatrices which are block tridiagonal and have the block size 4×4. The structure of the system is shown in fig. 2.

$$\left[\begin{array}{cccc|cccc|cccc}
\bar{E}_{2,2} & \bar{F}_{2,2} & & & \bar{H}_{2,2} & \bar{P}_{2,2} & & & & & & \\
\bar{D}_{2,3} & \bar{E}_{2,3} & \bar{F}_{2,3} & & \bar{G}_{2,3} & \bar{H}_{2,3} & \bar{P}_{2,3} & & & & & \\
& \bar{D}_{2,4} & \bar{E}_{2,4} & \bar{F}_{2,4} & & \bar{G}_{2,4} & \bar{H}_{2,4} & \bar{P}_{2,4} & & & & \\
& & \bar{D}_{2,5} & \bar{E}_{2,5} & & & \bar{G}_{2,5} & \bar{P}_{2,5} & & & & \\
\hline
\bar{B}_{3,2} & \bar{C}_{3,2} & & & \bar{E}_{3,2} & \bar{F}_{3,2} & & & \bar{H}_{3,2} & \bar{P}_{3,2} & & \\
\bar{A}_{3,3} & \bar{B}_{3,3} & \bar{C}_{3,3} & & \bar{D}_{3,3} & \bar{E}_{3,3} & \bar{F}_{3,3} & & \bar{G}_{3,3} & \bar{H}_{3,3} & \bar{P}_{3,3} & \\
& \bar{A}_{3,4} & \bar{B}_{3,4} & \bar{C}_{3,4} & & \bar{D}_{3,4} & \bar{E}_{3,4} & \bar{F}_{3,4} & & \bar{G}_{3,4} & \bar{H}_{3,4} & \bar{P}_{3,4} \\
& & \bar{A}_{3,5} & \bar{B}_{3,5} & & & \bar{D}_{3,5} & \bar{E}_{3,5} & & & \bar{G}_{3,5} & \bar{H}_{3,5} \\
\hline
& & & & \bar{B}_{4,2} & \bar{C}_{4,2} & & & \bar{E}_{4,2} & \bar{F}_{4,2} & & \\
& & & & \bar{A}_{4,3} & \bar{B}_{4,3} & \bar{C}_{4,3} & & \bar{D}_{4,3} & \bar{E}_{4,3} & \bar{F}_{4,3} & \\
& & & & & \bar{A}_{4,4} & \bar{B}_{4,4} & \bar{C}_{4,4} & & \bar{D}_{4,4} & \bar{E}_{4,4} & \bar{F}_{4,4} \\
& & & & & & \bar{A}_{4,5} & \bar{B}_{4,5} & & & \bar{D}_{4,5} & \bar{E}_{4,5}
\end{array}\right]
\times
\begin{bmatrix} S_{2,2} \\ S_{2,3} \\ S_{2,4} \\ S_{2,5} \\ S_{3,2} \\ S_{3,3} \\ S_{3,4} \\ S_{3,5} \\ S_{4,2} \\ S_{4,3} \\ S_{4,4} \\ S_{4,5} \end{bmatrix}
=
\begin{bmatrix} R_{2,2} \\ R_{2,3} \\ R_{2,4} \\ R_{2,5} \\ R_{3,2} \\ R_{3,3} \\ R_{3,4} \\ R_{3,5} \\ R_{4,2} \\ R_{4,3} \\ R_{4,4} \\ R_{4,5} \end{bmatrix}$$

Fig. 2. System of finite-difference equations [4] for $N = 5$, $M = 6$

In order to solve eq. (16) a direct solution procedure is applied. If a LU decomposition of the matrix Q is available then it is possible to solve eq. (1) for various right-hand sides by initializing the decomposition only once. This happens when linearization procedures of chord type are used. Hence, the first step consists of the block decomposition

$$Q = L \cdot U \tag{18}$$

where the lower and upper matrix are defined by

$$L = \begin{bmatrix} D_2 & & 0 \\ q_{3,1} & & \\ 0 & q_{N-1,1} & D_{N-1} \end{bmatrix}, \quad U = \begin{bmatrix} I & \Gamma_2 & 0 \\ & & \Gamma_{N-2} \\ 0 & & I \end{bmatrix} \tag{19}$$

I is the 4(M - 2) square identity matrix and the matrices D_i and Γ_i are given by:

$$D_2 = q_{2,2}$$

$$\left.\begin{aligned} \Gamma_{i-1} &= D_{i-1}^{-1} \, q_{i-1,3} \\ D_i &= q_{i,2} - q_{i,1} \, \Gamma_{i-1} \end{aligned}\right\} \quad i = 3(1)N-1 \tag{20}$$

Of course, we do not compute D_i^{-1} directly. In order to determine Γ_i for some i the system

$$D_i \, \Gamma_i = q_{i,3} \tag{21}$$

is used. The rewritten eq. (1)

$$L \, U \, S = R \tag{22}$$

can then be solved in two steps

$$L \cdot Z = R \tag{23}$$

$$U \cdot S = Z \, . \tag{24}$$

Assuming an appropriate partition of the vectors Z and R these steps are easily performed as the matrices L and U have a block triangular structure.

$$\begin{aligned} Z_2 &= D_2^{-1} \, R_2 \\ Z_i &= D_i^{-1} \, (R_i - q_{i,1} \cdot Z_{i-1}) \qquad i = 3(1)N-1 \end{aligned} \tag{25}$$

$$S_{N-1} = Z_{N-1}$$
$$S_i = Z_i - \Gamma_i \, S_{i+1} \qquad i = N-2(-1)2 \tag{26}$$

In this way we have obtained a direct solver for the complex system of Hermitian finite-difference equations.

Results and Discussion

The numerical results performed for air with $Pr = 0.71$ in the square enclosure are presented and discussed in this section. The investigations are considered with respect to the accuracy of the numerical method applied and the efficiency of the two iteration procedures for the nonlinear terms of the stream function equation.

A typical plot of the streamlines and isotherms for $Ra = 1000$ is shown in fig. 3 and 4, respectively. The streamlines are almost concentric whereas small departures occur near the

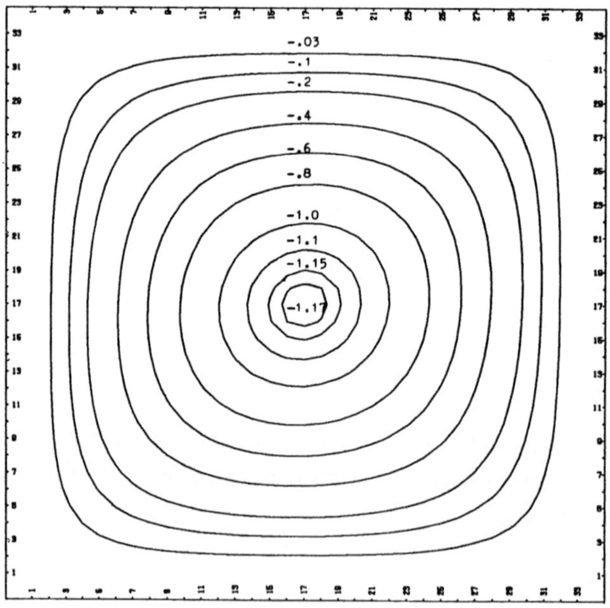

Fig. 3. Streamlines for $Ra = 1000$ and $Pr = 0.71$

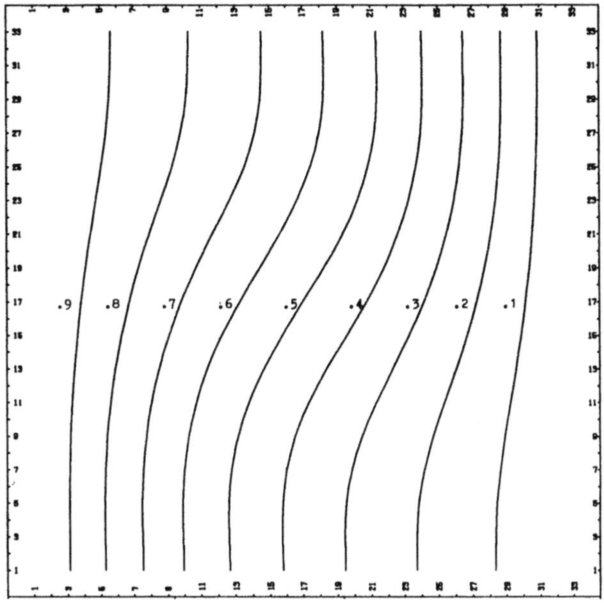

Fig. 4. Isotherms for Ra = 1000 and Pr = 0.71

corners. An appreciable convective heat transfer is indicated by the behaviour of the isotherms which depart from vertical lines for pure conduction.

For the comparison of the numerical results with the accurate calculation of de Vahl Davis [8] the local Nusselt number

$$Nu = - \frac{\partial \theta}{\partial x} dz \qquad (27)$$

and the average Nusselt number

$$Nu = - \int_0^1 \frac{\partial \theta}{\partial x} dz \qquad (28)$$

are computed. Table 1 presents the results for some characteristic properties of the flow field. It includes maximum and minimum local Nusselt numbers Nu_{max} and Nu_{min} on the hot wall and their locations z, maximum vertical velocity w_{max} on the horizontal mid-plane and its location x as well as the maximum horizontal velocity u_{max} on the vertical mid-plane and its location z. The values obtained indicate that accuracy increases with finer grid. Compared to the results of [8] the calculations for the 33 x 33 grid agree in three digits for the

grid Δx, Δz	9 x 9 0.135	17 x 17 0.0625	33 x 33 0.03125	Ref. [8]
Nu_{max} z	1.5156 0.125	1.5051 0.125	1.5095 0.0937	1.506 0.086
Nu_{min} z	0.7159 1.0	0.6961 1.0	0.6901 1.0	0.691 1.0
$\overline{Nu}$	1.1343	1.1223	1.1208	1.118
u_{max} z	3.4673 0.75	3.7151 0.8125	3.7121 0.8125	3.657 0.814
w_{max} x	3.5003 0.125	3.7447 0.1875	3.7398 0.1875	3.702 0.178
ψ_{max}	-1.1993	-1.1931	-1.1914	-1.174

Table 1. Results for Ra = 1000 and Pr = 0.71

Nusselt number. Concerning the values of velocity components and stream function agreement is only achieved up to the second digit. The reason for this behaviour might be the difference in the formulation of the Navier-Stokes equations.

The differential equations (4) and (5) have been solved by using the two methods NR and BID for the stream function equation. Table 2 presents the time required for one iteration step.

Method	Grid		
	9 x 9	17 x 17	33 x 33
NR	0.5	6.1	89.3
BID	0.1	0.5	2.7

Table 2. CPU time in [sec] on CYBER 175 per iteration

The results demonstrate the advantage of the BID method especially for the fine 33 x 33 grid. For the solution of the coupled equations (4) and (5) a relaxation procedure was applied at each iteration. Due to this procedure the overall iteration process does not converge quadratically even though the NR method was used for the stream function equation. Therefore the BID method converges as fast as the NR procedure as it is shown in table 3. The differences in the final values arise from the fact that the two methods are obtained by discretization of different differential equations (11) and (12).

Iteration number	NR method			BID method		
	error (ψ)	$\overline{Nu}$	ψ_{max}	error (ψ)	$\overline{Nu}$	ψ_{max}
1	1.2	1.000	-1.265	1.2	1.000	-1.265
2	3.0E-2	1.148	-1.284	3.E-2	1.146	-1.265
3	1.0E-1	1.139	-1.177	1.1E-1	1.135	-1.155
4	5.9E-3	1.117	-1.180	1.1E-2	1.113	-1.165
5	1.6E-2	1.119	-1.196	1.6E-2	1.117	-1.181
6	1.3E-3	1.123	-1.194	2.8E-3	1.120	-1.178
7	2.3E-3	1.122	-1.192	2.2E-3	1.119	-1.176
8	3.1E-4	1.122	-1.192	6.3E-4	1.119	-1.177
9	3.4E-4	1.122	-1.193	2.9E-4	1.119	-1.177
10	2.0E-4	1.122	-1.193	1.2E-4	1.119	-1.177

Table 3. Rate of convergence for Ra = 1000 and 17 x 17 grid

References

1. Jones, I.P.; Thompson, C.P.: Numerical solutions for a comparison problem on natural convection in an enclosed cavity. AERE-R 9955 (1981).

2. Cebeci, T.; Hirsh, R.S.; Keller, H.B.; Williams, P.G.: Studies of numerical methods for the plane Navier-Stokes equations. Computer Methods in Applied Mechanics and Engineering 27 (1981) 13-44.

3. Roache, P.J.; Ellis, M.A.: The BID method for the steady-state Navier-Stokes equations. Computers and Fluids 3 (1975) 305-320.

4. Thiele, F.; Wagner, H.: Solution of the Navier-Stokes equations using the finite-difference method of Hermitian

type. Notes on Numerical Fluid Mechanics 5 (1982) 326-336. Braunschweig/Wiesbaden: Vieweg & Sohn.

5. Hockney, R.W.: The potential calculation and some applications. Methods in Computational Physics 9 (1970) 135-154. New York: Academic Press.

6. Chandrasekhar, S.: Hydrodynamic and Hydromagnetic Stability. Oxford: Clarendon Press 1961.

7. Blottner, F.G.: Computational techniques for boundary layers. AGARD LS-73, 1975.

8. de Vahl Davis, G.: Natural convection of air in a square cavity: An accurate numerical solution. Rep. 1981/FMT/1 University of N.S.W., Kensington, Australia.

Computational Study of the Magnus Effect on Boattailed Shell at Supersonic Speeds

W. B. STUREK

U.S. Army Ballistic Research Laboratory/ARRADCOM
Aberdeen Proving Ground, Maryland 21005

Summary

The thin-layer Parabolized Navier-Stokes computational technique developed recently by Schiff and Steger has been shown to be a viable computational tool for predicting Magnus effects at small angle of attack for 2 < M < 4 for slender shell as verified by comparison to wind tunnel force measurements. This computational technique is applied here to examine the ability of the code to predict aerodynamic forces at angles of attack up to 10-degrees and to study the Magnus effect at small angle of attack for a parametric variation of boattail configuration. The model geometries in this study include boattail angles of 0°, 5°, 7-1/2°, and 10° for lengths of one and two calibers. This range of boattail angle and length effectively spans the range for practical shell application. Results are shown which quantize the significant effects of boattail length and angle on the Magnus force at supersonic velocity.

I. Introduction

Recent papers [1, 2, 3] have reported the development and application of the thin-layer Parabolized Navier-Stokes computational technique to predict the flow about slender bodies of revolution at supersonic velocities. Reference [3] showed the technique to be a viable computational tool for predicting Magnus effects for a six-caliber slender shell with a one-caliber, 7° boattail as verified by comparison to wind tunnel force measurements. The results of Reference [3] represent the first successful efforts to compute the Magnus effect for boattailed shell using sophisticated numerical computational techniques. The objective of this paper is to exploit this computational technique to study the flow over spinning shell for a parametric variation of boattail configurations. Additionally, results are shown illustrating the ability of the computational technique to predict Magnus and normal force coefficients for angles of attack up to ten degrees.

II. Computational Technique

The steady thin-layer Parabolized Navier-Stokes equations can be written for general spatial coordinates ξ, η, ζ as [1]

$$\frac{\partial \hat{E}_s}{\partial \xi} + \frac{\partial \hat{F}}{\partial \eta} + \frac{\partial \hat{G}}{\partial \zeta} = \frac{1}{\hat{Re}} \frac{\partial \hat{S}}{\partial \zeta} \tag{1}$$

where $\xi = \xi(x)$ is the streamwise (marching) coordinate
$\eta = \eta(x,y,z)$ is the circumferential coordinate
$\zeta = \zeta(x,y,z)$ is the normal coordinate

The inviscid flux vectors in Eq. (1) are represented by $\hat{E}_s$, $\hat{F}$, and $\hat{G}$ and contain all the terms for conservation of mass, momentum and energy in the three coordinate directions. The thin-layer viscous matrix, valid for high Reynolds No. flow, is represented by $\hat{S}$. Variations of body geometry are included in Eq. (1) through the presence of the metric terms η_x, η_y, η_z, etc., which appear in the flux vectors. $\hat{Re}$ is the Reynolds number.

Equation 1 is parabolic-like with respect to ξ and can thus be marched downstream in the ξ direction from an initial data plane (subject to appropriate body and free stream boundary conditions) under those conditions where the local flow is supersonic.

The numerical algorithm used to march Eq. (1) downstream is an approximately-factored, fully implicit, finite-difference scheme. The algorithm is conservative and of second-order accuracy in the marching direction. A two-layer, algebraic eddy viscosity model [4] is included for the computation of turbulent flows. Details of the Parabolized Navier-Stokes assumption and the derivation of the algorithm are included in Reference [1].

The computations are started from a converged conical solution near the tip of the projectile. The full solution is then obtained by marching over the body in the streamwise (axial) direction. This marching technique is depicted in Figure 1 along with the coordinate system. A logarithmic stretching is used to achieve adequate grid resolution of the turbulent viscous layer. The grid generator employs an adaptive capability which insures that adequate resolution of the viscous layer is maintained over the full length of the model. The streamwise marching step-size was adjusted to yield 500 to 800 computational steps for the full length of the shell. The computational grid consisted of 36 stations about the circumference of the model ($\Delta\phi = 10°$) and 50 points between the body and the outer boundary.

III. Results

Comparisons to Aerodynamic Force Measurements

In Reference [3], a series of results obtained for $\alpha = 2°$ were compared to experimental measurements [5] of aerodynamic forces for $2 \leq M \leq 4$. The sign convention for these aerodynamic forces on a spinning shell is indicated in Figure 2. The results in Reference [3] indicated that the thin-layer PNS computational technique achieved excellent agreement in comparison to experimental measurements of aerodynamic forces including the Magnus effect. Comparisons of the PNS computed results to experiment have now been carried out for $\alpha \leq 10°$. Computed results were obtained for $\alpha = 2°, 4°, 6°$, and $10°$.

A comparison between computation and experiment is shown as a function of angle of attack for normal force for $M = 3$ in Figure 3. Results for two boattail configurations are shown. The notation SOCBT depicts a model with a 7°, one caliber boattail model as shown in Figure 5. The notation SOC depicts a model with a 0° boattail. Excellent agreement is achieved for $\alpha \leq 10°$ for magnitude as well as the non-linear behavior of the normal force coefficient.

A similar comparison is shown for Magnus force for $M = 3$ in Figure 4. These results indicate very good agreement between computation and experiment for $\alpha \leq 6°$. However, the computed results for $\alpha = 10°$ are not in close agreement with the trend of the experimental data for $\alpha > 6°$. This discrepancy is not unexpected since, as discussed in Reference [2] where comparisons between computation and experiment for surface pressure were made, good agreement was achieved at $\alpha = 6.3°$ and significantly less satisfactory agreement was obtained at $\alpha = 10.4°$. The discrepancy at $\alpha = 10.4°$ was attributed to the inability of the computational technique to accurately model the significant development of lee-side vortical flow. The ability of the computational technique to accurately predict Magnus for $\alpha \leq 6°$ does represent a significant capability. This demonstrates that useful engineering results can be achieved for highly three-dimensional flow fields using a very simple turbulence model.

Parametric Results

Scope of Computational Effort

A series of computations have been accomplished for a parametric variation of boattail configuration. The geometries are shown in Figure 5 and include boattail lengths of one and two calibers for boattail angles of 0°, 5°, 7-1/2°, and 10°. This range of boattail length and angle effectively spans the range for practical shell application. The computations were accomplished for free stream standard atmospheric conditions and a constant wall temperature boundary condition ($T_W = T_\infty$).

Computational Results

The development of the Magnus force as a function of axial position is shown in Figure 6 illustrating the effects of boattail length. Figure 6 shows that the length of the boattail strongly affects the magnitude of the Magnus force.

Parametric comparisons for pitch and yaw plane aerodynamic parameters are shown in Figures 7 and 8. These examples illustrate the ability of the computational technique to develop data which reflect the effects of body configuration and Mach number.

The effect of boattail angle and length for the slope of the pitching moment (CG = 3.6 calibers) is shown in Figure 7. These results show that both boattail length and angle have a significant effect on the pitching moment. The trends indicate that C_{M_α} is increased for increasing boattail angle and boattail length. These trends are accentuated for flow velocities near Mach = 2.

Similar comparisons for the Magnus effect are shown in Figure 8. The slope of the Magnus moment coefficient (CG = 3.6 calibers) is shown in Figure 8 as a function of Mach number. The effect of boattail length is seen to strongly affect the Magnus moment coefficient. The trend illustrated is that $C_{M_{p\alpha}}$ is increased as boattail length and angle are increased.

IV. Summary

A computational aerodynamics study of the effects of boattail geometry has been described in which the thin-layer Parabolized Navier-Stokes numerical technique has been used at supersonic velocities. Results have been discussed which illustrate the ability of the computational technique to accurately predict Magnus and normal forces for angles of attack up to six degrees. Parametric results showing the effects of boattail length and boattail angle were discussed for pitching moment and Magnus moment. The parametric results illustrate the potential use of the computational code to provide shell design guidance for aeroballistics.

References

1. Schiff, L.B.; and Steger, J.L.: Numerical simulation of steady supersonic viscous flow. AIAA Journal, Vol. 18 No. 12 (1980) 1421-1430.

2. Schiff, L.B.; and Sturek, W.B.: Numerical simulation of steady supersonic flow over an ogive-cylinder-boattail body. AIAA Paper No. 80-0066, AIAA 18th Aerospace Sciences Meeting (1980).

3. Sturek, W.B.; and Schiff, L.B.: Computations of the magnus effect for slender bodies in supersonic flow. AIAA Paper No. 80-1586-CP, AIAA Atmospheric Flight Mechanics Conference (1980).

4. Baldwin, B.S.; and Lomax, H.: Thin layer approximation and algebraic model for separated turbulent flows. AIAA Paper No. 78-257 (1978).

5. Nietubicz, C.J.; and Opalka, K.: Supersonic wind tunnel measurements of static and magnus aerodynamic coefficients for projectile shapes with tangent and secant ogive noses. ARBRL-MR-02991, U.S. Army Ballistic Research Laboratory/ARRADCOM, Aberdeen Proving Ground, MD 21005 (1980).

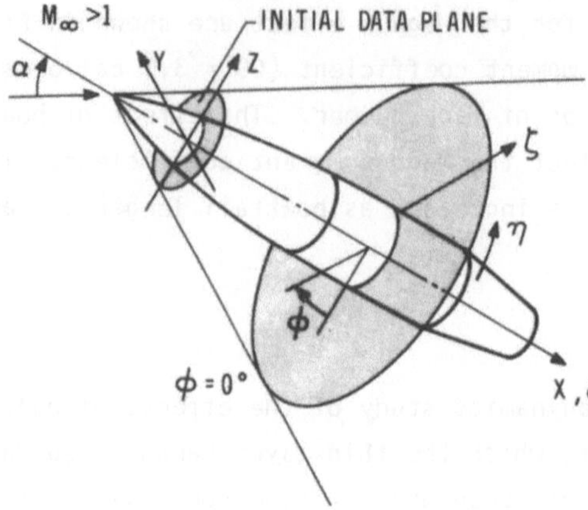

Figure 1. Coordinates and notation

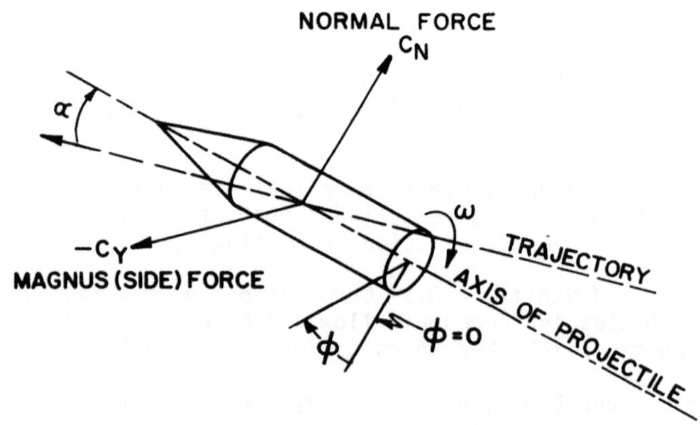

Figure 2. Aerodynamic forces on spinning shell

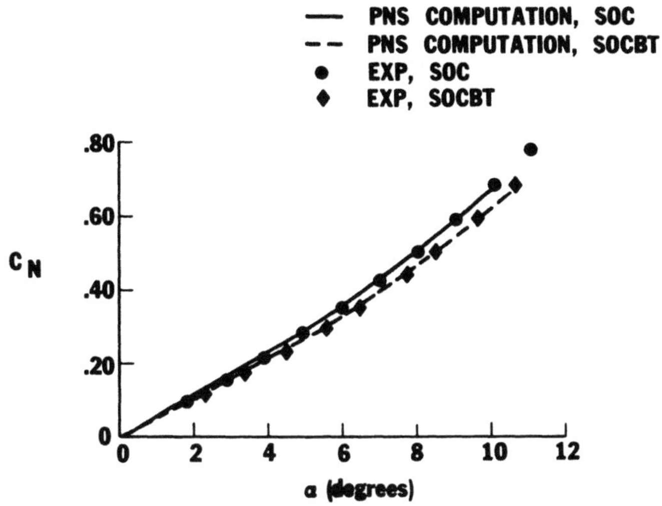

Figure 3. Normal force coefficient versus angle of attack, computation compared to wind tunnel force measurements, $M = 3$, $Re = 2.11 \times 10^7/m$

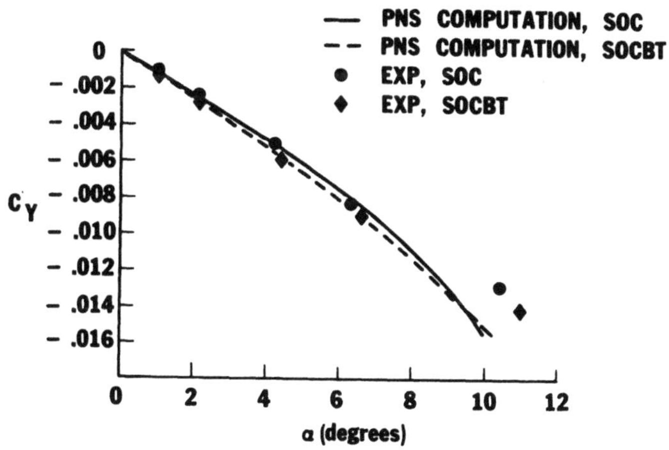

Figure 4. Magnus force coefficient versus angle of attack, computation compared to wind tunnel force measurements, $M = 3$, $Re = 2.11 \times 10^7/m$, $PD/V = .19$, $T_w = T_{aw}$

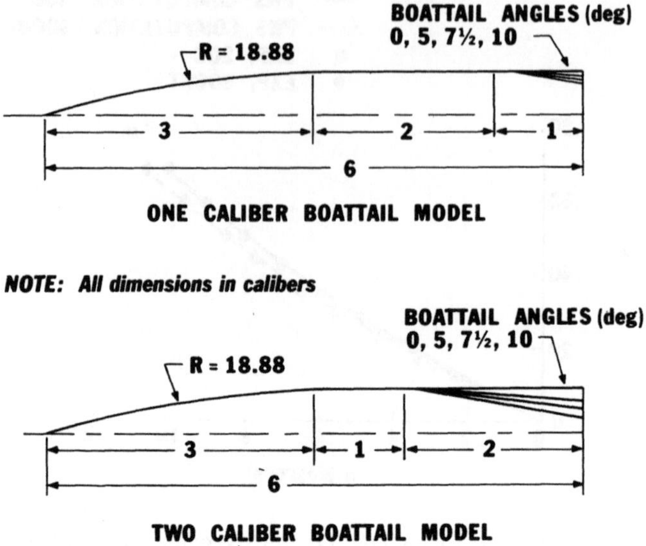

Figure 5. Model configurations for parametric study

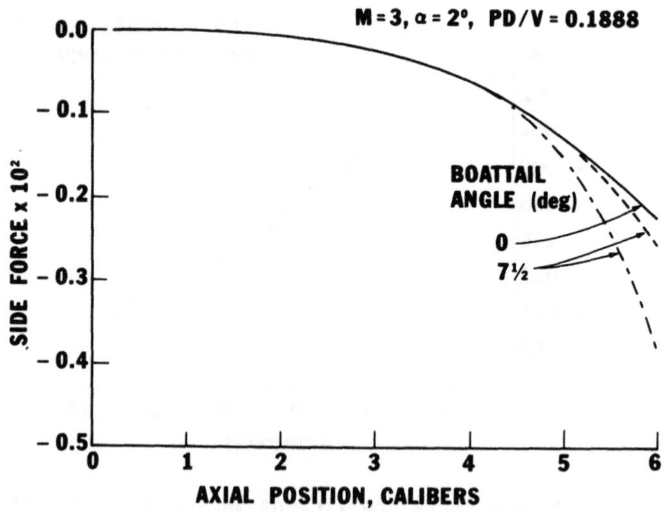

Figure 6. Magnus force coefficient versus axial position, parametric comparison, $M = 3$, $\alpha = 2°$, $PD/V = .19$, $T_w = T_{aw}$, $Re = 6.80 \times 10^7/m$

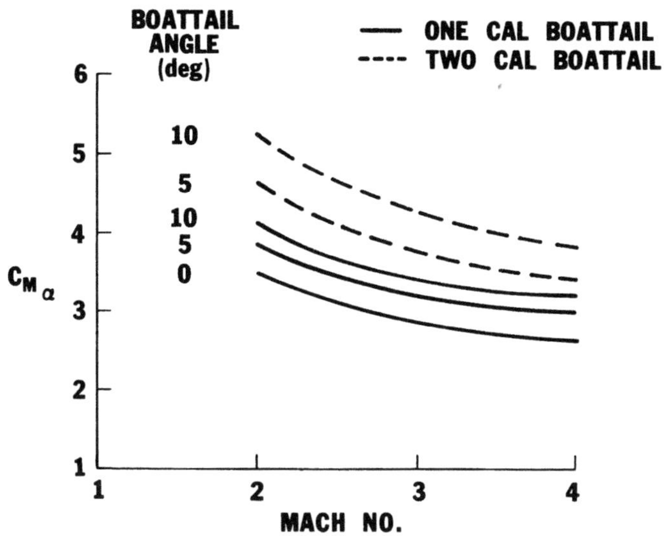

Figure 7. Slope of pitching moment coefficient versus Mach no., parametric comparison, T_∞ = 294K, T_W = 294K, CG = 3.6 calibers

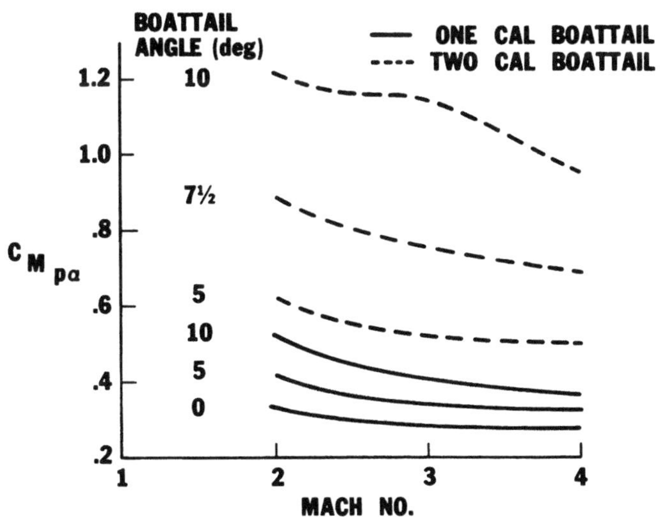

Figure 8. Slope of magnus moment coefficient versus Mach no., parametric comparison, T_∞ =294K, T_W = 294K, CG = 3.6 calibers

Viscid-Inviscid Interaction Analysis on Airfoils with an Inverse Boundary Layer Approach

P. THIEDE, G. DARGEL and E. ELSHOLZ

Messerschmitt-Bölkow-Blohm GmbH,
Unternehmensbereich Transportflugzeuge, Bremen

Summary

A non-hierarchical semi-inverse matching procedure is outlined, coupling an inverse boundary layer solution with a direct potential flow solution iteratively, to provide a highly efficient computation of strong viscid-inviscid interactions including separation.

Results obtained for separated wall flows encouraged the application of this technique to the trailing edge region of rear-loaded airfoils, where weak interaction methods usually fail. The agreement of both attached and separated subsonic airfoil flow predictions with experimental data confirm the reliability of the present approach.

Introduction

Significant progress in sub- and transonic airfoil flow predictions is closely related to an efficient analysis of strong viscid-inviscid interaction phenomena, occuring in the trailing edge and shock regions and intensified by flow separation.

In principle, for the iterative solution of viscid-inviscid interactions three different matching techniques can be applied, Fig. 1. The direct matching technique, which converges fast as long as the inviscid flow dominates over the boundary layer flow, is normally adopted for airfoil flow predictions. However, approaching the trailing edge, convergency difficulties arise due to the increasing interaction between the boundary layer and the inviscid flow. To avoid these difficulties, the inverse matching technique may be applied. Here, the inverse boundary layer approach removes the Goldstein singularities at the separation and reattachment point, so that separated flow regions can be included. Unfortunately, this technique has a very low convergency rate.

To overcome the drawbacks of both the direct and inverse matching, the semi-inverse technique has been suggested recently [1,2] , coupling an inverse boundary layer solution with a direct potential flow solution iteratively, which provides a highly efficient computation of strong viscid-inviscid interactions [3] .This technique is applied to attached and separated trailing edge flows of rear-loaded airfoils including the near wake, resulting in a substantial improvement of airfoil flow prediction methods.

Inverse boundary layer solution

The inverse boundary layer method [4,5] used solves the moment and the moment of momentum integral equations for steady 2-D compressible flows, Fig. 2.

To reduce the compressible quantities in the integral equations to incompressible ones, the van Driest transformation is used. The relations for laminar velocity profiles are deduced from the similar solutions of Stewartson [6] including the lower-branch solutions for reverse flow velocity profiles. In the turbulent case the relations of the Walz method [7] are incorporated in the attached flow regime, while for separated flow the similar solutions of Alber [8] are used as a basis. Deviations from the equilibrium condition are considered by an empirical dissipation law.

With these relations, the integral equations are solved with a prescribed displacement thickness by a standard Runge-Kutta integration.

Semi-inverse matching procedure

Fig. 3 shows the flow chart of the semi-inverse matching procedure in more detail. Here, from a given displacement thickness the boundary layer edge velocity is computed simultaneously by the direct potential flow method [9] and the inverse boundary layer method described above. The resulting edge velocities of both solutions are used to adjust the displacement thickness for the next iteration cycle by the matching condition of Carter [1] . To improve the overall convergency rate, a modification has been made by introducing an additional relaxation factor control.

The convergence behaviour of this procedure is demonstrated in Fig. 4 for a separated turbulent flow over a surface indentation. In these computations the initial viscous approximation was the flat plate solution. Fig. 4 shows on the right the iterative development of the pressure distribution in the strong interaction region, and on the left the pressure coefficient at two different stations as a function of the iteration number. Using the modified relaxation factor the solution converges significantly faster than with the original Carter condition.

Interaction analysis on airfoils

The results obtained encouraged the application of the semi-inverse matching procedure to the trailing edge region of rear-loaded airfoils, where conventional weak interaction methods usually fail.

For this purpose, the inverse boundary layer integral method was extended to wake flows, and boundary layer and wake curvature terms are considered in the trailing edge region [10,11,12] .

For the simulation of the viscous effects the equivalent surface mass flow concept is used, Fig. 5. For this reason the panel method [9] had to be modified to incorporate the appropriate viscous boundary conditions [10] , which are

- a normal velocity distribution, prescribed on the airfoil contour
- and normal and tangential velocity jump distributions, prescribed at the wake centerline.

Finally, a global method for subsonic airfoil flow predictions was developed, switching from a conventional weak interaction analysis to the present approach for strong trailing edge interactions.

Results

Fig. 6 shows viscid-inviscid flow predictions over a VFW rear-loaded airfoil at high incidence. For comparison with the experiments [13] three different calculations have been performed:

- inviscid mode
- viscous mode (with boundary layer and wake thickness terms only)
- full viscous mode (including curvature terms).

An excellent agreement between the predicted and measured pressure distribution is achieved with the full viscous solution. For the different calculation modes the convergence of the lift coefficient is also shown in Fig. 6. The viscous lift loss is significantly underpredicted, when curvature terms are not taken into consideration.

As a detail of this solution, Fig. 7 shows the streamline slope of the full viscous solution within the trailing edge region in comparison to the inviscid result. The discontinuity at the trailing edge obtained in the inviscid flow simulation is completely removed by the full viscous solution.

In Fig. 8 predicted lift, moment and drag polars of a NLR rear-loaded airfoil are compared with experimental data [14] . Beyond 6 degrees incidence the flow on the upper surface separates near the trailing edge. With exception of the drag coefficient at high incidence, where also large spanwise scatter were measured, the full viscous predictions agree remarkably with the measurements for both attached and separated flow cases.

In Fig. 9 compressible flow quantities (pressure distribution and boundary layer data) over a RAE rear-loaded airfoil are presented. This compressible test case too shows a generally good agreement between the predicted results and the experimental data of Cook et al [15] .

Concluding remarks

A semi-inverse matching procedure is presented,which couples an inverse boundary layer integral method with a direct panel method iteratively. With this technique applied highly efficient computations of strong viscid-inviscid interactions including separation have been performed.

The results obtained for subsonic airfoil flow cases show significant improvements in the trailing edge region. However, introducing a disymmetrical wake analysis would be advantageous at high incidence.

References

1. J.E. Carter: A new boundary-layer inviscid iteration technique for separated flow. AIAA-Paper 79 - 1450 (1979)
2. J.C. Le Balleur: Couplage visqueux - non visqueux: méthod numérique et applications aux écoulements bidimensionels transsoniques et supersoniques. La Recherche Aérospatiale 183 (1978)
3. P. Thiede, E. Elsholz, G. Dargel: Rationelle Berechnung partieller Ablösegebiete auf der Basis des Grenzschichtkonzepts. BMVg-FBWT RüFo IV T/RF 41/80050/81449, VFW (1981)
4. G. Dargel : Verbesserung eines Rechenverfahrens für Klappenprofile mit Grenzschichtablösung - Inverses Integralverfahren zur Berechnung abgelöster laminarer Grenzschichten. VFW-Report Ef 671 (1976), not published
5. P. Thiede, G. Dargel: Berechnung abgelöster turbulenter Grenzschichten bei transsonischer Profilströmung. ZTL-FAG 4 Final Report, VFW-Fokker 4.03/7/75 (1975)
6. K. Stewartson: Further solutions of the Falkner-Skan equation. Proc. Cambridge Phil. Soc. 50 (1954)
7. A. Walz: Strömungs- und Temperaturgrenzschichten. Verlag G. Braun, Karlsruhe, 1966
8. I.E. Alber: Similar solutions for a family of separated turbulent boundary layers. AIAA-Paper 71-203 (1971)
9. H. Jakob: Erweiterung eines 2-D Panelverfahrens auf Profile mit dicken Hinterkanten und Kopplung des Verfahrens mit einem Grenzschichtverfahren. VFW-Report Ef 612 (1976), not published
10. P. Thiede, G. Dargel: Hinterkantenströmung. Verbesserte Erfassung der Reibungseffekte im Hinterkantenbereich von überkritischen Profilen. BMVg-FBWT RüFo IV T/RF 41/70022/71421, VFW-Fokker (1979)
11. R.E. Melnik, R. Chow, H.R. Mead: Theory of viscous transonic flow over airfoils at high Reynolds number. AIAA-Paper 77-680 (1977)
12. R.C. Lock, M.C.P. Firmin: Survey of techniques for estimating viscous effects in external aerodynamics. RAE TM Aero 1900 (1981)
13. R.D. Boehe: Ergebnisse der Windkanalmessungen WX 77-1. VFW-Report Ef 2-12/77 (1977), not published
14. C.J.J. Joosen, C.G. Kho: 2-D low-speed wind tunnel investigation on a NLR 73-108-10 airfoil with fowler type flap, part 1. NLR TR 74058 C (1975)
15. P.H. Cook, M.A. Mc Donald, M.C.P. Firmin: Aerofoil RAE 2822-pressure distributions, and boundary layer and wake measurements. AGARD-AR-138 (1979)

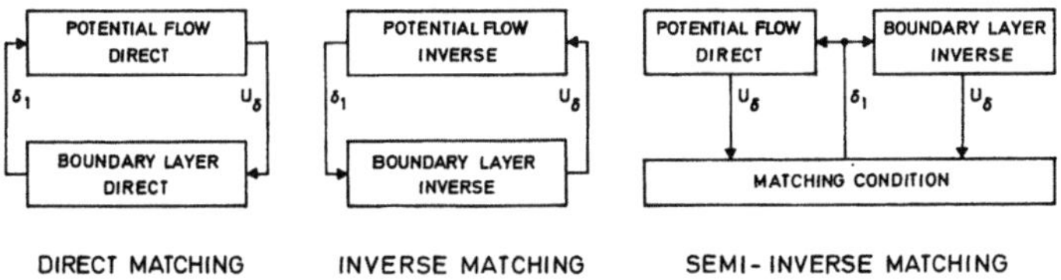

Fig. 1 Iterative solution of viscous/inviscid interaction

INTEGRAL EQUATIONS
FOR STEADY TWO-DIMENSIONAL COMPRESSIBLE B.L. FLOW

- **MOMENT**

$$\frac{dH_{21}^{*}}{ds} + \frac{1 + H_{21}^{*}(2 - M_e^2)}{1 + m_e} \; \frac{1}{M_e} \; \frac{dM_e}{ds} + \frac{H_{21}^{*}}{\delta_1} \; \frac{d\delta_1}{ds} = \frac{c_f}{2\delta_1}$$

- **MOMENT OF MOMENTUM**

$$\frac{dH_{31}^{*}}{dH_{21}^{*}} \; \frac{dH_{21}^{*}}{ds} + \frac{2H_{41}^{*} + H_{31}^{*}(3 - M_e^2)}{1 + m_e} \; \frac{1}{M_e} \; \frac{dM_e}{ds} + \frac{H_{31}^{*}}{\delta_1} \; \frac{d\delta_1}{ds} = \frac{c_D}{\delta_1}$$

RELATIONS

- COMPRESSIBILITY TRANSFORMATION: VAN DRIEST
- VELOCITY PROFILES
  - LAMINAR: STEWARTSON
  - TURBULENT: WALZ (ATTACHED B.L.) ALBER (SEPARATED B.L.)
- EMPIRICAL DISSIPATION LAW FOR NON-EQUILIBRIUM FLOWS

SOLUTION

- WITH PRESCRIBED DISPLACEMENT THICKNESS δ_1
- BY RUNGE-KUTTA INTEGRATION

Fig. 2 Inverse boundary layer integral method

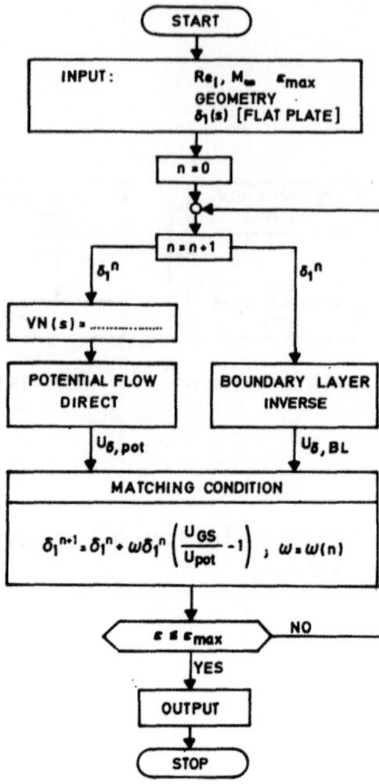

Fig. 3 Semi-inverse matching procedure

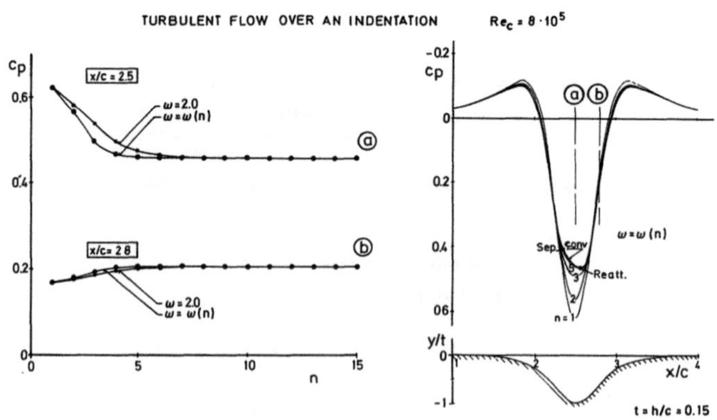

Fig. 4 Convergence of semi-inverse procedure. Turbulent flow

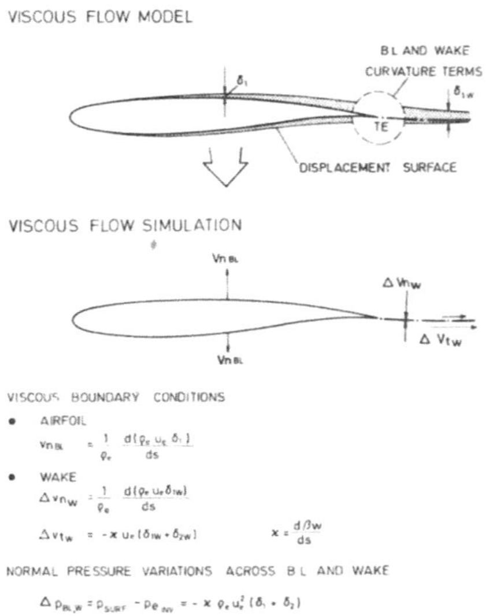

Fig. 5 Viscous effects simulation by surface massflow concept

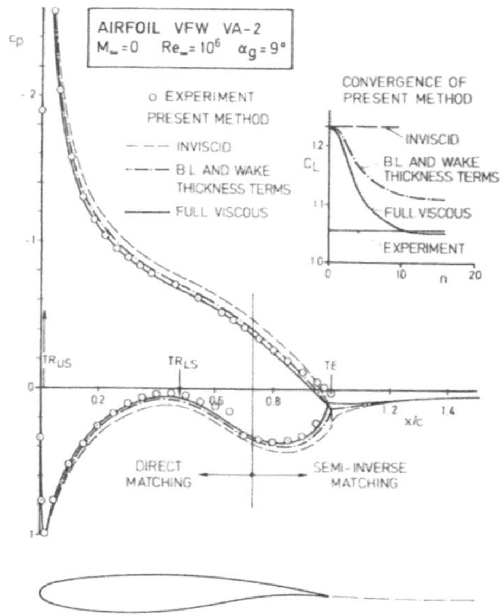

Fig. 6 Viscous/inviscid flow prediction over a rear-loaded airfoil

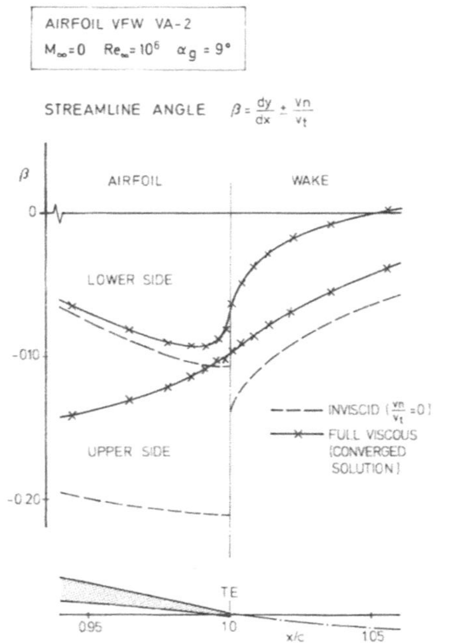

Fig. 7 Predicted streamline slope in the trailing edge region

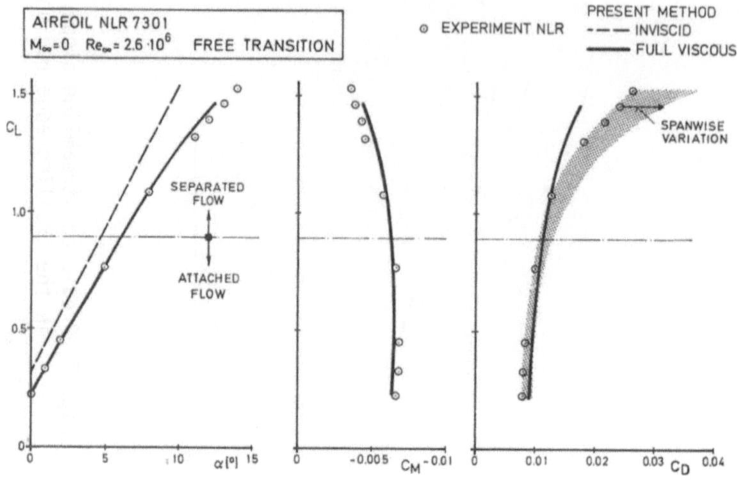

Fig. 8 Lift, moment and drag prediction of a rear-loaded airfoil

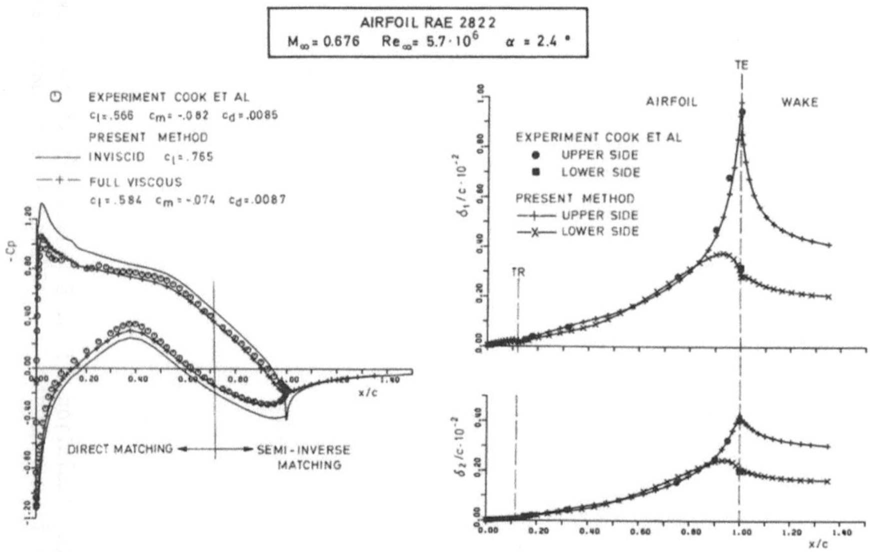

Fig. 9 Viscous/inviscid flow prediction over a rear-loaded airfoil

A Turbulent Boundary Layer Approaching Separation

B.E. THOMPSON AND J.H. WHITELAW

Fluid Mechanics, Heat Transfer and Combustion Section
Department of Mechanical Engineering
Imperial College of Science and Technology
London SW7 2BX, England

Summary

Measurements and calculations of the pressure and velocity characteristics of a two-dimensional boundary layer approaching separation are presented and compared. The velocity measurements were obtained with a combination of impact probes, stationary and flying hot wires and quantify, for example, the rapid growth of the boundary layer as separation is approached. The calculated results were obtained with a combination of potential flow and boundary-layer equations. The Laplace equation was solved, in finite-difference form, to obtain the locations of the iso-vorticity and iso-stream function lines throughout the flow; the potential velocity and pressure distributions were subsequently obtained and used as boundary conditions for a finite-difference solution of the boundary layer equations with a Reynolds stress turbulence model. The relative importance of conservation of momentum, normal to the surface, and of the interaction of the viscous and inviscid flow equations are examined.

Introduction

The recent measurements of references 1 and 2, made with a laser-Doppler anemometer and a flying wire anemometer respectively, have provided evidence of the advances which may be made with modern instrumentation and demonstrate that accurate measurement of velocity characteristics can be made in regions of recirculating flow. Similarly, predictions of separation have been reported in references 3, 4 and 5 using integral and differential equations whereas forward and inverse boundary-layer methods have been investigated in reference 6. It is clear that fully separated flows require the solution of the Navier-Stokes equations as reported, for example, in references 7 and 8 but that the flow approaching separation and perhaps in a short region downstream can be represented as a potential flow boundary layer interaction.

The present contribution is concerned with the flow on a flap located downstream of a flat plate and at an angle of 17.5 degrees. As a consequence of the adverse pressure gradient caused by the angle of the flap, the boundary layer separates and the static pressure distribution, together with profiles of the velocity characteristics, were obtained in the region of attached flow up to separation. Hot wire anemometry was employed and, in the region very close to separation, the flying wire arrangement of reference 9 was required to overcome the problems associated with regions of negative instantaneous velocity. The measured results provide boundary conditions for the solution of boundary-layer equations which have been solved, in finite-difference form, with the assumption of steady two-dimensional flow and with a turbulence model based on the solution of transport equations for the Reynolds stresses. The potential-flow equations have also been solved in finite-difference form to yield the wall-pressure distributions obtained before and after interaction with a boundary-layer solution and are compared with the measured distribution. Particular attention is paid to the region close to separation.

Flow Configuration, Instrumentation and Measured Results

A flat plate of 1 m length was located at 1 degree incidence in the wind tunnel previously described by Crabb et al (10). it was connected to a trailing flap of 250 mm length by a radius of 812.5 mm. The cross-section of the wind tunnel was 450 x 300 mm and the uniform flow velocity at exit from the 8 to 1 contraction was 26.3 m/s with a value of the rms of the corresponding fluctuations of 0.24 m/s. The boundary layer on the upper surface of the flap separated 200 mm upstream of the trailing edge of the flap and was investigated in the region up to detachment of the mean streamlines.

The region over which the mean velocity was two-dimensional was established by flow visualisation, wall pressure and local mean velocity measurements. Side-wall splitter plates, with

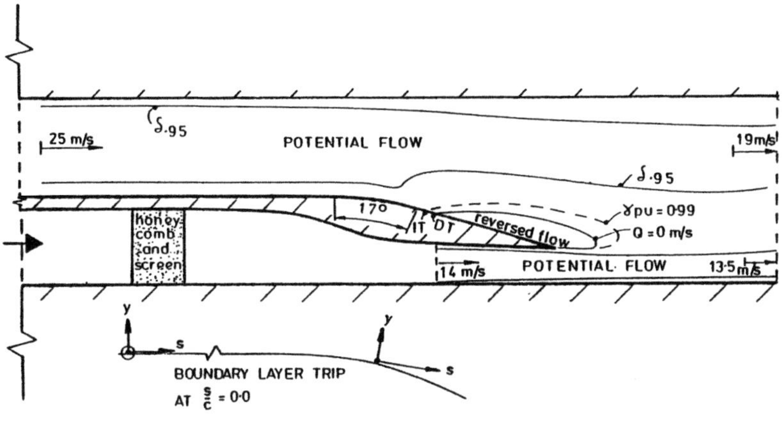

Fig.1. Flow configuration
DT - Mean streamline detachment
IT - Incipient separation (1per cent probability of reversed flow)
$Q = 0$ - Line at zero mean velocity
γ_{pu} - Fraction of time flow is in downstream direction
$\delta_{.95}$ - Boundary layer thickness ($U/U_G = .95$)

passive flow-control vanes, were used to reduce the growth of corner secondary flows on the curved surface and on the flap, and resulted in a two-dimensional region which narrowed to more than 100 mm width in the vicinity of separation. The extent of the region of two-dimensional flow at and downstream of boundary-layer separation was confirmed by mean velocity profiles, obtained with the flying-hot wire, and shown to be constant within 1% and also by off-centre two-dimensional momentum balances which were within 4% of those on the centre line.

Values of static pressure were obtained on the surface of the plate with taps which were 0.50 mm in diameter and were designed in accordance with the suggestions of reference 11. A pitot-static probe, of external diameter 2.2 mm was used to obtain values of static pressure in the flow and, except in the immediate vicinity of solid surfaces, provided results with an accuracy better than 2% of the local value. Flow angle was obtained, within approximately 0.5 degrees, using a calibrated five-hole probe of external diameter 2.5 mm. Velocity measurements were made with a combination of a pitot-static tube, a hot wire and a flying wire, depending on the region of flow.

In the attached boundary layer, mean-velocity values were obtained with the pitot-static tube and, in the near-wall region, with a pitot tube of 0.45 mm diameter and confirmed with a hot wire (DISA 55P11). The signals from the hot wire were linked to a DISA anemometer (55M10), and converted to velocity values with the equations of reference 12. The wire was calibrated in the potential core of the tunnel and digitally linearised using the method in reference 9. In the vicinity of separation, the flying-wire was preferred since turbulence intensities were large and negative velocities occurred.

Figure 1 shows a section of the plate and flap in the working section of the wind tunnel together with an indication, obtained from experiment, of the boundary layer and regions of flow separation. The measured variations of momentum thickness, shape factor and skin friction up to the location of separation are shown on figure 2 together with the free-stream velocity, the free-stream static pressure coefficient and the wall static-pressure coefficient based on the reference velocity and static-pressure at the boundary-layer trip. The values of skin friction were obtained from a Clauser chart and confirmed that the profiles conform to the logarithmic law of the wall, including that closest to the location of mean-streamline detachment where instantaneous flow reversals occurred close to the wall.

The location of streamline detachment is shown on figure 1, and can be deduced from figure 2 to have occurred at a distance from the location of the trip wire equal to 0.870 chord lengths (i.e. S/C = 0.870). The probability density distributions obtained closest to the wall suggested that intermittent back flow occurred some 30 to 35 mm upstream of this location. Profiles of static pressure, and the S and y com-

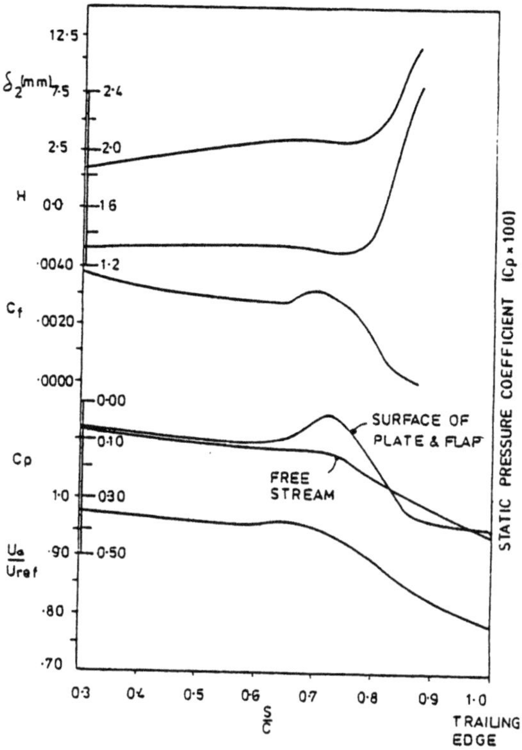

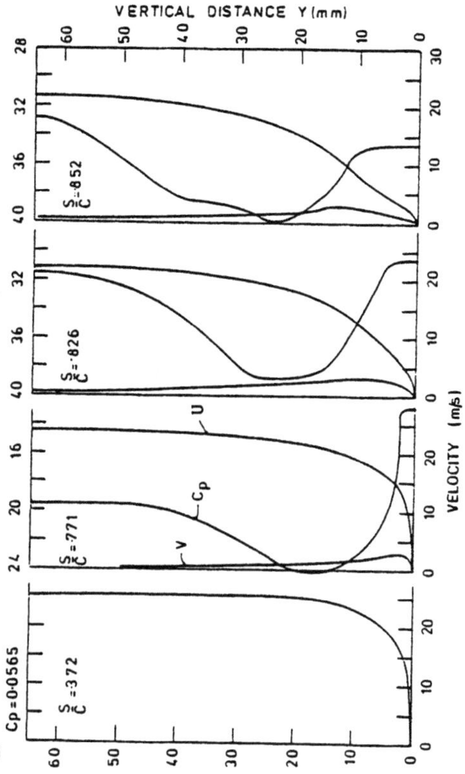

Fig.2. Momentum thickness, shape factor, skin friction, surface and free stream, static pressure coefficient and free stream velocity ratio on upper surface at plate and flap

Fig.3. Profiles of static pressure, longitudinal and cross stream mean velocity

ponents of mean velocity are shown on figure 3. The magnitude of cross-stream pressure gradient increases as detachment is approached and, at S/C of 0.826, achieves a value which is approximately 2.6 the value of the longitudinal pressure gradient of figure 4.

Calculation Methods, Result and Comparison with Measurements

The potential flow was determined by solving the Laplace equation to obtain the coordinates of lines of constant stream function and velocity potential. The finite-difference formulation of reference 13 was used and the pressure field obtain-

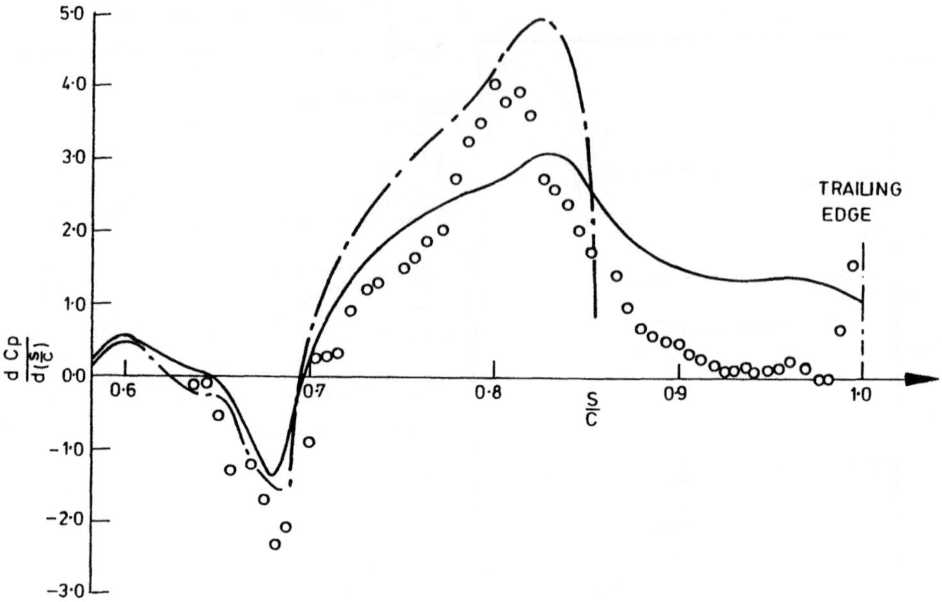

Fig.4. Comparison of measured and calculated non-dimensional longitudinal pressure gradient on upper surface of plate and flap
o Measurements
——— Calculated results based on wall coordinates
— — Calculated results based on wall coordinates plus displacement thickness up to calculated separation

ed by satisfying the equations of continuity and inviscid streamwise momentum.

The equations appropriate to the boundary layer were solved in the manner described in reference 14 with free-stream boundary conditions provided from measured wall pressure or from solution of the potential flow equations as described above. The cross-stream momentum equation was simplified to

$$\frac{\partial p}{\partial y} = \frac{\rho U^2}{R}$$

where R is the local radius of curvature calculated from the potential flow streamlines. A Reynolds-stress model of turbulence was used in the calculations although, as indicated in reference 15, it is likely to give similar results to appropriate eddy-viscosity assumptions.

The upstream boundary conditions for the boundary layer calculations included the measurements obtained at S/C = 0.352 shown on figure 3. Profiles of Reynolds stresses were measured with a stationary wire and agreed closely with those of Klebanoff (16); the required distribution of the rate of turbulence dissipation was obtained with a mixing length hypothesis.

Calculated results, obtained with the measured wall pressure distribution, were found to agree within 5% of the measured value of skin friction, shape factor and momentum thickness up to S/C = 0.6 and are not presented here. The results obtained further downstream are shown on figure 5 and are in reasonable agreement until separation is approached where, for example, the skin friction coefficient is under-predicted with its zero value extrapolated to S/C = 0.856 rather than to 0.870 as suggested by measurements. The calculated momentum thickness is in disaccord upstream of S/C = 0.845 which corresponds approximately to the extremity of instantaneous flow reversals in the near wall region. The cross-stream pressure gradient was calculated to be 50 per cent of the value suggested from measurements at S/C = 0.771 becoming only 12 per cent of the value measured immediately upstream of detachment: this error appears to stem from the simplified form of the cross-stream momentum equation. The numerical procedure failed to converge downstream of S/C = 0.853.

Wall coordinates were used to describe the boundary streamlines, without account of any effects of separation, and the potential equations solved to give one of the pressure distributions of figure 4. The agreement between this calculation and the measurements is adequate up to S/C of 0.78 where the process of separation begins to influence the results. Here the velocity component normal to the potential velocity direction reaches about 3 per cent of the local streamwise value in the near wall region but increases rapidly downstream to 10 per cent at S/C = .826 and 15 per cent at S/C = .852.

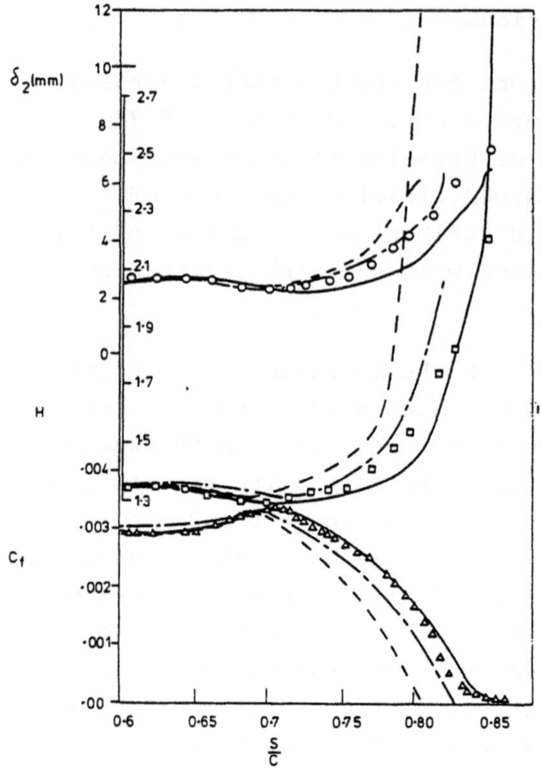

Fig.5. Calculated and measured momentum thickness, shape factor and skin friction coefficient in the vicinity of separation
—— Calculated results based on measured wall pressure
-·- Calculated results based on calculated wall pressure and wall coordinates
--- Calculated.results based on calculated wall pressure and wall coordinates plus displacement thickness

The influence of numerical approximations and of the downstream boundary condition are both important and, although preliminary testing has been carried out, further investigations are required. As shown on figures 2 and 4, the pressure gradient changes rapidly between S/C of 0.65 and 0.86 small streamwise step lengths are necessary in this region. They were, however, difficult to arrange since the solution of the Laplace equation tended to increase the streamwise spacing of nodes in regions of significant second derivatives in pressure. The calculated results of figure 4 were obtained with 40 nodes in the cross-stream direction between the plate and the wind tunnel roof and resulted in a maximum spacing of 5 mm. Initially 40 streamwise nodes were located between S/C of 0.352 and the downstream boundary of zero gradients at S/C of 1.00 and resulted in approximately 30 mm spacing. Longitudinal node spacing was subsequently reduced to about 10 mm between S/C of 0.65 and 0.75 and about 4 mm between S/C of 0.75 and 0.86 by shortening the solution domain using the velocity potential lines described from solution with the coarse grid as the boundaries.

Solution of the boundary layer equations with this calculated wall-pressure also agreed satisfactorily with measured values up to S/C = 0.68. Downstream, as shown on figure 5, the skin friction is under-predicted and the boundary layer grows more rapidly than is measured partly due to the incorrect stream-line curvature obtained from the potential flow solution which assumes the boundary layer remains attached.

The streamlines were corrected to account for separation by adding the calculated displacement thickness to the wall co-ordinates and recalculating the wall pressure. Downstream of the location of calculated separation at S/C = 0.853 the dis-placement thickness was not obtained and approximate treatment of the wall and outlet velocity potential line was required. The displacement thickness after separation was smoothly in-creased as suggested from measurements and a vertical velocity potential line positioned about 3 boundary layer thicknesses downstream from the separation point at S/C of 0.875. The streamwise pressure gradient was over-predicted near separa-tion, despite less than 4 mm between streamwise nodes.

The boundary layer equations were also solved with the pressu-re distribution obtained from the potential-flow equations with the assumption of a displacement thickness extrapolated into the separated flow. The location of the resulting boun-dary-layer separation was found to occur further upstream, figure 5. The agreement between measurements and calculations in the attached boundary layer is improved, away from the re-gion of separation, but significant boundary layer growth occurs too far upstream of separation with the skin friction coefficient under-predicted. The cross-stream pressure gra-dient is not significantly altered, remaining 84 per cent low-er than that measured immediately before separation.

The difference between the measured and calculated wall pres-sure distributions shows the need for further investigation of the coupling of the potential and boundary layer equations and, more particularly, of the downstream boundary with separation effects incorporated.

Concluding Remarks

The investigation and results described in the previous sections represent an early attempt to formulate and examine the problems of the viscous-inviscid interaction which occurs between the potential flow and a boundary layer separating on a trailing flap. The two sets of equations have been solved by finite-difference procedures and the results, when taken together with measurements performed to aid the evaluation of the calculation methods, allow the following conclusions:

1. The results of solutions to the boundary-layer equations, with the distributions of measured wall pressure and a simplified form of the cross-stream momentum equations providing the boundary conditions at the edge of the boundary layer, are in close accord with measurements up to the region where the influence of separation is significant. Thereafter, and up to a location very close to separation, the shear stress is well represented but integral quantities are less well calculated, perhaps due to the simplified form of the cross-stream momentum equation.

2. Solutions of the potential-flow equations have been obtained from the Laplace equation in finite difference form. The influence of truncation errors, associated with finite distances between nodes, is shown to be significant. The location and form of the downstream boundary conditions is also shown to be important and further work is necessary to determine the best combination of assumptions.

3. In addition to further testing of the present approach, it is desirable to evaluate the possibility of solving the steady, two-dimensional form of the Navier-Stokes equations in the region of and downstream of separation.

Acknowledgements

The authors are grateful to the Procurement Executive of the Ministry of Defence for financial support.

References

1. Simpson, R.K.; Chew, Y.-T; Shivaprasad, B.G.: The structure of a separating turbulent boundary layer. Part 1. Mean flow and Reynolds stresses. J. Fluid Mech. 113 (1981) 23.

2. Coles, D.; Wadcock, A.J.: Flying-hot-wire study of flow past a NACA 4412 airfoil at maximum lift. AIAA 17 (1979) 321.

3. Head, M.R.: Entrainment in the turbulent boundary layer. A.R.C. R & M. 3152 (1958).

4. Stratford, B.S.: The prediction of separation of the turbulent boundary layer. J. Fluid Mech. 5 (1959) 1.

5. Cebeci, T.; Mosinskis, G.J.; Smith, A.M.O.: Calculation separation points in turbulent flows. J. Aircraft 9 (1972) 618.

6. Cebeci, T., Keller, H.B.; Williams, P.G.: Separating boundary-layer calculations. J. Comp. Phys. 31 (1979) 363.

7. Briley, W.R.; McDonald, H.: Numerical predictions of incompressible separation bubbles. J. Fluid Mech. 69 (1974). 631.

8. Cebeci, T.; Kahlil, E.E.; Whitelaw, J.H.: The calculation of separated boundary layers. AIAA 17 (1979) 1291.

9. Thompson, B.E.: A preliminary evaluation of flying-hot-wire instrumentation. Imperial College, Mech. Eng. Dept, Fluids Section Report FS/81/23 (1981).

10. Crabb, D.; Durão, D.F.G.; Whitelaw, J.H.: A round jet normal to a cross-flow. J. Fluids Eng. 103 (1981) 142.

11. Shaw, R.: The influence of hole dimensions on static pressure measurements. J. Fluid Mech. 7 (1960) 550.

12. Ribeiro, M.M.C.; Whitelaw, J.H.: Statistical characteristics of a turbulent jet. J. Fluid Mech. 70 (1975) 1.

13. Johns, R.J.R.: Prediction of flow in Diesel engine cylinders. Ph.D. Thesis, University of London (1980).

14. Gibson, M.M.; Jones, W.P.; Younis, B.A.: Calculation of turbulent boundary layers on curved surfaces. Phys. Fluids 24 (1981) 386.

15. Cebeci, T.; Meier, H.U.: Modelling requirements for the calculations of the turbulent flow and airfoils, wings and bodies of revolution. AGARD CP271, Paper 16 (1979).

16. Klebanoff, P.S.: Characteristics of turbulence in a boundary with zero pressure gradient. NACA, TN 3178 (1954).

Considerations on the Transport-Theorem for Extensive Flow Properties

E. TRUCKENBRODT

Lehrstuhl für Strömungsmechanik
Technische Universität München
8000 München 2, Germany

1. Introduction

The transport theorem plays a significant roll in the representation of the balance equations in fluid mechanics. By this is meant the substantial derivatives (material time derivatives) of certain flow properties. These properties can be intensive as well as extensive. The latter group concerns, for example, volume or line properties. Volume properties consist of the volume itself, the mass, the momentum and moment of momentum, the energy in its different forms (kinetic, potential, internal energy, enthalply) as well as also the entropy. Line properties include particularly the circulation as the instantaneous line integral of the flow velocity. Several possibilities exist for the determination of the desired substantial derivatives.

The considerations presented here make a contribution to the uniform and analogous derivation of the transport theorem for volume and line properties, i.e. the Reynolds transport theorem and the Thomson circulation theorem, respectively. Two methods are pointed out, which can be characterized by the following:

1. Differentiation with respect to time of a function F(t) considering the definition of the differential quotient

$$\frac{dF}{dt} = \lim_{\delta t \to 0} \frac{1}{\delta t} [F(t + \delta t) - F(t)] \quad .$$

2. Exploitation of the fact that volume and line integrals are concerned with material quantities of constant mass m(t) = (m)

The english translation of this paper, originally in german, was made by K. Misegades.

= const thus the bounds of integration are independent of time.

2. Substantial Time Derivative

2.1 Field Property. The total time derivative of a scalar field function $\varepsilon(t,\vec{r})$ is given as

$$\frac{d\varepsilon}{dt} = \lim_{\delta t \to 0} \frac{1}{\delta t} [\varepsilon(t + \delta t,\ \vec{r} + \delta\vec{r}) - \varepsilon(t,\vec{r})] \quad (1)$$

whereby $\delta\vec{r} = \vec{v}\,\delta t$ is the distance covered in time δt at a velocity $\vec{v}$ [1]. From this follows

$$\frac{d\varepsilon}{dt} = \frac{\partial\varepsilon}{\partial t} + \vec{v}\cdot\mathrm{grad}\,\varepsilon \quad \text{(substantial derivative)} \quad (2)$$

with $\partial\varepsilon/\partial t$ as local and $\vec{v}\cdot\mathrm{grad}\,\varepsilon$ as convective derivative.

The change in density $\varrho(t,\vec{r})$ in a moving fluid is given by

$$\frac{d\varrho}{dt} + \varrho\ \mathrm{div}\,\vec{v} = 0 \quad \text{(continuity equation)}\ . \quad (3)$$

Here $d\varrho/dt$ represents the substantial derivative of density.

2.2 Volume Property. Considered is an arbitrarily chosen material volume $V(t)$ moving with the fluid. The amount of fluid mass $m(t) = (m) = \mathrm{const}$ contained in the volume is at all time the same. The integration of the field function $\varepsilon(t,\vec{r})$ in the material volume $V(t)$ gives the volume property

$$J(t) = \int_{V(t)} \varepsilon(t,\vec{r})\,dV = \int_{(m)} j(t,\vec{r})\,dm \quad \text{(material volume)}\ , \quad (4a,b)$$

with $dm = \varrho\ dV$ as mass of the material volume element dV, see fig. 1 , ε as the property density with respect to the volume (property/volume) and $j = \varepsilon/\varrho$ as the specific property with respect to the mass (property/mass). ε and j (and with it J) can be of scalar, vector or tensor form.

---

1) The substantial derivative is often given as $d/dt = D/Dt$.

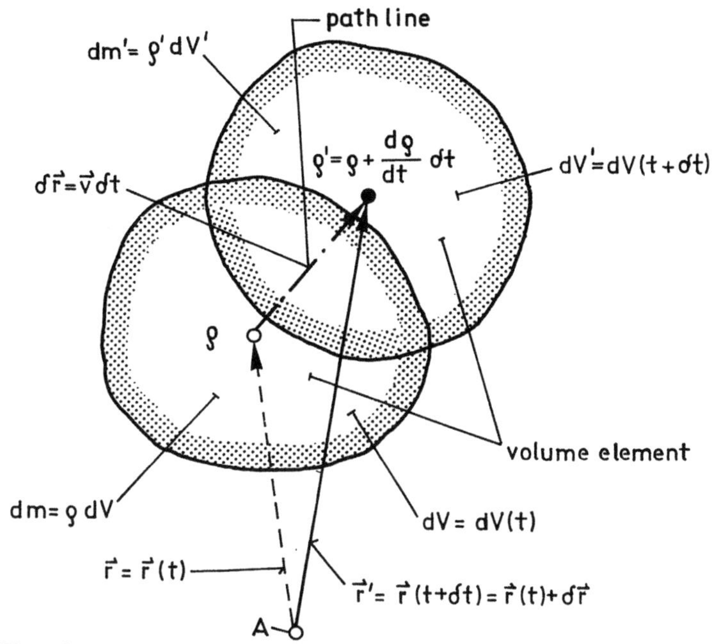

Fig.1

The volume properties J as well as the properties ε and j related to volume and mass, respectively, are listed below:

Property	J(t)	$\varepsilon(t,\vec{r})$	$j(t,\vec{r})$
Volume	V	1	$v = 1/\varrho$
mass	m	ϱ	1
momentum	$\vec{I}$	$\varrho\ \vec{v}$	$\vec{v}$
kinetic energy	E	$\varrho\ \vec{v}^2/2$	$\vec{v}^2/2$
internal energy	U	$\varrho\ u$	u
enthalpy	H	$\varrho\ h$	h
entropy	S	$\varrho\ s$	s

In order to arrive at the Reynolds transport theorem, the substantial derivative of J(t) must be developed,

$$\frac{dJ}{dt} = \frac{d}{dt}\int\limits_{V(t)} \varepsilon(t,\vec{r})\,dV = \int\limits_{(V)} \frac{d}{dt}(\varepsilon\ dV) = \int\limits_{V_o} \frac{d}{dt}(\varepsilon D)\,dV_o \qquad (5a,b,c)$$

In the differentiation of (5a) one must pay attention to the fact that after time t, not only the property under the integral but also the bounds on integration have changed with time. This leads to, among other things, the further forms (5b,c) [2]. Two other interpretations of (5a) will be discussed in sections 3.1 and 3.2.

2.3 Line Property. Considered is an arbitrarily chosen material line $\ell(t)$ that moves with the fluid and is always composed of the same fluid mass m(t) = (m) = const. Each point of the line moves with a flow velocity $\vec{v}(t,\vec{r})$. The line property L(t) is defined as a line integral of the velocity

$$L(t) = \int_{\ell(t)} \vec{v}(t,\vec{r})\cdot d\vec{\ell} \quad \text{(material line)} \; . \tag{6}$$

If this integration is carried out around a closed line, the line integral of the velocity represents the circulation as

$$\Gamma(t) = \oint_{\ell(t)} \vec{v}\cdot d\vec{\ell} = \oint_{(m)} \gamma(t,m)\,dm \quad \text{(circulation)} \; . \tag{7a,b}$$

In (7b) γ is the circulation related to mass and $m(\vec{r})$ is the variable of integration as depicted in fig. 2. With $\mu = dm/d\ell$ as line density,

$$\gamma(t,m) = \vec{v}\cdot\frac{d\vec{\ell}}{dm} \; , \quad m(\vec{r}) = \int_0^{\ell(\vec{r})} \mu(\ell)\,d\ell \; , \quad (m) = \oint_{\ell} \mu\,d\ell \; . \tag{8a,b,c}$$

In order to arrive at the Thomson circulation theorem, the substantial derivative of $\Gamma(t)$ must be developed. This is obtained from the differentiation of the line integral

$$\frac{dL}{dt} = \frac{d}{dt}\int_{\ell(t)} \vec{v}(t,\vec{r})\cdot d\vec{\ell} = \int_{\ell} \frac{d}{dt}(\vec{v}\cdot d\vec{\ell}) = \int_{\ell_0} \frac{d}{dt}(\vec{v}\cdot\vec{D})\,d\ell_0 \tag{9a,b,c}$$

---

2) Note,

$$\frac{d}{dt}(\varepsilon\, dV) = \frac{d\varepsilon}{dt}\,dV + \varepsilon\,\frac{d}{dt}(dV) \quad \text{and} \quad \frac{d}{dt}(\varepsilon D)\,dV_0 = \frac{d\varepsilon}{dt}\,dV + \varepsilon\,\frac{dD}{dt}\,dV_0$$

with V_0 = const, $D = dV/dV_0$ als Jacobian determinate and $d(dV)/dt = (dD/dt)dV_0 = \text{div}\,\vec{v}\,dV$ als volume derivative.

with subsequent transition to a closed integral. The time differentiation is obtained in a manner analogous to that mentioned for (5) [3]. Two other interpretations of (9a) will be discussed in sections 4.1 and 4.2.

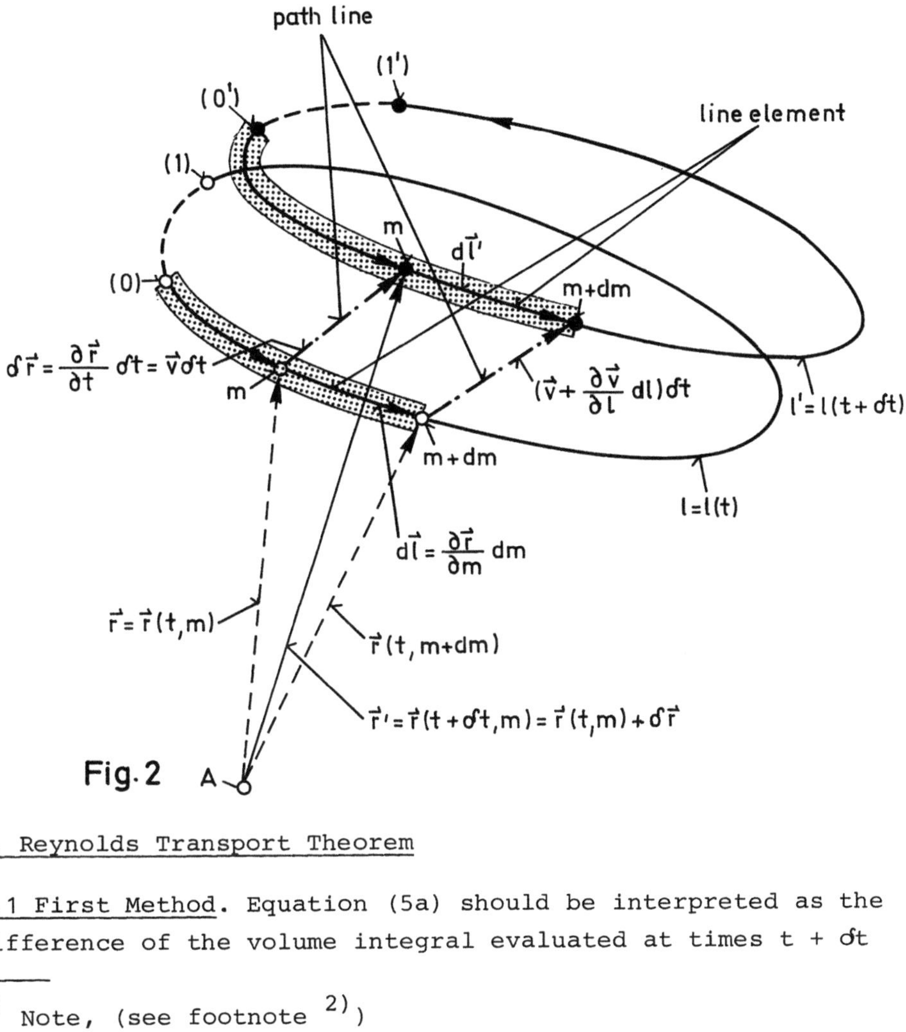

3. Reynolds Transport Theorem

3.1 First Method. Equation (5a) should be interpreted as the difference of the volume integral evaluated at times $t + \delta t$

---

[3] Note, (see footnote [2])

$\frac{d}{dt}(\vec{v}\cdot d\vec{\ell}) = \frac{d\vec{v}}{dt}\cdot d\vec{\ell} + \vec{v}\cdot\frac{d}{dt}(d\vec{\ell})$ and $\frac{d}{dt}(\vec{v}\cdot\vec{D})d\ell_o = \frac{d\vec{v}}{dt}\cdot d\vec{\ell} + \vec{v}\cdot\frac{d\vec{D}}{dt}\,d\ell_o$

with ℓ_o = const, $\vec{D} = d\vec{\ell}/d\ell_o$ and $d(d\vec{\ell})/dt = (d\vec{D}/dt)d\ell_o = \vec{e}_\ell\cdot\text{grad}\,\vec{v}\;d\ell = (\partial\vec{v}/\partial\ell)d\ell$ as length derivative.

and t, divided by δt as $\delta t \to 0$. This process is shown in the following:

$$\frac{dJ}{dt} = \frac{d}{dt}\int_{V(t)} \varepsilon\, dV = \lim_{\delta t \to 0} \frac{1}{\delta t}\,[J(t + \delta t) - J(t)] \qquad \text{(a)}$$

$$t' = t + \delta t,\ \vec{r}\,' = \vec{r} + \delta\vec{r}, \qquad V' = V(t + \delta t) \qquad \text{(b)}$$

$$\frac{dJ}{dt} = \lim_{\delta t \to 0} \frac{1}{\delta t}\,[\int_{V'} \varepsilon(t',\vec{r}\,')\,dV' - \int_{V} \varepsilon(t,\vec{r})\,dV] \qquad \text{(c)}$$

$$\varepsilon(t'\vec{r}\,') = \varepsilon(t,\vec{r}) + \frac{d\varepsilon(t,\vec{r})}{dt}\,\delta t, \qquad \varepsilon' = \varepsilon + \frac{d\varepsilon}{dt}\,\delta t \qquad \text{(d)}$$

$$dV' = (1 + \operatorname{div}\vec{v}\,\delta t)\,dV \qquad \text{(e)}$$

$$\frac{dJ}{dt} = \lim_{\delta t \to 0} \frac{1}{\delta t}[\int_{V} (\varepsilon + \frac{d\varepsilon}{dt}\,\delta t)\cdot(1 + \operatorname{div}\vec{v}\,\delta t)\,dV - \int_{V} \varepsilon\, dV] \qquad \text{(f)}$$

In rows (b) through (e) properties at time $t + \delta t$ are apostrophied. Row (d) follows from (1). The volume derivative in row (e) results from the fact that the motion of a non-changing element of mass $dm' = dm$ is considered. Paying attention to (3), we obtain next

$$\varrho\, dV = \varrho'\, dV' = (\varrho + \frac{d\varrho}{dt}\,\delta t)\,dV' = \varrho(1 - \operatorname{div}\vec{v}\,\delta t)\,dV'. \qquad (10)$$

Furthermore, we note that $\operatorname{div}\vec{v}\,\delta t \ll 1$. From row (f) the known relation for the substantial derivative of the volume property is obtained

$$\frac{dJ}{dt} = \int_{V} (\frac{d\varepsilon}{dt} + \varepsilon \operatorname{div}\vec{v})\,dV = \int_{V} [\frac{\partial\varepsilon}{\partial t} + \operatorname{div}(\varepsilon\vec{v})]\,dV\ , \qquad \text{(11a,b)}$$

where the second relation results from (2). Instead of integrating the material volume V(t) that moves with the fluid, the integration can be carried out for the volume (V) fixed in space that is just at time t filled with the material volume. Therefore, (V) as well as V(t) can be used as V in (11a,b). A further discussion of (11b), which leads to the Reynolds transport theorem, can not be made here; see, for example, ref. [3].

3.2 Second Method. For the integration of a volume property moving with the fluid, one sees that the included mass is m(t) = const and that integration is carried out over the mass (m), instead of over the volume. Since the bounds of the integration over mass are constant, the integration and differentiation can be exchanged. With (4b), the substantial derivative of the volume property can be written as

$$\frac{dJ}{dt} = \frac{d}{dt} \int_{(m)} j \; dm = \int_{(m)} \frac{dj}{dt} \; dm = \int_V \rho \frac{dj}{dt} \; dV \; . \qquad (12a,b,c)$$

The substantial derivative of the specific property $j = \mathcal{E}/\rho$ amounts to in connection with (3)

$$\frac{dj}{dt} = \frac{d}{dt}\left(\frac{\mathcal{E}}{\rho}\right) = \frac{1}{\rho}\left(\frac{d\mathcal{E}}{dt} - \frac{\mathcal{E}}{\rho^2}\frac{d\rho}{dt}\right) = \frac{1}{\rho}\left(\frac{d\mathcal{E}}{dt} + \mathcal{E} \; \text{div} \; \vec{v}\right) \; . \qquad (13)$$

Insertion of (13) into (12c) leads directly to (11a).

4. Thomson Circulation Theorem (Kelvin Theorem)

Using a method similar to that for the volume property J(t) in sections 3.1 and 3.2, the substantial derivative dL/dt is determined from the line integral of the velocity L(t).

4.1 First Method. Equation (9a) should be interpreted as the difference of the line integral evaluated at times $t + \delta t$ and t, divided by δt as $\delta t \to 0$. This process is shown in the following:

$$\frac{dL}{dt} = \frac{d}{dt} \int_{\ell(t)} \vec{v} \cdot d\vec{\ell} = \lim_{\delta t \to 0} \frac{1}{\delta t} \left[L(t + \delta t) - L(t) \right] \qquad (a)$$

$$t' = t + \delta t, \quad \vec{r}\,' = \vec{r} + \delta\vec{r}, \quad \ell' = \ell(t + \delta t) \qquad (b)$$

$$\frac{dL}{dt} = \lim_{\delta t \to 0} \frac{1}{\delta t} \left[\int_{\ell'} \vec{v}(t',\vec{r}\,') \cdot d\vec{\ell}\,' - \int_{\ell} \vec{v}(t,\vec{r}) \cdot d\ell \right] \qquad (c)$$

$$\vec{v}(t',\vec{r}\,') = \vec{v}(t,\vec{r}) + \frac{d\vec{v}(t,\vec{r})}{dt} \delta t, \quad \vec{v}\,' = \vec{v} + \frac{d\vec{v}}{dt} \delta t \qquad (d)$$

$$d\ell' = d\vec{\ell} + \frac{\partial v}{\partial \ell} d\ell \; \delta t \qquad (e)$$

$$\frac{dL}{dt} = \lim_{\delta t \to 0} \frac{1}{\delta t} \left[\int_{\ell} \left(\vec{v} + \frac{d\vec{v}}{dt} \delta t\right) \cdot \left(d\vec{\ell} + \frac{\partial \vec{v}}{\partial \ell} d\ell \; \delta t\right) - \int_{\ell} \vec{v} \cdot d\vec{\ell} \right] \qquad (f)$$

In rows (b) through (e) properties at time $t + \delta t$ are apostrophied. Row (d) follows from (1) if $\mathcal{E}$ is replaced by $\dot{\vec{v}}$. The length derivative in row (e) is seen in fig. 2 as the total distance:

$$\vec{v}\,\delta t + d\vec{\ell}\,' = d\vec{\ell} + (\vec{v} + \frac{\partial \vec{v}}{\partial \ell}\,d\ell)\,\delta t \quad . \tag{14}$$

The interpretation of row (f) is the known relation for the substantial derivative of the line property

$$\frac{dL}{dt} = \int_\ell \frac{d\vec{v}}{dt}\cdot d\vec{\ell} + \int_\ell \vec{v}\cdot\frac{\partial \vec{v}}{\partial \ell}\,d\ell = \int_\ell \frac{d\vec{v}}{dt}\cdot d\vec{\ell} + \frac{1}{2}(\vec{v}_1^{\,2} - \vec{v}_0^{\,2}) \quad . \tag{15a,b}$$

If a closed line integral is considered, the points (O) and (1) are the same, i.e., $v_1 = v_o$. In this case, the line integral $L(t)$ represents the circulation $\Gamma(t)$. Consequently,

$$\frac{d\Gamma}{dt} = \oint_\ell \frac{d\vec{v}}{dt}\cdot d\vec{\ell} = \oint_\ell \mathrm{grad}(u_B + i)\cdot d\vec{\ell} = 0 \quad \text{(inviscid)} \tag{16a,b,c}$$

is valid for the substantial derivative of the circulation along a closed material line. The second relation is valid for the inviscid flow of a barometric fluid with $\varrho = \varrho(p)$ as density, u_B as specific potential of inertia and $i = i(p)$ as specific potential of pressure forces. This is the well-known Thomson circulation theorem, often also known as the Kelvin theorem. It states that for the given assumptions, the circulation is independent of time, i.e., $\Gamma(t)$ = constant. A further discussion of (16a) can not be made here; see for example, ref. [3].

4.2 Second Method. For the differentiation of the line property moving with the fluid it is assumed that the mass of the material line $m(t)$ = const and that the integration is carried out over the mass (m) instead of over the length. From (7b), the substantial derivative of the line integral can be written as

$$\frac{dL}{dt} = \frac{d}{dt}\int_{(m)} \gamma\,dm = \int_{(m)} \frac{d\gamma}{dt}\,dm \quad \text{with} \quad L(t) = \int_{(m)} \gamma\,dm \quad . \tag{17a,b,c}$$

In order to obtain $d\gamma/dt$, the following must be considered: On the material lines $\ell(t)$ and $\ell(t + \delta t)$ are located in the sense of circulation respectfully two neighboring points that are from starting point (O) as masses m and m + dm denoted. The distance between the two ponts is $d\ell$. As seen from an origin A,

$$d\vec{\ell} = \frac{\partial \vec{r}}{\partial m}\, dm \; , \qquad \frac{d\vec{\ell}}{dm} = \frac{\partial \vec{r}}{\partial m} \tag{18a,b}$$

is also valid. Under consideration of the definition from (8a), one obtains first

$$\gamma = \vec{v}\cdot\frac{d\vec{\ell}}{dm} = \vec{v}\cdot\frac{\partial \vec{r}}{\partial m} \tag{19a,b}$$

and from this the desired time derivative

$$\frac{d\gamma}{dt} = \frac{d\vec{v}}{dt}\cdot\frac{\partial \vec{r}}{\partial m} + \vec{v}\cdot\frac{d}{dt}\left(\frac{\partial \vec{r}}{\partial m}\right) = \frac{d\vec{v}}{dt}\cdot\frac{d\vec{\ell}}{dm} + \vec{v}\cdot\frac{\partial \vec{v}}{\partial m} \; . \tag{20a,b}$$

The flow velocity $\vec{v} = d\vec{r}/dt$ has been introduced. After substitution in (17b) we obtain for the substantial derivative of the line property

$$\frac{dL}{dt} = \int\limits_{\ell} \frac{d\vec{v}}{dt}\cdot d\vec{\ell} + \int\limits_{(m)} \vec{v}\cdot\frac{\partial \vec{v}}{\partial m}\, dm = \int\limits_{\ell} \frac{d\vec{v}}{dt}\cdot d\ell + \frac{1}{2}(\vec{v}_1^{\,2} - \vec{v}_0^{\,2}) \, , \tag{21a,b}$$

which is in agreement with (15b). The further comments made in connection with (15) are also valid here.

References

1. Becker, E. and Bürger, W.: Kontinuumsmechanik, chap. 1.3, Stuttgart: Teubner, 1975.

2. Schade, H.: Kontinuumstheorie strömender Medien, chap. 1.2.4, Berlin, Heidelberg, New York: Springer, 1970.

3. Truckenbrodt, E.: Fluidmechanik, Berlin, Heidelberg, New York: Springer, 1980.

The Structure of Flows with Unsteady Boundary Conditions

HERMANN VIETS

College of Engineering
West Virginia University
Morgantown, WV 26506
USA

RICHARD J. BETHKE

Engineering Department
Wright State University
Dayton, Ohio 45435
USA

Summary

In order to control unsteady flows, it is necessary to understand the details of their production and the effect that the unsteady component has on the mean flow. It is the purpose of this paper to describe several forced unsteady flows in terms of the source of the time dependency and the dynamics of the vortical motions produced in the flowfield. Three techniques are described briefly to locate the coherent structures and some possible applications for driven unsteady flows are discussed.

Background

Unsteady flows have generally been viewed as undesirable. Substantial efforts have been made to eliminate them, as in the cases of inlet buzz and wing flutter. More recently, with the development of rapid response instrumentation, unsteady flows are being quantitatively studied and thus better understood. With the improved understanding comes the potential for control and useful applications. The following overview describes an ongoing research program aimed at the understanding of forced unsteady flows.

The imposition of unsteady boundary conditions generally leads to large scale coherent structures in the flow. In order to identify these convected structures, they must be separated from the mean flow on which they are superimposed. In relatively simple flows the vortex structure can be identified by essentially a recognition of a typical flow pattern, even in the convected flow. In more complicated flows, more formal techniques are required.

The Unsteady Driver

In the present discussion, two different techniques are employed to generate the unsteady flow. Both are cyclical flows so that the desired flowfields can be replicated often. The first technique shown in Figure 1, is chosen because of its potential for application, since it is fluidically controlled and produces a jet which oscillates from side to side without the need for moving parts. The details of its operation are presented in [1]. A photograph illustrating the vortex structure of the flow is shown in Figure 2 employing smoke flow visualization. The flow is from the left and the vortices may be seen on each side of the jet flow.

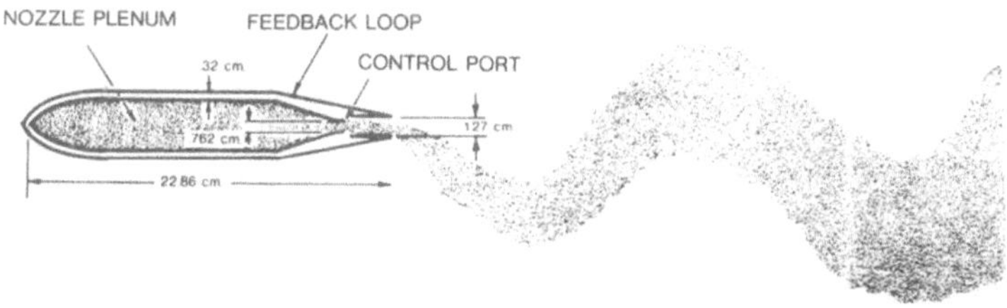

Figure 1. Fluidically controlled unsteady jet.

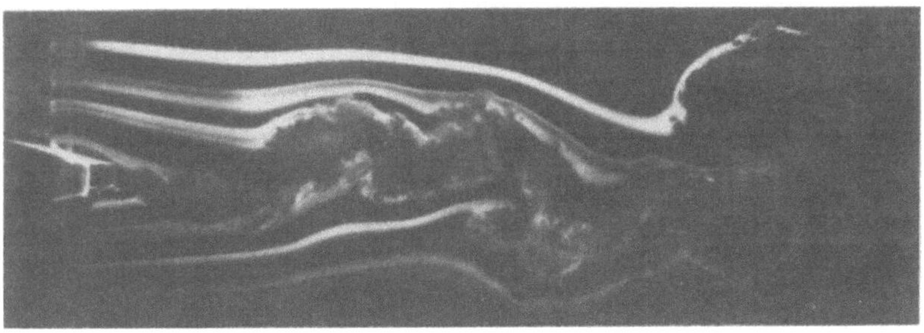

Figure 2. Unsteady jet flowfield photograph.

The second unsteady flow generator is the rotor device shown in Figure 3. The flow is from the left and the rotor turns in a counter clockwise direction. Each turn of the rotor produces one vortex structure which is then convected downstream. In the photograph of Figure 4, the flowfield generated by the rotor is visualized by smoke addition and the vortex structures are readily identifiable, one near the rotor and the other further downstream. Both the unsteady jet [1-4] and the rotor [4-9] have been investigated on a fundamental and applied level.

The Unsteady Flowfield

In order to identify quantitatively the large scale structures in the flowfields produced by the unsteady drivers, it is necessary to separate the structure from the general background velocities on which it is superimposed. For example, the

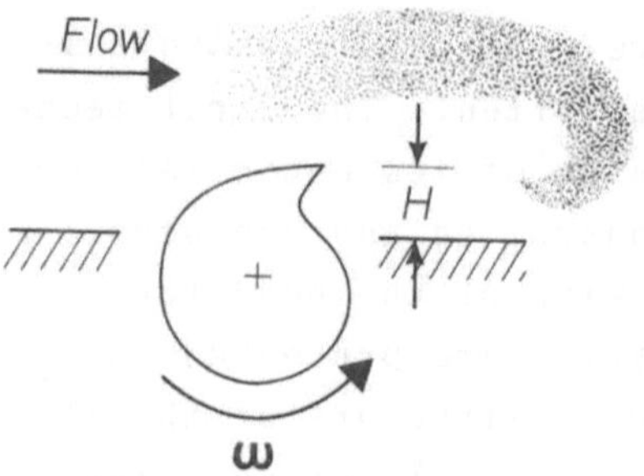

Figure 3. Rotor vortex generator.

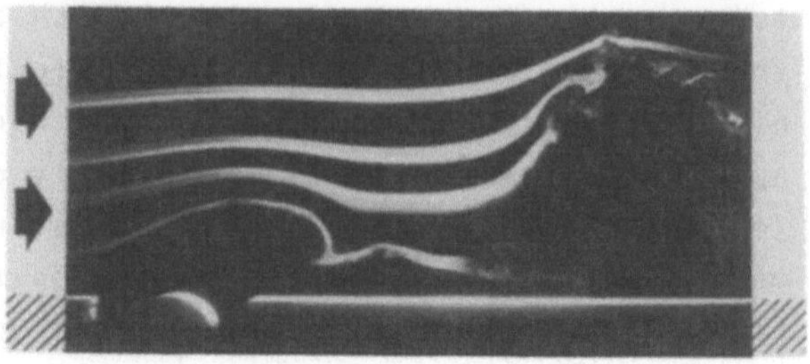

Figure 4. Rotor flowfield photograph.

instantaneous jet flowfield shown in Fig. 5 shows no clear sign of the large scale vortex structures visible in the flow visualization results of Figure 2. In order to observe these structures, it is necessary for the observer to be moving with them in the same frame of reference. This frame can be found by examining the field for a telltale pattern of a flow curvature with a decreasing velocity as one approaches the apparent center of the curvature. When the streamwise velocity at the center position is subtracted from the entire field, the observer is then in a frame of reference moving with the large scale structure [3]. The result is that the flowfield of Figure 5 becomes the vortex structure of Figure 6.

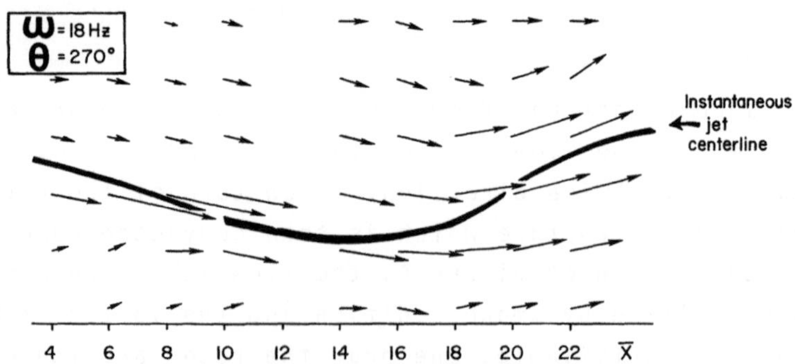

Figure 5. Quantitative jet flowfield.

The same technique has been applied to the flowfield produced by the rotor of Figures 3 and 4. By plotting the positions of the identified large scale structures with time, the trajectory of the vortex produced by the rotor can be found [9] as shown in Figure 7.

Vortex Identification

Although the technique of identifying the large scale structures by visual pattern recognition has been shown to be useful, a more precise method is required to allow the definition of smaller and more diffuse patterns. In order to accomplish this, a more mathematically based technique is required. Two such techniques have been developed.

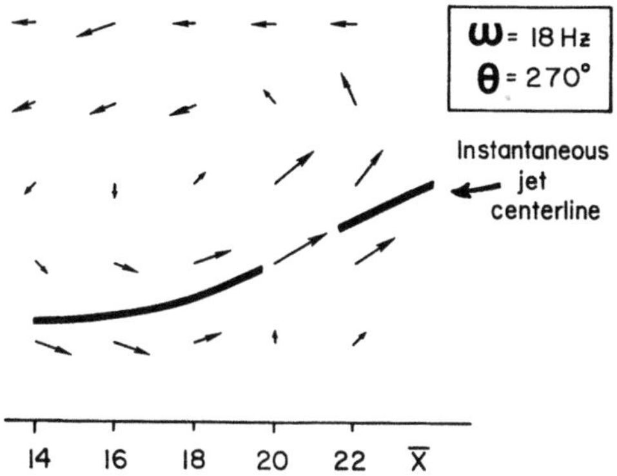

Figure 6. Coherent vortex structure in jet flowfield.

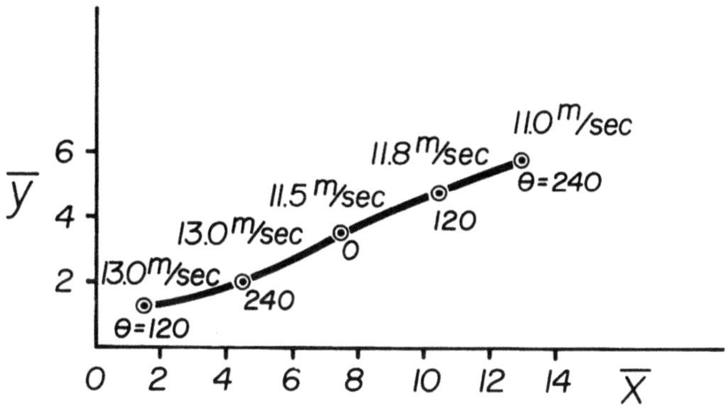

Figure 7. Trajectory of vortex near wall.

A method based on the fourier transformation of the flowfield was developed in [10] and achieved reasonably good results. The technique allowed the search process to be automated and hence eliminated the need for visual pattern recognition.

An improved technique has been developed based on the concept of decomposing the velocity field into its component parts including rotation, two translation components, two stretching components and a shear component. If a rectangular data cell is chosen consisting of four spacial positions, each with two velocities, then the total of eight velocities could be written as a function of the components described above plus an error which is necessarily present in the experiment as in equation 1.

$$\begin{bmatrix} u_1 \\ v_1 \\ u_2 \\ v_2 \\ u_3 \\ v_3 \\ u_4 \\ v_4 \end{bmatrix} = \begin{bmatrix} -A & 1 & 0 & -1 & 0 & B \\ -B & 0 & 1 & 0 & 1 & -A \\ -A & 1 & 0 & 1 & 0 & B \\ B & 0 & 1 & 0 & 1 & A \\ A & 1 & 0 & 1 & 0 & -B \\ -B & 0 & 1 & 0 & -1 & -A \\ A & 1 & 0 & 1 & 0 & -B \\ B & 0 & 1 & 0 & -1 & A \end{bmatrix} \begin{bmatrix} \beta_1 \\ \beta_2 \\ \beta_3 \\ \beta_4 \\ \beta_5 \\ \beta_6 \end{bmatrix} + \begin{bmatrix} \varepsilon_1 \\ \varepsilon_2 \\ \varepsilon_3 \\ \varepsilon_4 \\ \varepsilon_5 \\ \varepsilon_6 \\ \varepsilon_7 \\ \varepsilon_8 \end{bmatrix} \quad (1)$$

The velocities (u,v) are related to the decomposed components (β) by the matrix produced by determining the velocity contributions of each of the components. The errors (ε) must be present but must be small if the components (β) are sufficient to describe the flowfield.

Various flowfields have been examined by the decomposition technique. Any of the components could be employed to identify the location of the vortex structures. Perhaps it is not surprising that the rotational component seems to most successfully locate these structures. An example of the plotted contours is shown in Figure 8, where the locations of the vortices are signified by the closed contours. The technique describes the vortices very well as may be seen by comparison with the photograph of Figure 4. In addition, the technique identifies a third vortex in the field which is rotating in the opposite (i.e. CCW) direction as compared to the two main structures. Examining Figure 4 shows some evidence of the existence of this additional vortex which had not been identified earlier but which is evident in other photographs. Thus the decomposition technique is useful for small and diffuse vortices as well as the larger driven vortices.

Further Examples

A flow which combines the effects of an unsteady jet with the vortex generation properties of the rotor is shown in Figure 9. Rotors are symmetrically located on either side of the nozzle near the exit. As in the other examples, the rotors turn to produce the unsteady flow. They can have any phase relationship to each other. In Figure 9, they are out of phase and hence produce a jet which oscillates from side to side in a manner similar to that of References [1-4]. As in the jet results of Figure 5, the large scale structures of Figure 9 are obscured by the superimposed translational velocity field. Removing that translational field yields the results of Figure 10 where the structures themselves are evident. Additional work is currently underway to define the trajectories of these structures.

Figure 8. Rotor flowfield by decomposition.

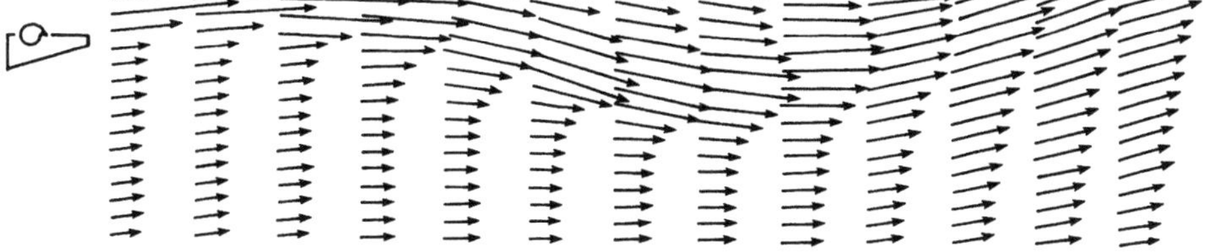

Figure 9. Quantitative rotor jet flowfield.

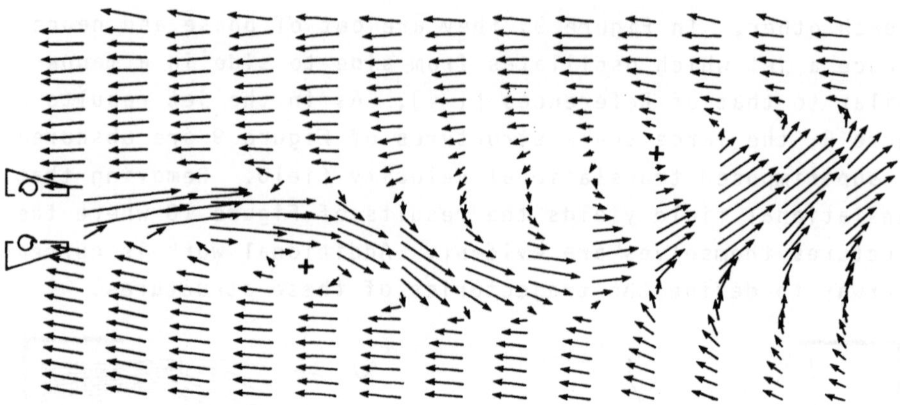

Figure 10. Coherent vortex structures in a rotor jet flowfield.

The final example is that of the vortex structure near the rear of a bluff body vehicle shown in Figure 11. The objective is to employ the rotor to reduce the size of the separated flow region behind the vehicle. It was shown in [5] that the use of the rotor could allow the flow to turn through a larger angle without steady separation. In this case by employing the techniques described above, the structure of the vortex may be clearly seen. Additional study is underway to define the trajectories of the vortices and their effect on the body.

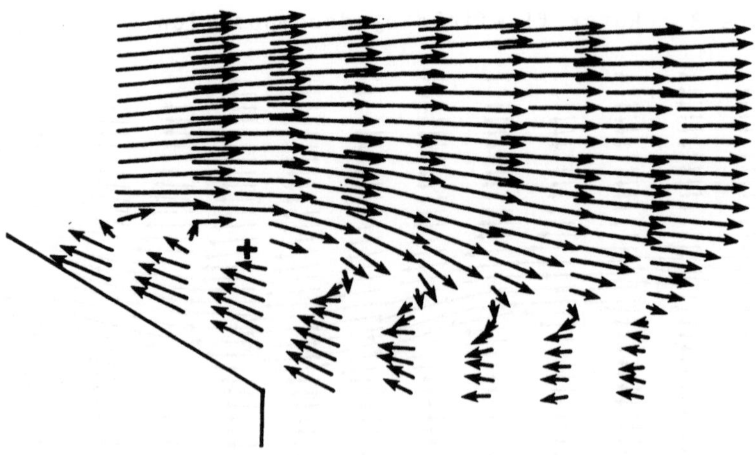

Figure 11. Vortex structure behind vehicle with rotor.

Conclusions

Forced unsteady velocities have a very significant effect on the overall flowfield. In order to take advantage of these effects and control them the unsteady flows must be understood and successfully modeled. A first step to that end is to identify the unsteady structure of the flows as in the present results.

Acknowledgment

The support of the U.S. Air Force Office of Scientific Research under Grant No. 81-0025 monitored by Capt. (Dr.) M. S. Francis is much appreciated. In addition, the able assistance of Mont Ball in the construction and testing phases is gratefully appreciated.

References

1. Piatt, M.; and Viets, H.: Conditioned Sampling in an Unsteady Jet, AIAA Paper No. 79-1857, AIAA Aircraft Systems and Technology Meeting, New York, August 1979.

2. Viets, H.: Unsteady Ejectors, Proc. NATO/AGARD Meeting on Jet Flows with Application to V/STOL, NATO/AGARD CP-308, Nov. 1981.

3. Viets, H.; Ball, M; and Piatt, M.: Experiments in a Sub-scale Pilot Gust Tunnel, AIAA Journal, Vol. 19, No. 6, June 1981, pp. 820-822.

4. Viets, H.: Coherent Structures in Time Dependent Flow, - Proceedings of the NATO/AGARD Specialists Meeting on Turbulent Boundary Layers, The Hague, Netherlands, September 1979.

5. Viets, H.; Piatt, M.; and Ball, M.: Boundary Layer Control by Unsteady Vortex Generation, J. Wind Eng. & Ind. Aero., Vol. 7, 1981, pp. 135-144.

6. Viets, H.; Piatt, M.; and Ball, M.: Unsteady Wing Boundary Layer Energization, AIAA Paper No. 79-1631, Aug. 1979.

7. Viets, H.; and Piatt, M.: Induced Unsteady Flow in a Dump Combustor, AIAA Progress in Astro. & Aero., Vol. 76, 1981, pp. 611-624.

8. Viets, H.; Bethke, R. J.; and Bougine, D.: Three Dimensional Vortex Dynamics Near a Wall, to be issued.

9. Viets, H.; Piatt, M; and Ball, M.: Forced Vortices Near a Wall, AIAA Paper No. 81-0256, Jan. 1981.

10. Bethke, R. J.; and Viets, H.: Data Analysis to Identify Coherent Flow Structures, AIAA Paper No. 80-1561, Aug. 1980.

11. Bethke, R. J.; and Viets, H.: Identification of Convected Flow Structures by Decomposition Techniques, to be issued.

12. Viets, H.; Ball, M; and Smith, J.: The Rotor Jet Nozzle, to be issued.

13. Catalano, G. D.; Viets, H.; and Bougine, D.: Reduction of the Turbulent Wake of a Road Vehicle by use of Unsteady Vortex Shedding, to be issued.

Boundary Layer Calculations for Cryogenic Wind Tunnel Flows *

B. WAGNER

Dornier GmbH, Friedrichshafen

Summary

In order to estimate theoretically the simulation accuracy in planned cryogenic wind tunnels some turbulent boundary calculations have been carried out taking into account real gas effects and heat transfer. Cebeci's finite difference method has been used including real gas equations of state for representation of the gas behaviour close to its liquefaction boundary and the assumption of isothermal walls for simulation of not correctly cooled down model surfaces. From thermodynamics and boundary layer assumptions a special formulation for the Prandtl number in real gas boundary layers has been derived which shows this number to be a function of the actual position within the boundary layer. But the results achieved with different constant Prandtl numbers indicate that the influence of this Prandtl number variation through the boundary layer is negligible compared with the direct effects introduced by use of the real gas equations of state. In general, the calculated results confirm some fundamental viscous effects predicted earlier by Navier-Stokes calculations for supersonic shock wave-laminar boundary layer interaction in real nitrogen. Although systematic deviations in skin friction always occur as a consequence of real gas effects and small temperature steps, the influence of these differences on the separation behaviour are found to be small under cryogenic wind tunnel conditions.

1. Introduction

Cryogenic Wind Tunnels achieving high Reynolds number flow operate at very low temperatures close to the liquefaction boundary of the test gas which is usually envisaged to be nitrogen. In this operating range deviations from the behaviour of ideal gases must be expected and measurement errors are

* This research was sponsored by the German Ministry of Research and Technology under support number LVW 7901 and by the Technical Group ETW, Amsterdam, under contract ETW/79/01/68730/143963

possible if the model surface is not correctly cooled down to recovery temperature. While the pure inviscid real gas effects for adiabatic walls have been shown to be insignificant (see e.g. ref. [1,2]), a sensitivity to these effects has been recognized for viscous phenomena, especially for the skin friction [3]. The systematic differences in the skin friction coefficient appeared to result from two opposing effects, namely from a change of the wall temperature due to thermodynamic real gas effects and from a slight slope change of the viscosity characteristic. This conjecture was confirmed by a thoroughly detailed investigation of laminar and turbulent boundary layers [4]. However, it was still suggested that the mechanism of flow separation could be affected by those effects [5] but this possibility was called in question in the light of the above mentioned boundary layer results [6]. Hence, the present investigation was carried out with special emphasis on possible changes in separation behaviour. Furthermore, nonadiabatic wall effects have been included in the study in order to estimate the influence of model surfaces not having the correct recovery temperature. Knowledge on such effects is important since the adjustment of the wall temperatures in consequence of inaccurate precooling of the model surfaces or of temperature steps during a single run of the tunnel proceeds slowly [7]. While the corresponding effects are already known in terms of Reynolds number corrections [8], the present study aims to provide calculation examples which are characteristic for judging the influence on the separation behaviour in high Reynolds number transonic flow.

2. Analysis

Correct account of heat transfer effects requires the energy equation of the flow to be retained which is neglected in many adiabatic boundary layer methods. The usual boundary layer assumptions reduce the momentum equation in the direction normal to the wall to the condition of constant pressure and lead to the following set of equations for conservation of mass, tangential momentum, and energy (see e.g. [9])

$$\frac{\partial(\rho u)}{\partial x} + \frac{\partial(\rho v)}{\partial y} = 0, \tag{1}$$

$$\rho \left(u \frac{\partial u}{\partial x} + v \frac{\partial u}{\partial y}\right) = -\frac{dp}{dx} + \frac{\partial}{\partial y}\left(\eta \frac{\partial u}{\partial y}\right), \tag{2}$$

$$\rho \left(u \frac{\partial h_t}{\partial x} + v \frac{\partial h_t}{\partial y}\right) = \frac{\partial}{\partial y}\left(\eta u \frac{\partial u}{\partial y} + k \frac{\partial T}{\partial y}\right) \tag{3}$$

with h_t being the total enthalpy

$$h_t = h + \frac{u^2}{2} . \tag{4}$$

u and v are the velocity components in the x-direction parallel to the wall and in the y-direction normal to the wall, respectively. ρ is the density, p the pressure, T the temperature, and h the enthalpy. The molecular viscosity η and the heat conductivity k can be described by a Sutherland formula for each with the constants given in ref. [3].

Concerning the thermodynamics of the real gas it will be assumed now that the thermal equation of state expresses the pressure as a function of density and temperature

$$p = p(\rho, T). \tag{5}$$

For the present purpose the Beattie-Bridgeman equation is used which contains five constants for the description of the real gas behaviour [10]. By the usual assumption that a real gas behaves like an ideal one in the case of very low pressures, caloric equations of state can be derived from Eq. (5), e.g. for the internal energy $e(\rho,T)$ which leads immediately to a corresponding formulation for the enthalpy

$$h = h(\rho,T) = e(\rho,T) + p(\rho,T)/\rho . \tag{6}$$

For a real gas the temperature gradient in Eq. (3) can be replaced by an enthalpy gradient in the following manner. From eq. (6) we get

$$\frac{\partial h}{\partial y} = \left(\frac{\partial h}{\partial \rho}\right)_T \frac{\partial \rho}{\partial y} + \left(\frac{\partial h}{\partial T}\right)_\rho \frac{\partial T}{\partial y} . \tag{7}$$

Because of the pressure constancy normal to the wall Eq. (5) yields

$$dp = \left(\frac{\partial p}{\partial \rho}\right)_T d\rho + \left(\frac{\partial p}{\partial T}\right)_\rho dT = 0$$

or

$$\frac{\partial \rho}{\partial y} = - \frac{\partial T}{\partial y} \frac{(\partial p/\partial T)\rho}{(\partial p/\partial \rho)_T} . \tag{8}$$

Eq. (8) introduced into Eq. (7)

$$\frac{\partial h}{\partial y} = \left[\left(\frac{\partial h}{\partial T}\right)_\rho - \left(\frac{\partial h}{\partial \rho}\right)_T \frac{(\partial p/\partial T)\rho}{(\partial p/\partial \rho)_T}\right] \frac{\partial T}{\partial y} \tag{9}$$

makes it possible to replace the temperature gradient in Eq. (3). Hence, an adiabatic wall ($\partial T/\partial y = 0$ at the wall) forces the enthalpy gradient to vanish at the wall as a consequence of the usual boundary layer simplifications. With a new definition of the Prandtl number for real gas cases

$$Pr^* = \frac{\eta(T)}{k(T)} \left[\left(\frac{\partial h}{\partial T}\right)_\rho - \left(\frac{\partial h}{\partial \rho}\right)_T \frac{(\partial p/\partial T)\rho}{(\partial p/\partial \rho)_T}\right] \tag{10}$$

the energy equation (3) becomes

$$\rho \left(u \frac{\partial h_t}{\partial x} + v \frac{\partial h_t}{\partial y}\right) = \frac{\partial}{\partial y} \left[\eta \frac{\partial (\frac{u^2}{2})}{\partial y} + \frac{\eta}{Pr^*} \frac{\partial h}{\partial y}\right].$$

Using Eq. (4) we get

$$\rho \left(u \frac{\partial h_t}{\partial x} + v \frac{\partial h_t}{\partial y}\right) = \frac{\partial}{\partial y} \left[\eta \left(1 - \frac{1}{Pr^*}\right) u \frac{\partial u}{\partial y}\right] + \frac{\partial}{\partial y} \left(\frac{\eta}{Pr^*} \frac{\partial h_t}{\partial y}\right) \tag{11}$$

where also the Prandtl number (10) can be written in another form by the aid of Eq. (6)

$$Pr^* = \frac{\eta(T)}{k(T)} \left\{C_v (\rho,T) + \left[\frac{p}{\rho^2} - \left(\frac{\partial e}{\partial \rho}\right)_T\right] \frac{(\partial p/\partial T)\rho}{(\partial p/\partial \rho)_T}\right\} \tag{12}$$

which reduces immediately to the usual expression in the ideal gas case (C_v = specific heat at constant volume).

Eqs. (1), (2), (11) and (12) represent the basic equations for a proper boundary layer method. The finite difference method of Cebeci [11] and [12] seems to be well suited since it provides the possibility to introduce real gas equations of state and contains the treatment of Eq. (11). The only disadvantage of that method is the presumption of constant Prandtl number. Therefore, if this method is used for real gas calculations without changes, it must be proved thoroughly that this restriction does not affect the results essentially because after Eq. (12) the Prandtl number is a function of temperature and density in reality. For the present purpose the calculations have been carried out using Cebeci's method with the real gas equations (5) and (6) implemented and constant Prandtl number Pr = 0.72 (turbulent Pr_t = 0.9) but in each numerical step also the real Prandtl number (12) has been recalculated from the results. This real Prandtl number varied for the cryogenic cases in the range from 0.74 to 0.8. Hence, for further examination additional calculations have been performed with the value of the Prandtl number changed to 0.8 (and similar changes in the turbulent Prandtl number). The results show only an extremely small influence on the pressure and skin friction distributions which is negligible compared with the effects due to the direct introduction of the real gas equations of state into the boundary layer method. Consequently, we may conclude that consideration of changes in the Prandtl number is generally unneccessary for thermodynamic states corresponding to cryogenic wind tunnel conditions.

For turbulence modeling the two-layer eddy viscosity model of Cebeci and Smith resulting from a mass weighted time averaging procedure has been chosen which is well proven in applications and has recently been described in a very sophisticated form [13]. However, no special account can be given for the cryogenic cases since no information is available on the turbulence behaviour in a real gas.

For non-adiabatic walls which can also be treated by Cebeci's method, the isothermal wall (presently, more correctly speaking: "isoenergetic wall", see below) has been chosen as a sample case. In all calculations the boundary layer was spanned by more than 30 points in the y-direction and the transition from laminar to turbulent flow has been assumed at about a local Reynolds number of $Re_x = 2.5 \cdot 10^6$.

3. Results

In order to verify the calculation method and to analyse the principal effects without pressure gradient, flat plate calculations have been carried out in supersonic external flow. The flow conditions correspond to those of some Navier-Stokes solutions for transonic shock wave - turbulent boundary layer interaction [14].

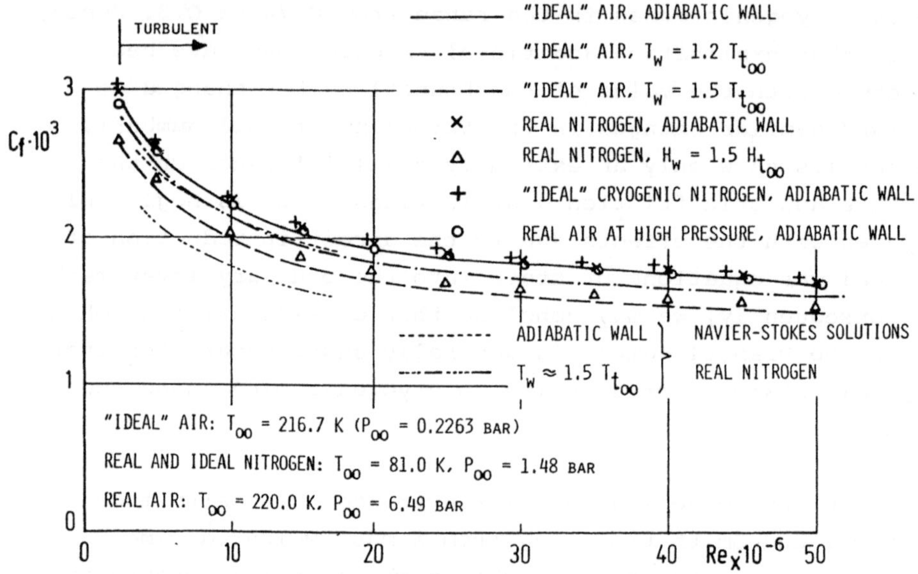

Fig. 1: Skin friction coefficient in a flat plate boundary layer at M_∞ = 1.35

For Mach number M = 1.35 the results are plotted in Fig. 1 up to an ETW-relevant local Reynolds number of 50 million. As already pointed out in section 2 transition is assumed at about 2.5 million local Reynolds number. That explains to a great extent the differences in C_f observed in Fig. 1 between boundary layer and Navier-Stokes results, the latter being calculated by assuming transition at the leading edge.

The differences between adiabatic and isothermal wall cases exhibit identical trends but the resulting step is slightly higher for the Navier-Stokes solutions although the temperature step is smaller since it is based on a percentage of the recovery temperature for the Navier-Stokes cases instead of a percentage of the total temperature for the boundary layer calculations. However, the essential non-adiabatic effects seem to be represented very similarly in both methods.

The real gas state corresponds to typical ETW-conditions where the thermodynamic state at maximum local Mach number (immediately upstream of the shock terminating the supersonic region) is assumed to represent a point just on the vapor pressure curve of nitrogen. For real nitrogen the results for the skin friction coefficient lie slightly above the ideal gas solutions for an adiabatic wall as well as for an isothermal wall with the wall enthalpy being 50 percent higher than the total enthalpy of the undisturbed flow. Additionally, in order to show the favourable opposition of the both effects mentioned in the introduction, a result is plotted in Fig. 1 where ideal gas equations are used under cryogenic conditions. This curve, denoted as "ideal" cryogenic nitrogen, exhibits correctly that the effect of steeper slope of the viscosity characteristics alone leads to higher deviations from the "ideal" air solution than the combined effects. Furthermore, a high pressure solution for air is contained in Fig. 1 where at ambient stagnation temperature the same Reynolds number with the same chord length as in the cryogenic case is achieved by increasing the pressure to a sufficiently high level. Obviously, the use of real gas equations does not produce significant deviations from ideal gas solutions although in inviscid flows such deviations have been indicated [15].

In order to judge the separation behaviour some boundary layer calculations have been carried out on airfoil sections in transonic flow. The well known profile NACA 0012 has been chosen in transonic flow with Mach number M_∞ = 0.8 and at angle of attack α = 0.37^{O} resulting in supersonic regions on both sides of the airfoil. The inviscid pressure distributions shown in Fig. 2 are gained by means of the Dornier finite volume method for solving the Euler equations [16]. The undis-

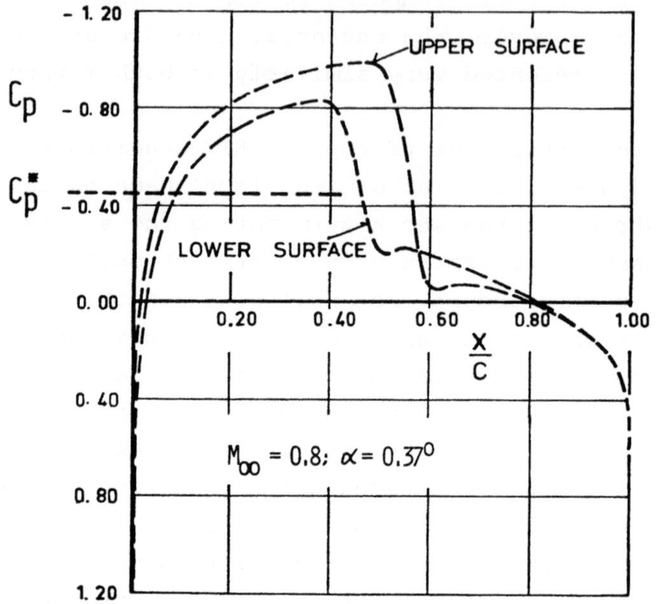

Fig. 2: Euler solution for the pressure coefficient on the surface of NACA 0012 airfoil in transonic flow at small angle of attack

turbed state has been chosen, so that at the highest ETW stagnation pressure of $p_{t\infty}$ = 4.5 bar and a typical Reynolds number Re = $50 \cdot 10^6$ the state at the highest local Mach number is just on the saturation boundary giving $T_{t\infty}$ = 110 K. The ideal gas solutions correspond to air in real flight at the lower edge of the stratosphere.

In Fig. 3 the skin friction coefficient is plotted versus the local Reynolds number based on the arc length s which is measured on the airfoil surface from the stagnation point. Fig. 3a presents results for the upper surface of the NACA 0012 which are primarily shaped by the strong shock leading always to separation at about 60 percent of the chord length. A hot wall diminishes C_f by proper amounts but shifts the separation point only slightly forward. (The case T_W = 1.5 $T_{t\infty}$

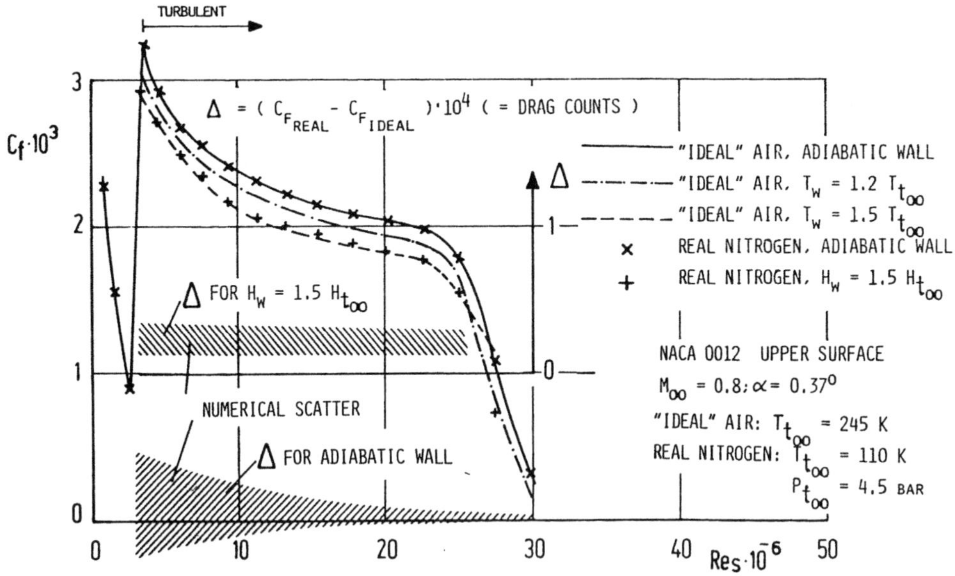

a) Upper surface

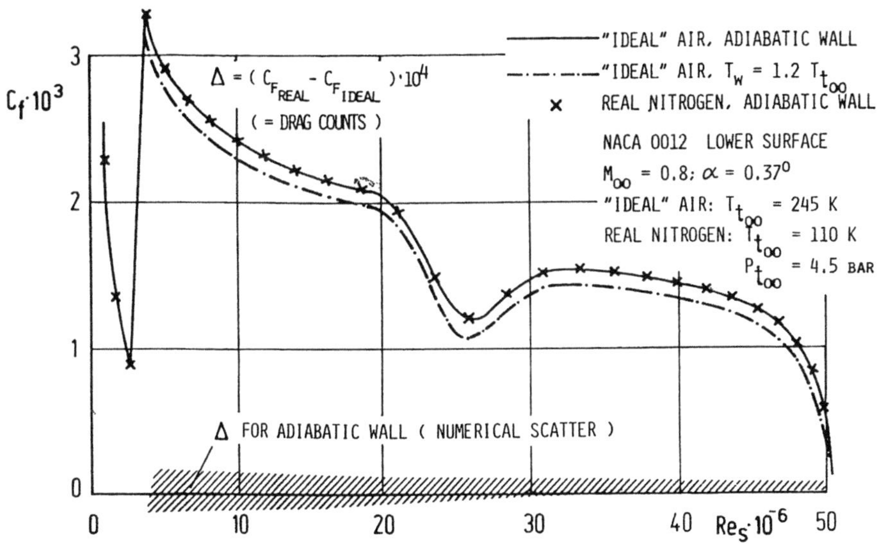

b) Lower surface

Fig. 3: Skin friction coefficient on NACA 0012 airfoil in transonic flow

may be numerically unreliable beyond $Re_S = 25 \cdot 10^6$). The deviations in C_f for real nitrogen are almost an order of magnitude lower than the desired ETW-accuracy of one drag count (0.0001) and no change in separation behaviour can be detected. On the lower surface (Fig. 3b) separation does not occur because of the weaker shock but all other trends are similar.

4. Conclusion

It can be stated that the boundary layer method used is well suited to accurately describe the non-adiabatic and real gas effects in cryogenic wind tunnels. It reproduces correctly the favourably opposing effects with respect to skin friction shown in previous investigations. The assumption of constant Prandtl number has no significant influence on the results. The separation characteristics are not visibly affected by real gas behaviour and the influences on C_f remain well below the intended accuracy of the experiments in the ETW. Deviations of the wall temperature from the recovery temperature should not exceed a few percent because a 20 percent temperature step changes the friction drag by approximately one count although the separation pattern does not alter essentially.

For high pressure wind tunnels at ambient temperature the real gas influences seem not to affect the viscous behaviour noticeably.

References

1. Hall, R.M.: Real gas effects I - Simulation of ideal gas flow by cryogenic nitrogen and other selected gases. In: AGARD Lecture Series 111 on "Cryogenic Wind Tunnels", 1980

2. Adcock, J.B.: Real-gas effects associated with one-dimensional transonic flow of cryogenic nitrogen. NASA TN D-8274, 1976. Also in: Adcock, J.B.; Kilgore, R.A.; Ray, E.J.: Cryogenic nitrogen as a transonic wind-tunnel test gas. AIAA-Paper 75-143, 1975

3. Wagner, B.; Schmidt, W.: Theoretische Untersuchungen zur Stoß-Grenzschicht-Wechselwirkung in kryogenem Stickstoff. In: Beiträge zu Transportphänomenen in der Strömungsmechanik und anverwandten Gebieten (Prof.Dr. A. Walz zum 70. Geburtstag gewidmet). DLR-FB 77-16 (1977). Also in: ZFW 2, pp. 81-88, 1978

4. Adcock, J.B.; Johnson, C.B.: A theoretical analysis of simulated transonic boundary layers in cryogenic-nitrogen wind tunnels. NASA Technical Paper 1631, 1980

5. Inger, G.R.: Transonic shock-boundary-layer interactions in cryogenic wind tunnels. Journal of Aircraft 16, pp. 284-287, 1979

6. Adcock, J.: Simulation of flat-plate turbulent boundary layers in cryogenic tunnels. Journal of Aircraft 17, pp. 284-285, 1980

7. Johnson, C.B.: A study of non-adiabatic boundary-layer stabilization time in a cryogenic tunnel for typical wing and fuselage models. AIAA-Paper 80-0417, 1980

8. Green, J.E.; Weeks, D.J.; Pugh, P.G.: Heat transfer as a source of spurious scale effects in subsonic and transonic wind tunnels. In: Proc. First Int. Symp. on Cryogenic Wind Tunnels, Southampton, 1979

9. Walz, A.: Strömungs- und Temperaturgrenzschichten. Verlag A. Braun, Karlsruhe, 1966

10. Landolt-Börnstein: Zahlenwerte und Funktionen, IV. Band, 4. Teil, Bandteil a. 6. Auflage (herausgegeben von H. Hausen), Springer-Verlag, Berlin/Heidelberg/New York, 1967

11. Cebeci, T.; Smith, A.M.O.: Analysis of turbulent boundary layers. Academic Press, New York, 1974

12. Cebeci, T.: A brief description of a computer program for solving the two-dimensional boundary-layer equations for standard and inverse compressible laminar and turbulent flows. Calif. State Univ., Long Beach, Ca., 1974

13. Reynolds, W.C.; Cebeci, T.: Calculation of turbulent flows. In: Turbulence (edited by P. Bradshaw), 2nd edition, Topics in Applied Physics, Vol. 12, Springer-Verlag, Berlin/Heidelberg/New York, 1978

14. Wagner, B.: Estimation of Simulation Errors in the European Transonic Wind Tunnel (ETW). Proc. of the 13th ICAS Congress, Seattle, 1982

15. Wagner, B.; Schmidt, W.: Theoretical investigations of real gas effects in cryogenic wind tunnels. AIAA-Journal 16, pp. 580-586, 1978

16. Schmidt, W.; Jameson, A.; Whitfield, D.: Finite volume solutions for the Euler equations for transonic flow over airfoils and wings including viscous effects. AIAA-Paper 81-1265, 1981

An Inverse Integral Computational Method for Compressible Turbulent Boundary Layers *

D. L. Whitfield†, T. W. Swafford††, and T. L. Donegan†††

Department of Aerospace Engineering
Mississippi State University
Mississippi State, MS 39762
USA

Introduction

An inviscid flow solver used in conjunction with a viscous flow solver through a patching or matching approach [1] has proved to be a powerful technique for computing viscous-inviscid interacting flow, as pointed out in the reviews of Lock [2], Melnik [3], and Le Balleur [4]. This computational technique can be extended to separated flow if an inverse boundary-layer calculation method (where the pressure distribution is obtained as part of the solution) is used in place of a direct boundary-layer calculation method (where the pressure distribution is specified). A review of computational methods capable of handling separated flow is given by Le Balleur [4].

The purpose of this paper is to present a compressible turbulent momentum and mean-flow kinetic energy inverse integral boundary-layer equation calculation method capable of computing attached or separated flow over an adiabatic surface. The method is similar to that reported earlier [5]. A singularity that occurs in the method described in [5] for certain attached flow conditions is, however, removed by a rearrangement of the inverse equations. Analytical expressions for auxiliary relations are

---

*Computer support for this work was kindly and efficiently provided by United Information Services, Inc., United Telecom Computer Group, Atlanta, GA (USA).

†Professor

††Ph.D. student on leave from Sverdrup Technology, Inc., Arnold Air Force Station, TN (USA)

†††Calspan, Arnold Air Force Station, TN (USA)

presented that are valid for boundary-layer edge Mach numbers up to about three. Transonic viscous-inviscid interaction calculations using this inverse method are discussed to illustrate the types of problems that can be solved.

Inverse Boundary-Layer Formulation

The direct momentum and mean-flow kinetic energy compressible integral boundary-layer equations are [6]

$$\frac{1}{r_w^k \rho_e u_e^2} \frac{d}{dx} (r_w^k \rho_e u_e^2 \theta) + H_{\delta^*} \frac{\theta}{u_e} \frac{du_e}{dx} = \frac{c_f}{2} \tag{1}$$

$$\frac{1}{2 r_w^k \rho_e u_e^3} \frac{d}{dx} (r_w^k \rho_e u_e^3 \theta H_{\theta^*}) + H_{\delta^{**}} \frac{\theta}{u_e} \frac{du_e}{dx} = \frac{c_f}{2} D \tag{2}$$

where

$$\theta = \int_0^\infty \frac{\rho u}{\rho_e u_e} \left(1 - \frac{u}{u_e}\right) dy \tag{3}$$

$$\delta^* = \int_0^\infty \left(1 - \frac{\rho u}{\rho_e u_e}\right) dy \tag{4}$$

$$\theta^* = \int_0^\infty \frac{\rho u}{\rho_e u_e} \left(1 - \frac{u^2}{u_e^2}\right) dy \tag{5}$$

$$\delta^{**} = \int_0^\infty \frac{u}{u_e} \left(1 - \frac{\rho}{\rho_e}\right) dy \tag{6}$$

$$D = \int_0^\infty \frac{\tau}{\tau_w} \frac{\partial \left(\frac{u}{u_e}\right)}{\partial y} dy \tag{7}$$

$$H_\xi = \frac{\xi}{\theta}, \quad \xi = \delta^*, \theta^*, \text{ and } \delta^{**} \tag{8}$$

$$\frac{c_f}{2} = \frac{\tau_w}{\rho_e u_e^2} \tag{9}$$

and r_w^k is the local body radius with k = 0 for two-dimensional planar flow and k = 1 for axisymmetric flow. Equations (1) and (2) can be arranged to obtain

$$(1 + H_{\delta*}) \frac{d}{dx} (\ln u_e) - \frac{d}{dx} (\ln H_{\delta*}) = \frac{c_f}{2\theta} - \frac{d}{dx} [\ln(r_w^k \rho_e u_e \delta^*)] \quad (10)$$

$$(1 + 2 \frac{H_{\delta**}}{H_{\theta*}} - H_{\delta*}) \frac{d}{dx} (\ln u_e) + \frac{d}{dx} (\ln H_{\theta*}) = \frac{c_f D}{\theta H_{\theta*}} - \frac{c_f}{2\theta} \quad (11)$$

Note that the right hand side of Eq. (10) contains a term involving u_e which could be grouped with the term involving u_e appearing on the left hand side of Eq. (10). It is, however, the arrangement given by Eq. (10) that avoids a singularity that occurs for certain attached flow conditions. Moreover, the last term on the right hand side of Eq. (10) can be specified for viscous-inviscid interaction calculations. For example, for transonic interaction, δ^* can be specified using Carter's method [7] as explained in [5]; and because the product of density and velocity (ρq) computed from the viscous solution converges to that computed from the inviscid solution, the quantity ρq from the last inviscid solution can be used for $\rho_e u_e$ in Eq. (10).

For a gas with constant specific heat ratio, γ, and constant total enthalpy at the boundary-layer edge (not necessarily isentropic)

$$\frac{d}{dx} (\ln u_e) = \frac{1}{F_c^2} \frac{d}{dx} (\ln M_e) \quad (12)$$

where

$$F_c^2 = 1 + \frac{\gamma - 1}{2} M_e^2 \quad (13)$$

and M_e is the boundary-layer edge Mach number. By expressing $H_{\delta*}$, $H_{\theta*}$, $H_{\delta**}$, c_f, and D as functions of $\bar{H}$ and M_e where

$$\bar{H} = \frac{\bar{\delta}^*}{\bar{\theta}} = \frac{\int_0^\infty (1 - \frac{u}{u_e})\, dy}{\int_0^\infty \frac{u}{u_e} (1 - \frac{u}{u_e})\, dy} \quad (14)$$

and

$$\frac{\theta}{\bar{\theta}} = 1 - \frac{0.92\, M_e^2}{7.09 + M_e^2} \tanh\,[1.49\,(\bar{H} - 0.9)] \qquad (15)$$

Eqs. (10) and (11) can be written with $\bar{H}$ and M_e as dependent variables and solved numerically. For example, by using Eq. (12) in Eqs. (10) and (11), and noting that $H_\xi = H_\xi\,[\bar{H}(x), M_e(x)]$ where $\xi = \delta^*$ and θ^*, Eqs. (10) and (11) can be solved for dM_e/dx and $d\bar{H}/dx$ to obtain the following two equations to be solved simultaneously:

$$\frac{dM_e}{dx} = \frac{a_{22}b_1 - a_{12}b_2}{\Delta} \qquad (16)$$

$$\frac{d\bar{H}}{dx} = \frac{a_{11}b_2 - a_{21}b_1}{\Delta} \qquad (17)$$

where

$$a_{11} = \frac{1 + H_{\delta^*}}{M_e F_c^2} - \frac{1}{H_{\delta^*}} \frac{\partial H_{\delta^*}}{\partial M_e} \qquad (18)$$

$$a_{12} = - \frac{1}{H_{\delta^*}} \frac{\partial H_{\delta^*}}{\partial \bar{H}} \qquad (19)$$

$$a_{21} = \frac{1 + 2\,\dfrac{H_{\delta^{**}}}{H_{\theta^*}} - H_{\delta^*}}{M_e F_c^2} + \frac{1}{H_{\theta^*}} \frac{\partial H_{\theta^*}}{\partial M_e} \qquad (20)$$

$$a_{22} = \frac{1}{H_{\theta^*}} \frac{\partial H_{\theta^*}}{\partial \bar{H}} \qquad (21)$$

$$b_1 = \frac{c_f}{2\theta} - \frac{d}{dx}\,[\ln(r_w^k \rho_e u_e \delta^*)] \qquad (22)$$

$$b_2 = \frac{c_f D}{\theta H_{\theta^*}} - \frac{c_f}{2\theta} \qquad (23)$$

and

$$\Delta = a_{11}a_{22} - a_{12}a_{21} \qquad (24)$$

Equations (16) and (17) are easily solved using a fourth-order, four-stage, explicit Runge-Kutta scheme provided the a_{ij} and b_i (where i = 1 or 2 and j = 1 or 2) are known. In this regard, the present integral method is attractive because essentially all this information is obtained from the velocity profile correlation. For attached flows the velocity profile for $0 \leq y < \infty$ is reasonably well established [8]. An extension of the attached flow velocity profile to separated flow is given in [5] and [9]. The following correlations of the auxiliary relations for attached or separated flow have provided extremely useful solutions for complex turbulent flows [5, 10, 11]. (The correlations should be improved, however, as new or improved experimental data for compressible separated flow become available.)

$$H_{\delta*} = \bar{H}\,(1 + 0.113\, M_e^{\,2}) + 0.29\, M_e^{\,2} \tag{25}$$

$$(H_{\theta*})_{M_e = 0} = 1.48061 + 3.83781 e^{-2\bar{H}} + 0.33 - \frac{1}{8.5484} \tan^{-1}\left(\frac{10^{7-\bar{H}} - 1}{1.23}\right)$$

$$- \left(0.33 - \frac{\pi}{17.1}\right) \tanh^{\frac{1}{2}}\left[(1.2874 \times 10^{-6})(10^{7-\bar{H}})^{1.45761}\right] \tag{26}$$

$$H_{\theta*} = \frac{(H_{\theta*})_{M_e = 0} + 0.028\, M_e^{\,2}}{1 + 0.014\, M_e^2} \tag{27}$$

$$H_{\delta**} = \left[\frac{0.064}{\bar{H} - 0.8} + 0.251\right] M_e^{\,2} \tag{28}$$

$$\frac{c_f}{\bar{c}_f} = \frac{\bar{Re}_\theta}{Re_\theta} = \frac{1}{F_c} \tag{29}$$

$$\bar{c}_f = \frac{0.3 e^{-1.33\bar{H}}}{(\log_{10}\bar{Re}_\theta)^{1.74+0.31\bar{H}}} + (1.1 \times 10^{-4})\left[\tanh\left(4 - \frac{\bar{H}}{0.875}\right) - 1\right] \tag{30}$$

$$\frac{c_f D}{2} = \frac{0.009 - 0.011e^{-0.15\bar{H}^{2.1}} + 3.0\mathrm{X}10^{-5}e^{0.117\bar{H}^2} + A(\bar{R}e_\theta)^{-0.574}}{F_c(1 + 0.05\, M_e^{1.4})} \tag{31}$$

where

$$A = 0.438 - 0.280\,\bar{H} \quad \text{for } \bar{H} \leq 3.5$$

$$A = 0.160(\bar{H}-3.5) - 0.550 \quad \text{for } \bar{H} > 3.5$$

Results and Discussion

The singularity that occurs in the inverse formulation of Eqs. (1) and (2) as given in [5] is a consequence of the system coefficient matrix becoming singular for certain attached flow conditions. A comparison of the determinants as computed from the formulation given in [5] and from Eq. (24) is illustrated in Figure 1. For low to moderate M_e, both formulations exhibit similar behavior. However, for $M_e = 2$ and $M_e = 3$, it is seen that the formulation as given in [5] becomes singular at $\bar{H} \simeq 1.35$ and $\bar{H} \simeq 1.85$, respectively, whereas the present formulation remains non-singular. Although such singular conditions do not exist in transonic flow, these conditions have occurred in the early cycles of high Mach number transonic interactive calculations.

Implementing the present method involves solving the system (16) and (17) by specifying $r_w^k \rho_e u_e \delta^*$ as a function of axial distance. For viscous-inviscid interaction calculations, $\rho_e u_e$ is obtained from an inviscid solver, r_w^k from geometrical considerations, and δ^* from Carter's method [7]. Viscous-inviscid interaction can be achieved using the surface-source model as described in [5].

An example of transonic viscous-inviscid interactive calculations using the present inverse boundary-layer method and an Euler equation solver [10] is given in Figure 2. Note that the comparisons of theoretical and experimental velocity profiles [12] are from those regions most difficult to compute accurately, namely the shock and trailing edge regions of the upper surface. The agreement in Figure 2 is considered good. Examples of other solutions are included in [5], [10], and [11].

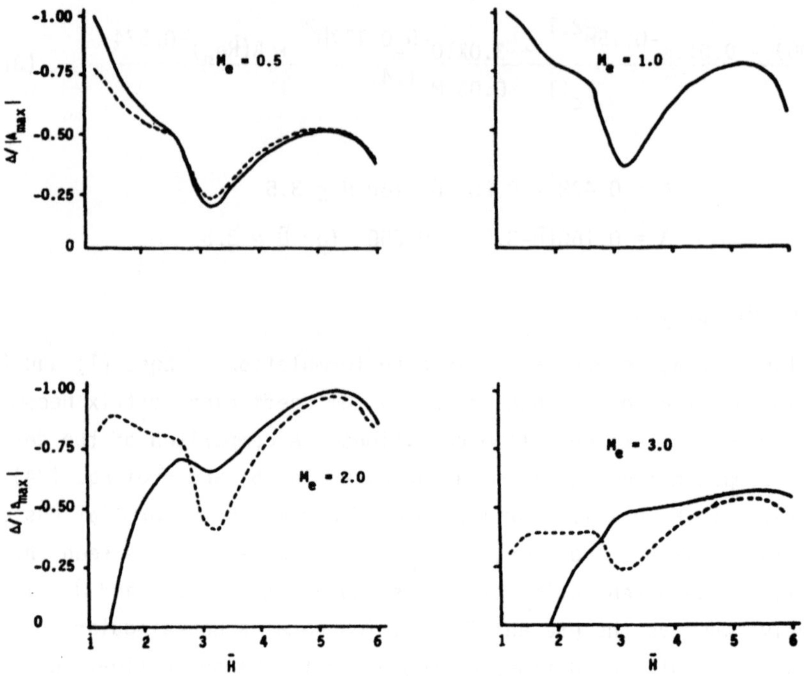

Figure 1. Determinant of System Coefficient Matrix [solid lines represent determinant as given in [5]; dashed lines represent determinant as given by Eq. (24)]

Concluding Remarks

An inverse integral method suitable for calculating attached and separated compressible turbulent boundary layers on adiabatic surfaces has been presented. The inverse-boundary-layer equations were formulated such that a singularity that occurs in inverse methods for certain attached flow conditions is avoided thus enabling viscous-inviscid interaction calculations to be made for flows containing both attached and separated boundary layers without switching between direct and inverse methods. Correlations of auxiliary relations necessary for implementing the present method were also presented.

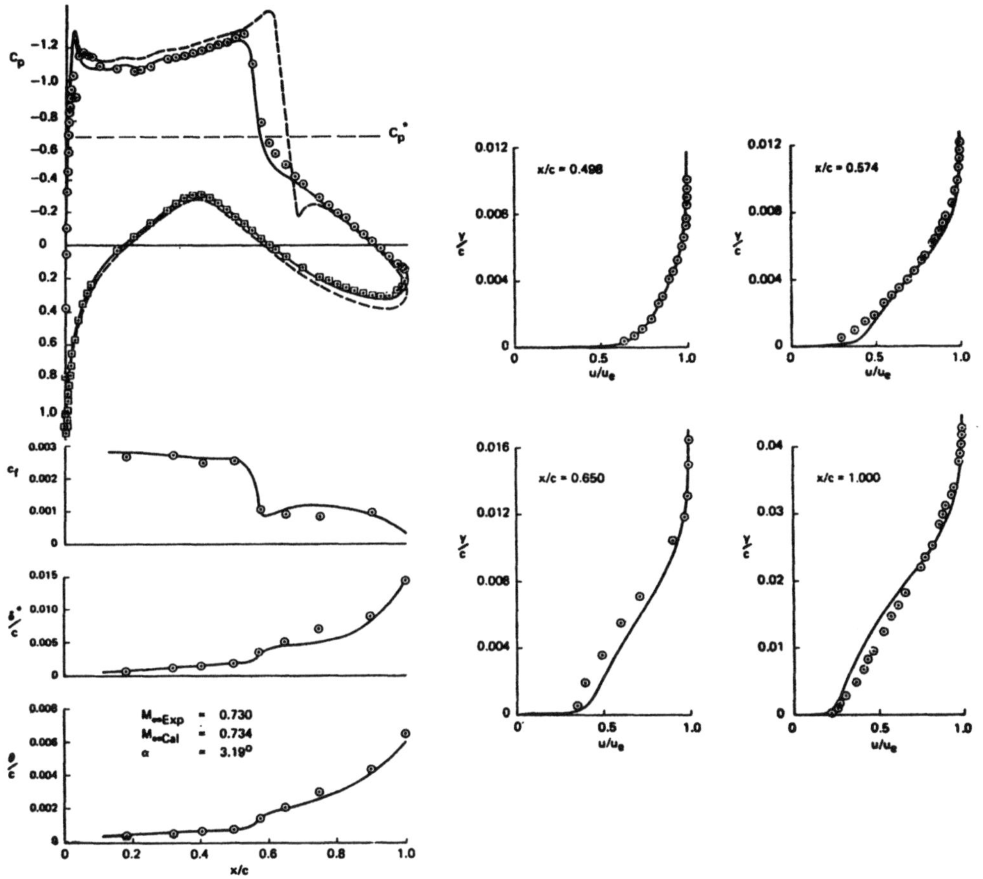

Figure 2. Boundary-layer Characteristics of RAE 2822 Aerofoil-Case 9 (solid lines represent interactive calculations using present method, dashed line represents inviscid solution, symbols represent measured data [12])

References

1. Le Balleur, J. C., Peyrett, R. and Viviand, H., "Numerical Studies in High Reynolds Number Aerodynamics," Computers and Fluids, Vol. 8, No. 3, March 1980, pp. 1-30.

2. Lock, R. C., "A Review of Methods for Predicting Viscous Effects on Aerofoils and Wings at Transonic Speeds," AGARD Conference Preprint No. 291, 29 Sept. - 1 Oct., 1980.

3. Melnik, R. E., "Turbulent Interactions on Airfoils at Transonic Speeds-Recent Developments," AGARD Conference Preprint No. 291, 29 Sept. - 1 Oct., 1980.

4. Le Balleur, J. C., "Viscous-Inviscid Coupling Calculations for Two and Three Dimensional Flows," von Karman Institute for Fluid Dynamics, Lecture Series 1982-04, March 29-April 2, 1982.

5. Whitfield, D. L., Swafford, T. W., and Jacocks, J. L., "Calculation of Turbulent Boundary Layers with Separation and Viscous-Inviscid Interaction," AIAA Journal, Vol. 19, No. 10, Oct. 1981, pp. 1315-1322.

6. Whitfield, D. L., "Integral Solution of Compressible Turbulent Boundary Layers Using Improved Velocity Profiles," Arnold Air Force Station, TN, AEDC-TR-78-42, Dec. 1978.

7. Carter, J. E., "A New Boundary-Layer Inviscid Iteration Technique for Separated Flow," AIAA Paper No. 79-1450, July 1979.

8. Whitfield, D. L., "Analytical Description of the Complete Turbulent Boundary-Layer Velocity Profile," AIAA Journal, Vol. 17, No. 10, Oct. 1979, pp. 1145-1147.

9. Swafford, T. W., "Analytical Approximation of Two-Dimensional Separated Turbulent Boundary-Layer Velocity Profiles," Arnold Air Force Station, TN, AEDC-TR-79-99, Oct. 1980.

10. Schmidt, W., Jameson, A., and Whitfield, D., "Finite Volume Solutions to the Euler Equations in Transonic Flow," AIAA Journal, Vol. 20, 1982.

11. Whitfield, D., Jameson, A., and Schmidt, W., 1980-81 AFOSR-HTTM-Stanford Conference on Complex Turbulent Flows: Comparison of Computation and Experiment, Stanford, CA, Sept. 1981.

12. Cook, P. H., McDonald, M. A., and Firmin, M. C. P., "Aerofoil RAE 2822 --Pressure Distributions, Boundary Layer and Wake Measurements," Royal Aircraft Establishment, UK, AGARD WG 04, July 1977.

Transition of Rotational Flow into Outer Irrotational Flow

K. WIEGHARDT
Institut für Schiffbau
der Universität Hamburg

Summary

From a physical point of view it seems almost obvious to suppose that - outside a boundary layer, wake or jet at great Reynolds number - vorticity $\vec{\omega}$ should not only fall off, yet even faster than deformation rate d so that $W_K(\eta \to \infty) \to 0$ with $W_K = |\vec{\omega}|/d$. Yet, strictly speaking, so far this is only confirmed by two exact solutions of Navier-Stokes equation and by tests in a turbulent wake.
In particular, in the near wake behind a body where all the separated rotational fluid particles of the former boundary layer are mixed up with the irrotational fluid of the outer flow entering the wake, this mixing is certainly an important process. To describe it correctly a mathematical model should approximate not only the field of mean velocity but also that of the velocity gradients, or - at least - the kinematical condition $W_K(\eta \to \infty) \to 0$.

Introduction

Boundary layer theory assumes that the vorticity $\vec{\omega}$ of the velocity $\vec{v}$ ($\vec{\omega}$ = rot $\vec{v}$) dies away at the edge to the outer potential flow where there is only a deformation rate d but no longer vorticity. Hence, one would expect that Truesdell's kinematic vorticity number [1] $W_K = |\vec{\omega}|/d$ also dies away for great wall distances η . For incompressible flow d is defined by $d^2 = \vec{\omega}^2 + 2 \operatorname{div} \vec{a}$ with a = acceleration and with div $\vec{v} = 0$ $2 \operatorname{div} \vec{a} = 4[v_x u_y - u_x v_y + w_y v_z - v_y w_z + u_z w_x - w_z u_x]$. Yet, when checking this conjecture for the simplest boundary layer, i.e. the Blasius flow along a plate, one finds $W_K \gg 1$ for a n y wall distance η.
If Prandtl had been told of this he might have answered:" Of course, boundary layer simplifications such as e.g. $u_y \gg v_x$ are no longer true at the edge and so equilibrium of velocity

gradients cannot be preserved there. - Yet, these are just mathematical niceties without any physical significance."
And indeed, for all the essential properties of a boundary layer viz. displacement, wall shear and separation, yet also for momentum and energy loss, a good description of the flow near the wall is quite sufficient.
However, near and after the hind part of a body the convective mixing of the separated rotational boundary layer material with the outer irrotational flow is surely a physically important process taking place in a large part of the near wake and not in an insignificant edge region. That the real wake flow usually is turbulent does not matter. The kinematical condition $W_K(\eta \to \infty) \to 0$ is to be expected also here for the mean velocity field at high Reynolds number.
Indeed, this is veryfied by threedimensional measurements near the stern of the double model of a ship in a windtunnel (Fig.1) at $Re = 5 \cdot 10^6$ [2]. Fig.2 shows lines W_K= const in the cross plane x = -19 mm (see Fig.1). The line W_K = 2 denotes a longitudinal vortex roll which is again to be seen in Fig.3 in the wake at x = 50 mm(behind the here not existing propeller). Elsewhere, velocity and rotation are mostly nearly perpendicular to each other, - as in the upstream boundary layer before separation. In both figures 2 and 3 W_K falls off to 0.5 resp. to 0.25 although only the inner part of the wake is shown; half midship breadth was y = 202 mm. (The distortions of the lines W_K = 0.75 and 0.5 in the lower half of Fig.2 and 3 is due to the vortex wake of a span wire of 1 mm about 400 mm upstream the plane x = 50 mm.)

Known similarity solutions

How far is the decay of W_K described by known solutions for steady, plane or axisymmetric boundary layers, laminar or turbulent jets or wakes [3, 4] ? In two special cases these are even exact solutions of Navier-Stokes equation.

Plane flow

With x in main flow direction and y cross coordinate the similarity assumption for the stream function Ψ becomes:

$\psi = C\,x^n F(\eta)$ resp. $\psi = U\,y - C\,x^n F(\eta)$ for wakes; $\eta = \sigma\,y\,x^k$ $= y/b(x)$ with $b(x) = 1/(\sigma\,x^k)$. The deformation rate consists of rates of extension and shears : $d^2 = 2(u_x{}^2 + v_y{}^2) + (v_x + u_y)^2$. (Incidentally, in plane potential flow with complex function $F(z) = \phi + i\psi$ and $z = x + iy$ the deformation rate is $d = 2|F''(z)|$ and this is a remarkably simple interpretation of the magnitude of the second derivative of F(z).) Furtheron,

$$2\,\mathrm{div}\,\vec{a} = 4(u_y v_x - u_x v_y) = 4(\psi_{xy}^2 - \psi_{xx}\psi_{yy})$$
$$= 4C^2\sigma^2 x^{2(n+k-1)} . \left\{ (n+k)^2 F'^2 + k(k+1)\eta F'F'' + n(1-n)F\,F'' \right\}$$

$$\vec{\omega}^2 = C^2\sigma^4\,x^{2(n+2k)} . \left\{ F'' + \frac{1}{\sigma^2 x^{2(k+1)}} \left[n(n-1)F + k(2n+k-1)\eta F' + k^2\eta^2 F'' \right] \right\}$$

$$W_K = |\vec{\omega}|/d = 1/\sqrt{1 + 2\,(\mathrm{div}\,\vec{a})/\vec{\omega}^2} .$$

In boundary layers, jets and wakes for $\eta \to \infty$ F' and $F'' \to 0$, but $F(\eta \to \infty)$ remains finite. Hence, $\vec{\omega}^2(\eta \to \infty) \to 0$ is only possible if n = 0 or n = 1. Yet, good approximate solutions can be expected for other Values of n too, since the factor $b^2/x^2 = \sigma^{-2} x^{-2(k+1)}$ is often small.

Examples:

a) n = 1, k = -1: Smoothing out of a velocity discontinuity in cross direction. Here, $\mathrm{div}\,\vec{a} = 0$ and $W_K = 1$ everywhere and for a n y velocity profile $F'(\eta)$; i.e. for this assumed similarity only $\vec{\omega}^2(\eta \to \infty) \to 0$ can be enforced, but not $W_K \to 0$. Yet, this is just the special case - as with Blasius flow - when the deformation rate in the outer flow vanishes in zeroth approximation.

b) n = 0, k = -1/2: Plane wake, laminar or turbulent (with constant eddy viscosity). The approximation $F' = \exp(-\eta^2/4)$ gives for W_K values between 0.99 and 1 for greater wall distances and $W_K(\eta \to \infty) \to 1$. Only very near to the wake centre W_K falls off to zero because $\omega = v_x - u_y$ changes sign at $\eta = 0$ for symmetry reasons. - To enforce $W_K \to 0$ monotonically one could assume $F' \sim \eta^{-3}$; yet, then e.g. $W_K = 0.5$ is reached only at $\eta/\sigma x = y/x = \sqrt{27/4} = 2.6$ or at a cross distance of the order of 100 half-value-breadths.

c) $n - k = 1$: Similar boundary layers. Main stream $U(x) \sim x^m$ and $n = (m+1)/2$, $k = (m-1)/2$, see Fig.4. The case $m = 1$ or $n = 1$ and $k = 0$ is the exact solution of the Navier-Stokes equation for the flow near a stagnation point. Here, as expected, W_K falls off monotonically from 1 at the wall to 0 as shown in Fig.4 for $Re = 10^5$. - For $0 < m < 1$, e g. for $m = 1/3$ (flow along a 2×45° wedge), at first $W_K \rightarrow 0$. Yet then, for large values of η W_K increases again to 1 since ω^2 remains finite.

The case $m = 0$ gives the Blasius flow mentioned above with $\mathrm{div}\,\vec{a} \leqslant 0$ and so $W_K \gg 1$ everywhere. For any Reynolds number there is a wall distance η_o where $2\,\mathrm{div}\,\vec{a} = -\omega^2$ or $d^2 = 0$ and $W_K \rightarrow \infty$. This singularity is found only for a very small range of $m \leqslant 0$ and $|m| \leqslant 2\lambda\eta_o/Re$ with $\lambda(m) \approx 1.2$.

As shown in Fig.4 for $Re = 10^5$ and $m = -0.07$ and -0.09 at first W_K falls off to a small value, yet, lateron it increases again to 1.

d) $n = 1/3$, $k = -2/3$: Laminar plane jet. The analytic solution of the boundary layer equation $F' = 1 - \tanh^2\eta$ gives $W_K \gg 1$ for $\eta \geqslant 0.794$, and for $\eta \rightarrow \infty$ $W_K \rightarrow 1$ with ω^2 remaining finite.

e) $n = 1/2$, $k = -1$: Turbulent plane jet. The approximations for the velocity profile $F'_I = 1 - \tanh^2\eta$ resp. $F'_{II} = \exp(-0.892\,\eta^2)$ give $W_K \gg 1$ for $\eta \geqslant 0.658$ resp. 0.571, and again for $\eta \rightarrow \infty$ $W_K \rightarrow 1$ and $\omega^2 > 0$.

Axisymmetric flow

Also the corresponding similarity solutions for axisymmetric flow don't yield the expected $W_K(\eta \rightarrow \infty) \rightarrow 0$ with the only exception of the two exact solutions: stagnation point flow and round laminar jet. (The latter is also a good approximation for the turbulent jet with constant eddy viscosity.)

For the round jet the boundary layer solution is also the asymptotic of the exact solution for $Re \rightarrow \infty$ or $\varepsilon \rightarrow 0$. In spherical coordinates $\Psi = 2\nu R\,(1-\eta^2)/(\varepsilon + 1 - \eta)$ with $\eta = \cos\vartheta$ (ϑ angle between jet axis and rdius vector R from origin) and $\varepsilon \rightarrow 32\pi\varrho\nu^2/3M$ for $\varepsilon \ll 1$; ϱ = density, ν = kinematic viscosity and M = longitudinal momentum of jet [4]. Fig.5 shows velocity and W_K for $\varepsilon = 0.01$ vs $r/b_{0.5}$ with r= axial distance and $b_{0.5}$ = half breadth at half velocity. Within design accuracy exact and boundary layer solution give same results.

Other exact solutions

It is, by no means, selfevident that the two exact solutions above meet the ecpectation $W_K(\eta \to \infty) \to 0$ for $Re \to \infty$. The simplest counter example is the Rayleigh flow when an infinite flat plate is accelerated to a constant speed in its own direction. In this pure shear flow there is again everywhere $\operatorname{div} \vec{a} = 0$ and $W_K = 1$. Only at the half infinite plate at great Reynolds number a Blasius flow would be established very soon, except near the plate edge [5]. Taking into account the displacement of this boundary layer, i.e. assuming parabolic stream lines outside, the deformation rate at great wall distance becomes finite; and, since ω decreases exponentially, $W_K(\eta \to \infty) \to 0$.

One might even construct exact solutions with $W_K \to \infty$ as e.g. the instationary, plane, viscous vortex flow when the velocity on circles changes sign: $v = r\, t^{-3}(1 - \eta^2/2)\exp(-\eta^2)$ with $\eta = r/(2\sqrt{\nu t})$ and $\omega = t^{-3}(2 - 4\eta^2 + \eta^4)\exp(-\eta^2)$ and $W_K = |\eta^4 - 4\eta^2 + 2|/|\eta^2(\eta^2 - 3)|$.

Yet, apart from such paper experiments one would still expect for any real flow at high Reynolds number around a finite body in unbounded fluid $W_K(\eta \to \infty) \to 0$.

References

1. Truesdell, C. Two Measures of Vorticity. J.Rat.Mech.Anal. 2, 1953, 173
2. Wieghardt, K.; Kux, J. Nomineller Nachstrom auf Grund von Windkanalversuchen. Jahrbuch Schiffbautechn.Ges.74, 1980
3. Schlichting, H. Boundary Layer Theory. Pergamon 1980
4. Rosenhead, L. Ed.Laminar Boundary Layers. Oxford 1963
5. Hall, M.G. Proc.Roy.Soc.A 310, 1969, 401.

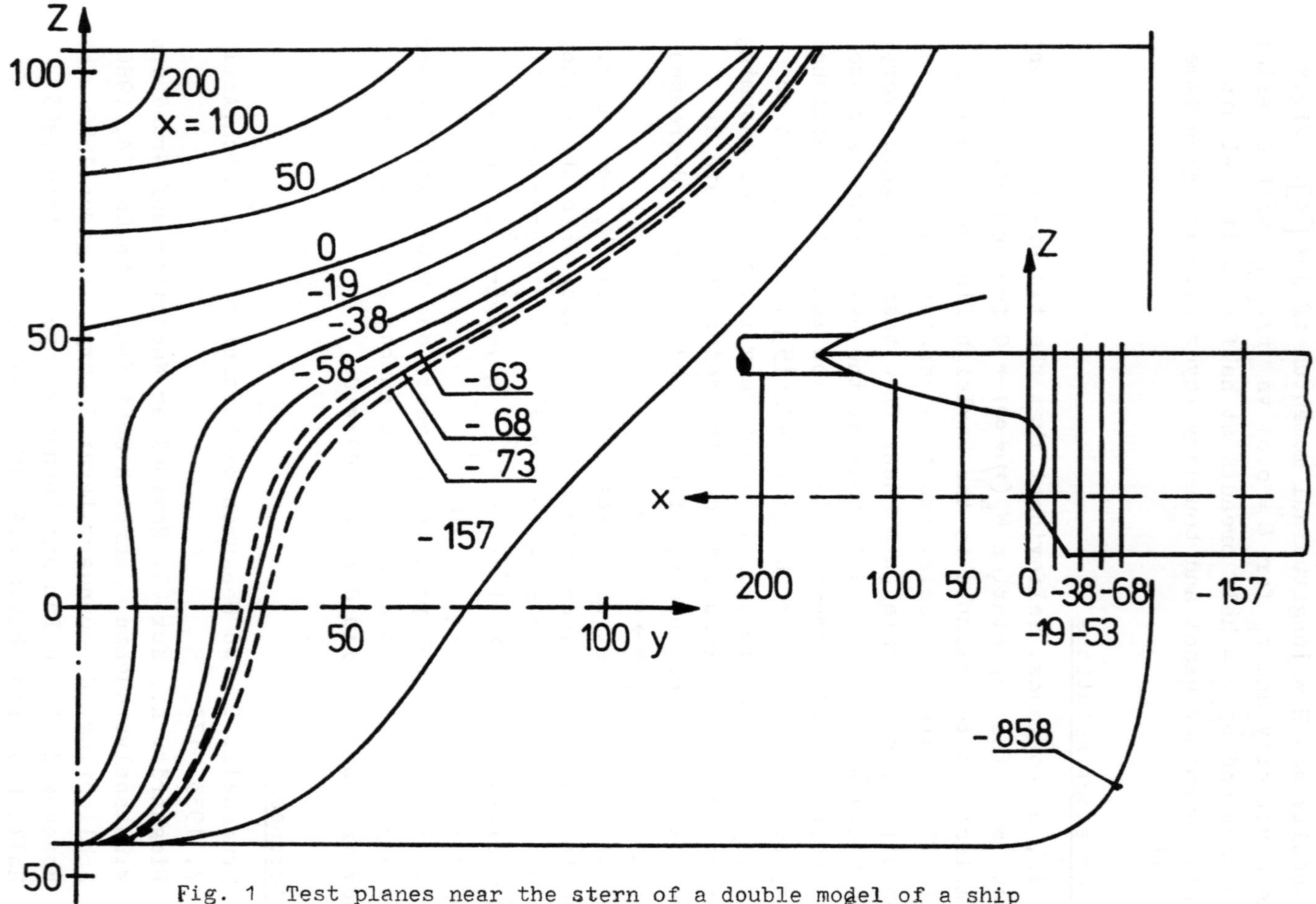

Fig. 1 Test planes near the stern of a double model of a ship in a windtunnel (length 2.74 m, Re = 5 10^6, half breadth 0.202 m, depth 0.148 m)

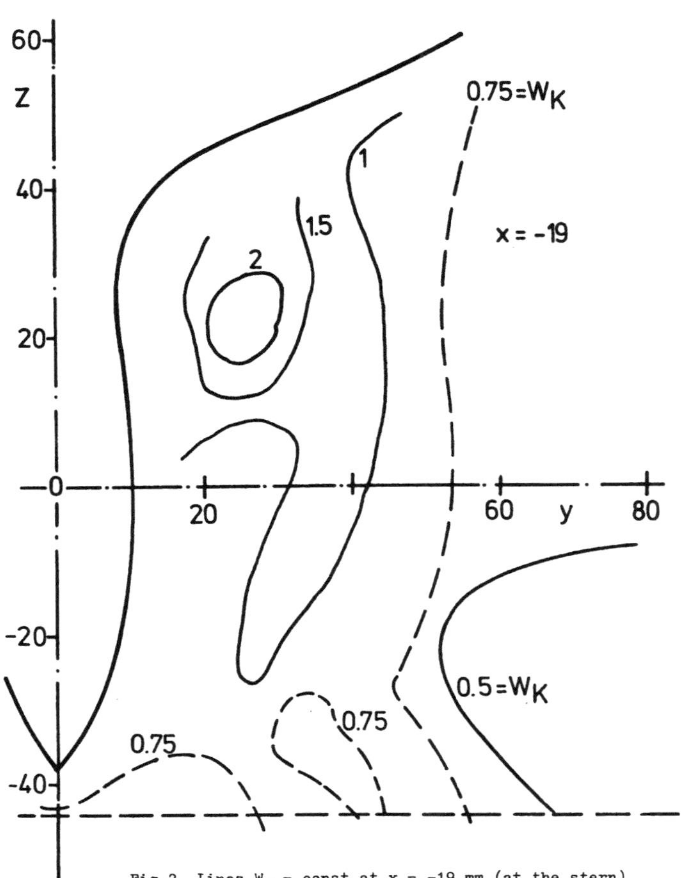

Fig.2 Lines W_K = const at x = -19 mm (at the stern)

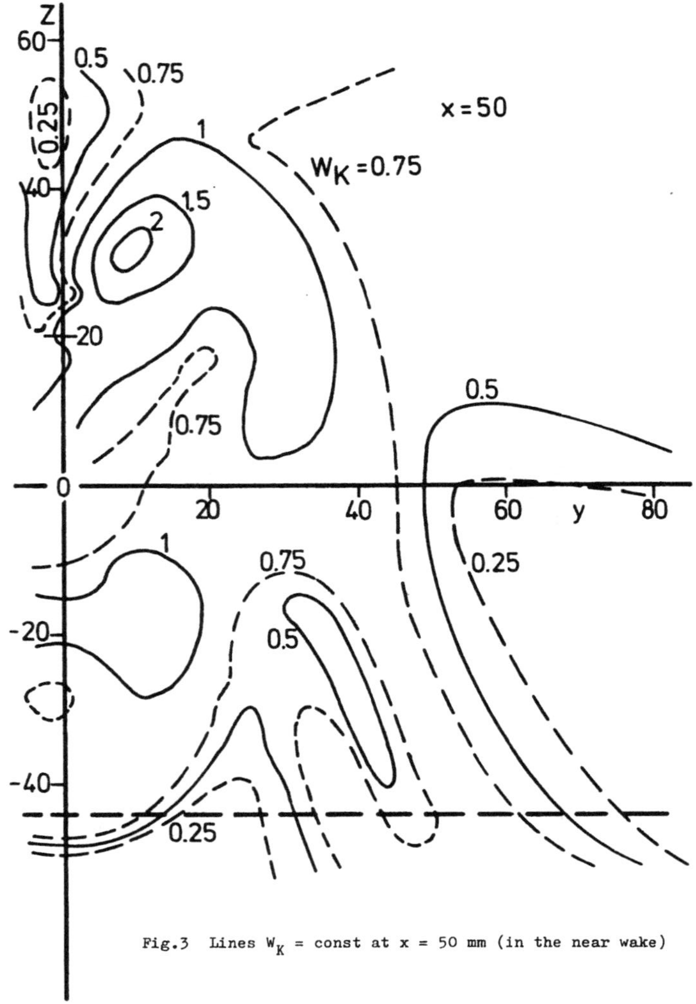

Fig.3 Lines W_K = const at x = 50 mm (in the near wake)

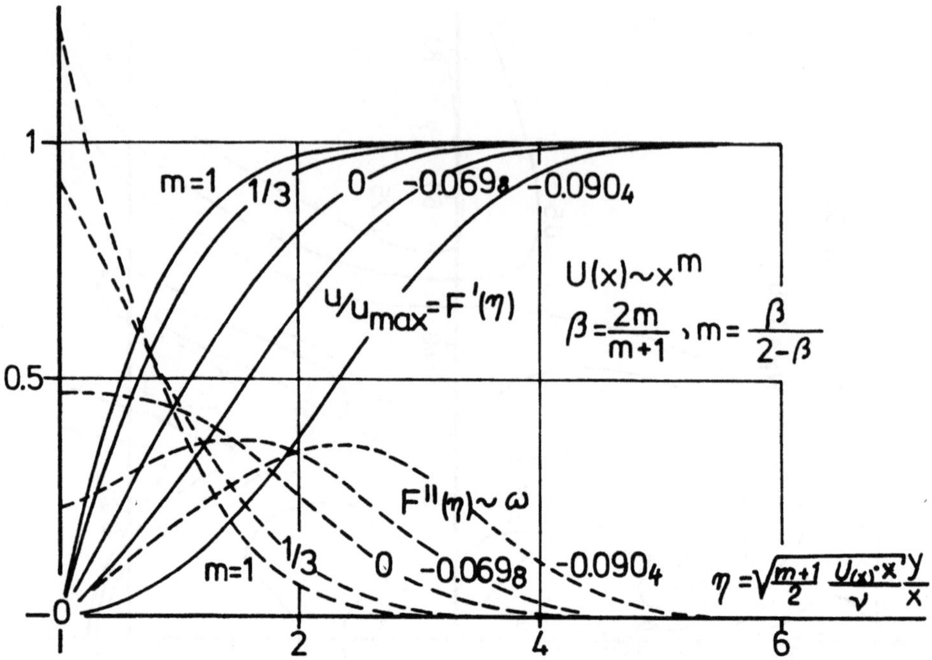

Fig. 4 Similarity Boundary Layer Profiles

m = 1	Stagnation Point	β =	1
1/3	2 x 45° Wedge		0.5
0	Blasius Flat Plate		0
- 0.0904	Separation Profile		- 0.1988

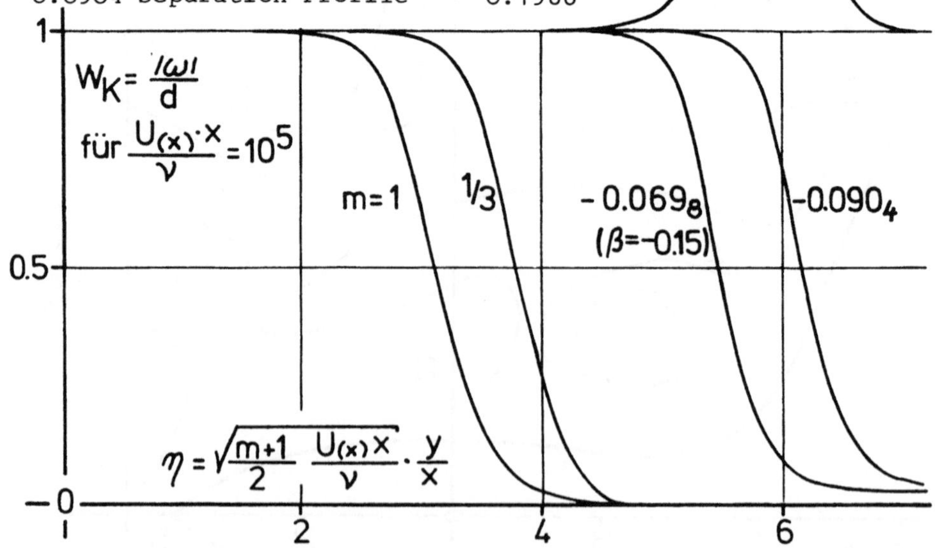

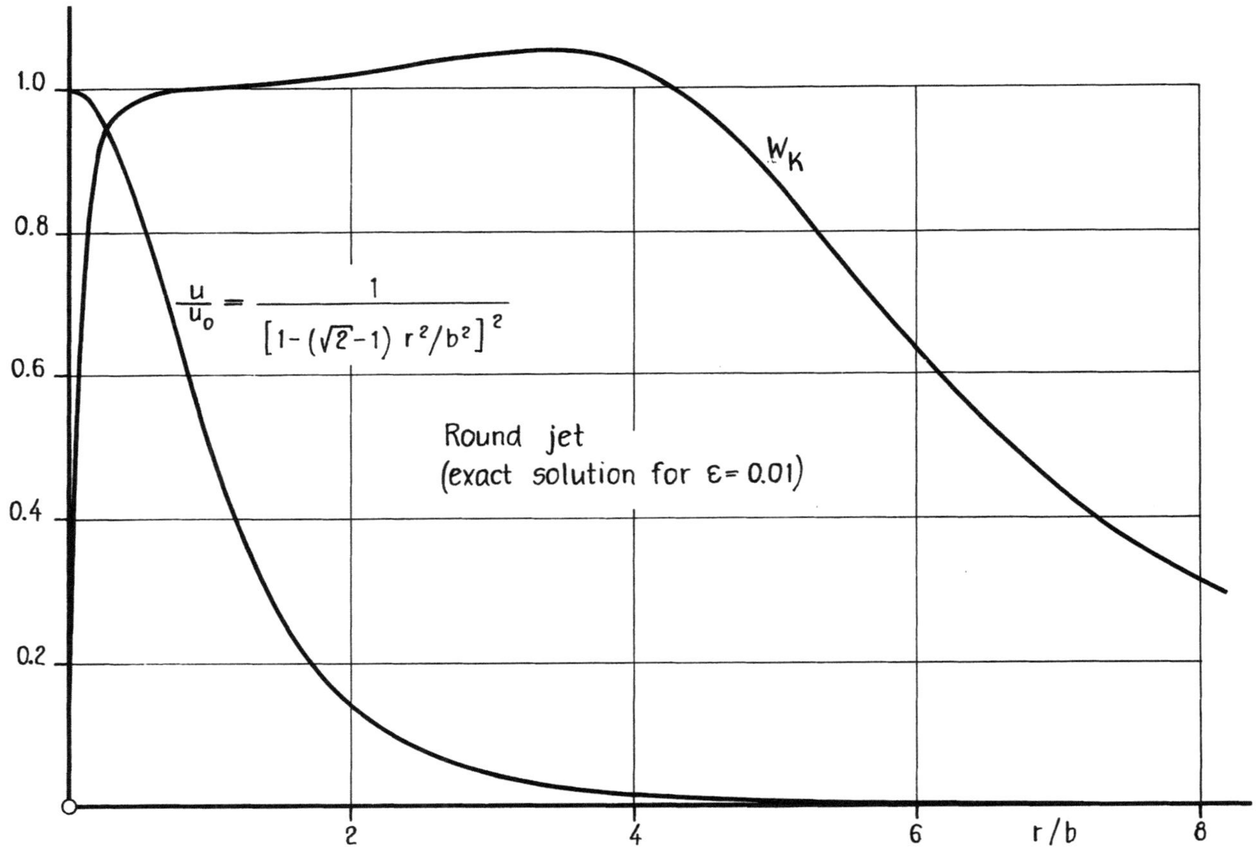

Fig. 5 Round laminar jet: velocity and W_K vs axial distance

The Design of Optimum Diffusers for Incompressible Flow

K. G. WINTER and L. F. EAST

Aerodynamics Department
Royal Aircraft Establishment

Summary

For diffusers which are sufficiently short the flow will not become fully developed and can be treated as a boundary layer surrounding an inviscid core. In most practical applications the flow will be turbulent and the calculation of the boundary layer will depend in some measure on empirical information. In the paper use is made of the empirically-derived properties of equilibrium turbulent boundary layers to derive an optimum form for the pressure distribution along a diffuser. It is shown that the condition is given by a velocity distribution of the form $U \propto x^m$ where m has the value of about -0.3. For a diffuser with this velocity distribution the loss of mechanical energy is about half that for a flow with incipient separation. A method of designing diffusers to achieve optimum performance is indicated and an example of a design given. The gain in efficiency compared with a conical diffuser of similar effective expansion angle is not large but is just significant.

1 Introduction

The flow in a diffuser is regarded as being composed of two regions. The main part of the flow is taken to have a velocity distribution such that the boundary layer is in equilibrium. The flow is further idealized to have constant Reynolds number. The major part of the pressure rise is assumed to take place in this equilibrium region. Thus, by considering equilibrium flows only, an assessment can be made of desirable forms of the pressure distribution in order to achieve high efficiency. The assumption of a distinct boundary layer on the walls of the diffuser limits the application to diffusers which are sufficiently short for a fully-developed 'pipe-flow' not to occur. In a practical diffuser of the type envisaged, the part having equilibrium flow needs to be preceded by an entry region. In this region the pressure distribution is so chosen that the boundary layer on the walls develops rapidly from its initial state to a condition corresponding to the desired equilibrium flow.

2 Optimum Equilibrium Flow

It is assumed that the flow over any section of radius R of the major part of the diffuser can be represented by an inviscid core, of radius r and velocity U, uniform over the cross section, surrounded by a boundary layer of displacement thickness δ^* so that

$$R = r + \delta^* \quad . \tag{1}$$

It can then be shown[1] that a good approximation to the momentum equation for the boundary layer flow is given by

$$\frac{1}{r}\frac{d(r\theta)}{dx} + (H+2)\frac{\theta}{U}\frac{dU}{dx} = \frac{C_f}{2} \quad , \tag{2}$$

where the symbols have their usual meaning as for a two-dimensional flow. Combination of equation (2) and the condition of continuity ($r \propto U^{-\frac{1}{2}}$) leads to

$$\frac{d\theta}{dx} + (H+1.5)\frac{\theta}{U}\frac{dU}{dx} = \frac{C_f}{2} \quad . \tag{3}$$

If the flow is in equilibrium as defined by H, C_f being constant, equation (3) integrates to

$$\frac{\theta}{x} = \frac{C_{f/2}}{1+(H+1.5)m} \quad . \tag{4}$$

The origin of x is chosen to be the effective start of the boundary layer, and m is given by $m = \frac{x}{U}\frac{dU}{dx}$. It is interesting to note that the linear growth characteristic of two-dimensional equilibrium boundary layers is still retained despite the convergence or divergence effects in an axisymmetric flow.

The model of Ref 2 is now used for the equilibrium flow which leads to the following relationsgip between m and H

$$\frac{1}{m} = \frac{0.8H}{1-K} - (H+1.5) \quad , \tag{5}$$

where $K = 1.037028 \left(\frac{H-1}{H}\right)^2 \left(\frac{H_0}{H_0-1}\right)^2 \frac{H - 0.4\,H_0}{1.1\,H_0-0.5H}$, and H_0 is the shape parameter for a flat-plate boundary layer at the given value of Re_θ. The variation of H with m is shown in Fig 1 for three values of Reynolds number. The feature to be noted is the existence of a minimum for m, indicating the possible existence of two different flows for a given pressure gradient and Reynolds number.

An approximate expression for the efficiency, η, of a diffuser in terms of the ratio of pressure energy recovered to the reduction in kinetic energy was derived in Ref 1 as

$$\frac{1}{\eta} = 1-2n\,\frac{\left(\frac{U_2}{U_1}\right)^{2.5+1/m}-1}{\left(\frac{U_2}{U_1}\right)^{2}-1} \qquad (6)$$

where n is the ratio of the energy thickness to the core radius at the start of the equilibrium flow ($n = \delta_{E1}/r_1$). The maximum value of efficiency will occur for the largest value of $2.5+\frac{1}{m}$ and, m being negative, this will occur for the largest negative value of m, that is, for the points of minimum m typically illustrated in Fig 1. It is also interesting to note that, since $U_2 < U_1$, $2.5+\frac{1}{m} < 0$, in order that the flow does not gain mechanical energy, and thus m cannot be less than -0.4. For a two-dimensional flow the equivalent result is that m cannot be less than -1/3.

The geometry of such a diffuser is given by

$$\frac{R}{R_1} = \frac{\frac{r}{r_1} + \frac{nH}{H_{32}}\left(1+\frac{x}{\theta_1}\frac{d\theta}{dx}\right)}{1+\frac{nH}{H_{32}}}\,, \qquad (7)$$

and area ratio

$$\frac{A_2}{A_1} = \left\{\frac{\frac{r_2}{r_1} + \frac{nH}{H_{32}}\left(1+\frac{1}{\theta_1}\frac{d\theta}{dx}\right)}{1+\frac{nH}{H_{32}}}\right\}^2 . \qquad (8)$$

The above equations for the diffuser characteristics can be evaluated for given Re_θ and H by making use of the equations for equilibrium flow given in Ref 2, and an additional expression for H_{32} taken from the empirical result of Ref 3 as

$$H_{32} = \frac{2}{1+0.5885\left(\frac{H-1}{H} - 0.036\right)} . \qquad (9)$$

Details are given in Ref 1.

The analysis is in a somewhat idealistic form in respect of the practical design of diffusers because, firstly, it relates only to the region where the boundary layer is of equilibrium type and, secondly, because the Reynolds number for a given flow is effectively taken to be independent of the streamwise development. Nevertheless, some broad conclusions can be drawn. The fact that there is a minimum value for m implies that for values greater than this there exist two possible solutions for the boundary-layer flow. The rate of growth of the boundary layer for the two flows will be grossly different, that for the upper branch of the curves in Fig 1 being considerably in excess of that for the lower. For a diffuser with sufficiently large angle the only possible physical solution will involve flow appropriate to the higher value of H over at least some portion of the circumference, in order to satisfy the continuity equation. It may therefore not be physically realistic to consider axisymmetrical flows which imply boundary layers with characteristics corresponding to the upper branches of the curves. However, in order to illustrate the low values of diffuser efficiency obtained for diffusers with incipient separation calculations have been made for such flows. The results in terms of efficiency and diffuser angle as a function of area ratio are compared with optimum flows in Fig 2 for one value of Reynolds number $Re_\theta = 10^4$, and a range of values of the ratio, n, of the initial boundary layer energy thickness to the inlet radius. Roughly the fractional loss of mechanical energy $(1-\eta)$ for the diffusers with incipient separation is twice that for optimum diffusers and the angles are also roughly twice as great.

3 Design of Entry Region

The problem posed is to design a wall shape for the entry to the diffuser which starts as a cylindrical pipe having a boundary layer on its wall with shape parameter appropriate to zero pressure gradient, and then expands to join, at some match point with continuity of slope, the wall shape corresponding to the required equilibrium flow.

The adverse pressure gradient in the entry region will grow very rapidly so that a plausible assumption is that the entrainment into the boundary layer may be neglected. Thus

$$rU(\delta-\delta^*) = \text{constant} \quad . \tag{10}$$

By assuming power law velocity profiles and eliminating U by means the of continuity equation we can rewrite (10) as

$$\frac{2\theta}{r}\,\frac{H}{H-1} = \text{const} = \lambda,\ \text{say}, \tag{11}$$

which, differentiated with respect to x, gives

$$\frac{d\theta}{dx} = \lambda r\,\frac{H-1}{2H}\left[\frac{1}{r}\frac{dr}{dx} + \frac{1}{H(H-1)}\frac{dH}{dx}\right] \quad . \tag{12}$$

Substitution of these equations in the momentum equation (2) results in a relationship between r and H

$$\frac{1}{r}\frac{dr}{dx} = \frac{1}{2H(H^2-1)}\frac{dH}{dx} - \frac{C_f}{2\lambda r} \quad . \tag{13}$$

This can be written approximately as

$$\frac{\Delta r}{r_1+\frac{\Delta r}{2}} = \frac{\Delta H}{2\hat{H}(\hat{H}^2-1)} - \frac{C_f}{2\lambda r}\Delta x \quad , \tag{14}$$

where r_1 is the radius of the cylindrical entry, Δr is the change in r appropriate to the required change ΔH in H, and $\hat{H}$ is a mean value. The second term on the right can be shown to be small compared with the first and has been neglected.

For a diffuser of length l with velocity distribution defined by the parameter m (ie $U \propto x^m$), and having initial and final velocities U_1 and U_2 the velocity distribution and radius of the core flow are given by

$$\frac{U}{U_1} = \left(\frac{r_1}{r}\right)^2 = \left\{1 + \frac{x}{l}\left[\left(\frac{U_2}{U_1}\right)^{1/m} - 1\right]\right\}^m \quad . \tag{15}$$

Hence, given a "match point" with radius $r_1+\Delta r$ from equation (14), the location of the point, and the wall slope at the point, can be determined from equation (15). A suitable curve which can match the ordinate and slope of the wall at the match point and have zero slope at entry is of cubic form. The length of the entry region, l_e, is taken as a parameter for the curve leading to

$$\frac{r}{r_1} = 1 + \left[3\left(\frac{r_3-1}{r_1}\right) - r_3'\frac{l_e}{r_1}\right]\left(\frac{x_e}{l_e}\right)^2 + \left[r_3'\frac{l_e}{r_1} - 2\left(\frac{r_3-1}{r_1}\right)\right]\left(\frac{x_e}{l_e}\right)^3 \tag{16}$$

where $r_3 = r_1 + \Delta r$, $r_3' = \left(\frac{dr}{dx}\right)_{r=r_3}$ and x_e is th axial coordinate with origin at the entry.

As an example the design of a diffuser based on an optimum equilibrium shape for an area ratio of 2.5, effective angle 5°, with boundary layer shape parameter H = 1.684 has been investigated. The boundary-layer development was calculated by the method of Ref 2.

Specifically the conditions taken for the calculation were as follows. For the initial boundary layer n = 0.02, equivalent to $\frac{\theta_1}{r_1} = 0.011$, H = 1.313, $Re_{\theta_1} = 10^4$, and for the equilibrium flow m = -0.289. In principle the jump condition on H was $\Delta H = 0.371$, ie an increase from the initial value to 1.684, the optimum equilibrium value for $Re_\theta = 10^4$. However, there were two reasons for reducing ΔH to the value taken of 0.3, namely, that for an actual flow, as compared with the idealized equilibrium flow at constant Reynolds number, H would decrease with distance downstream, and also preliminary calculations showed that some overshoot in H would in fact occur.

The results of the calculation are shown in Figs 3, 4 and 5. Fig 3 shows the velocity distribution along the diffuser and Fig 4 the variation of H. The calculation method of Ref 2 enables approximate allowance to be made for the effects of streamwise curvature of the flow, which as shown in Fig 4, has an important effect. Because of the overshoot beyond the match point, the datum equilibrium value of H was very nearly achieved with a subsequent decrease in H with distance downstream. This decrease is to be expected, as is illustrated in Fig 5, where the variation of H with Re_θ is shown. The calculation is made for a given value of m and the equilibrium value of H for this given m decreases with increase of Re_θ as shown by the chain-dotted line. The calculated H tends slowly towards this line. The flow therefore does not have the optimum variation of properties which would require the variation of H indicated by the line labelled "Optimum" H in Fig 5, together with the appropriate variation of m.

For comparison with this "optimum" design a calculation was also made for a diffuser of the same length and velocity ratio, but with a conically-shaped region of inviscid flow, utilizing the same inlet displacement surface. The distribution of H is shown in Fig 6 and the shape of the diffuser is compared in Fig 7 with that for the "optimun" design. The effective diffuser angles are 5.2 and 4.8 degrees respectively for the 'conical' and optimum design and the efficiencies are 0.959 and 0.956. These values are apparently very close together but nevertheless the losses for the conical core flow design are 7% greater than for the optimum design. However, the difference between the optimum design and a diffuser with a conical wall shape would be even less since the displacement thickness growth would tend to lead to a displacement surface of tulip shape.

4 Concluding Remarks

It is shown that in an idealization of the flow in a diffuser with velocity distributions of the form $U \propto x^m$ the optimum efficiency is obtained for values of m of about -0.3. Such flows are far from the incipient separation often assumed to lead to high efficiency and, in fact, the losses for the optimum flows derived are about half those for incipient separation. Because of the nature of the equilibrium locus for a turbulent boundary layer the latter type of flow is also likely to lead to asymmetry and consequent unsteadiness.

Within the limits of the assumption of uniformity of the inviscid core flow over sections of the diffuser, calculations have been made for one example of a practical design. The loss of mechanical energy is shown to be slightly less than for a diffuser of approximately conical shape, having the same length and velocity ratio.

References

1. East, L.F.; Winter, K.G.: On efficient pressure recovery. RAE TR 79143 (1979).

2. Green, J.E.; Weeks, D.J.; Brooman, J.W.G.: Prediction of turbulent boundary layers and wakes in compressible flow by a lag-entrainment methods. ARC R&M 3791 (1973).

3. East, L.F.; Sawyer, W.G.; Nash, C.R.: An investigation of the structure of equilibrium turbulent boundary layers. RAE TR 79040 (1979).

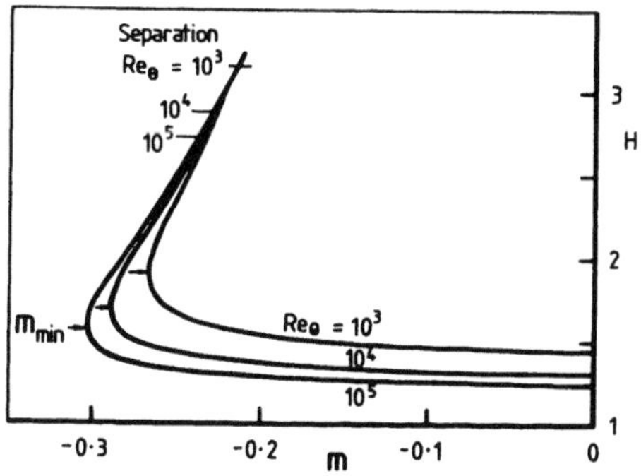

Fig 1. Shape parameter variation for equilibrium boundary layers in axially-symmetric diffusing flows.

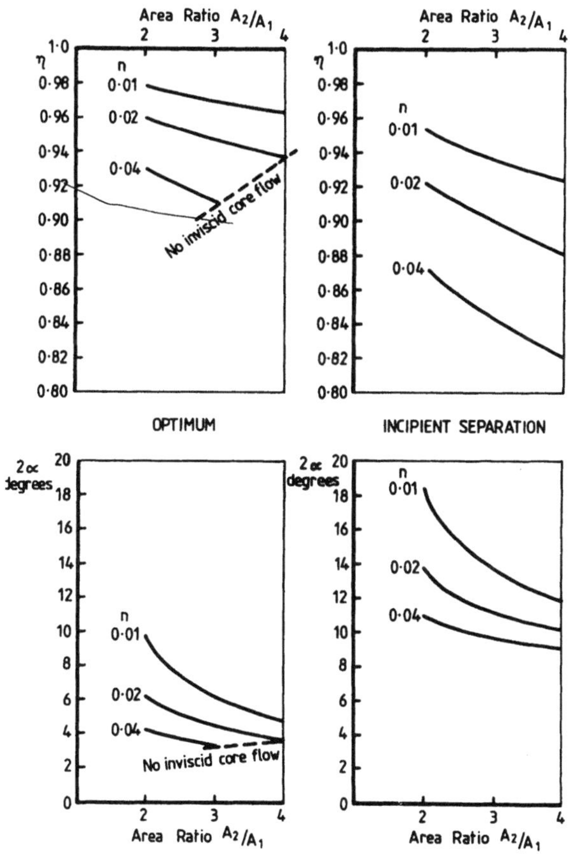

Fig 2. Comparison of characteristics of diffusers for optimum conditions and conditions of incipient separation $Re_\theta = 10^4$.

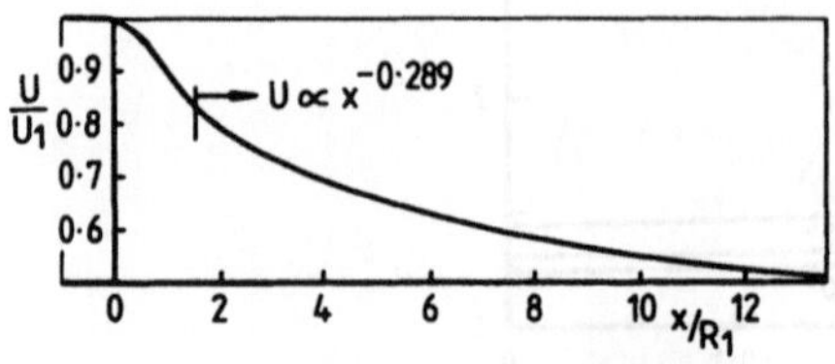

Fig 3. Design velocity distribution.

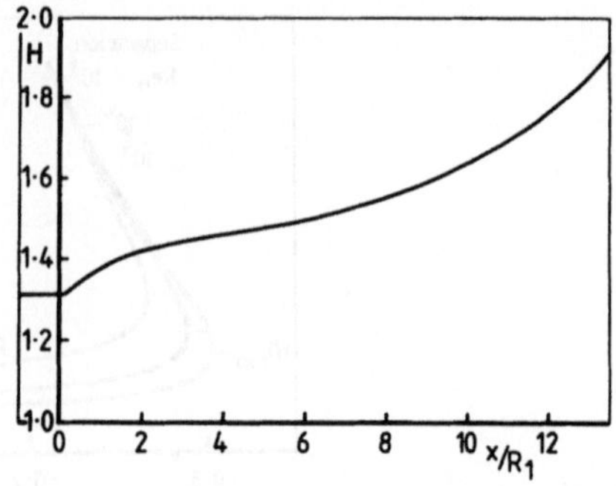

Fig 6. Calculated H development for conical core flow.

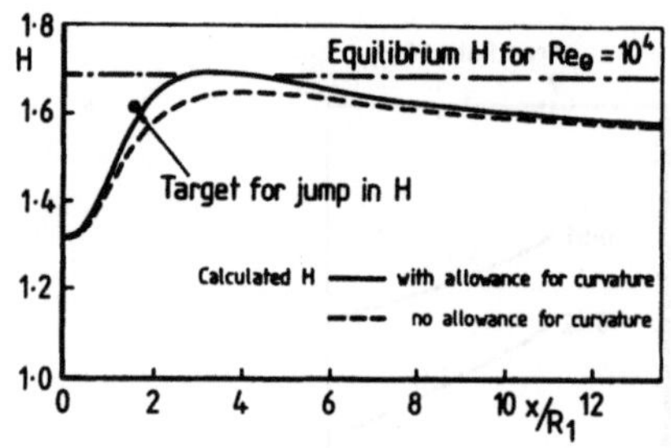

Fig 4. Calculated H development

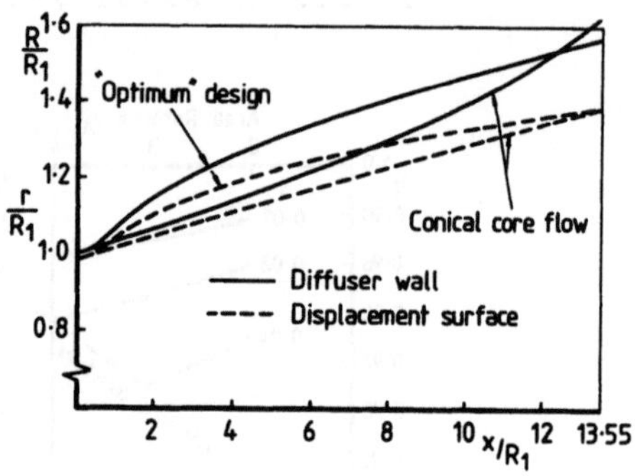

Fig 7. Diffuser shapes.

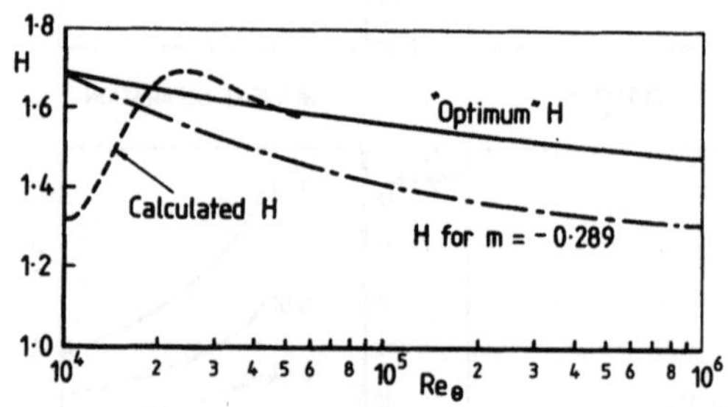

Fig 5. Optimum H development.

Relaxing Evolution Systems for the Solution of an Ill-Posed Design Problem

H.J. WIRZ

Westinghouse Nuclear International
Brussels-Belgium

Summary

For a relatively simple design problem, which is shown to be ill-posed in the usual sense, new artificial relaxing evolution systems (ARES) are considered, which provide approximate (regularized) solutions in an extremely efficient way.

1. Introduction

Design problems, as opposed to the more conventional analysis cases, are typically man made having no counterpart in observable nature. A well known experience is that such problems (sometimes also called inverse problems) are by far more difficult to solve, if a solution exists at all. This, in part, could explain why the development of efficient design methods, though of considerable practical importance, seems not to be as advanced as one would wish.

Another reason could be that engineers who are highly interested in developing design tools often lack the necessary mathematical background while mathematicians seem to be less interested in solving engineering problems. Intentionally, we shall treat here a rather simple design problem, namely that of finding the shape of an upper flexible low speed wind tunnel wall, see fig. 1, where the (inviscid) velocity distribution at the tunnel floor is assumed to be known. This design problem exhibits some of the fundamental problems (ill-posedness) which are typical for this class. The efficient solution methods we are presenting appear to be new since general "regularization" techniques for the treatment of such ill-posed systems seem not be known. Due to the limited space available, the solution idea based on artificial and relaxing evolution equations can only be outlined.

2. The ill-posed design problem

The local microclimate around buildings is often studied in stationary low speed wind tunnels with appropriate scaled models, see fig. 1. The upper, flexible tunnel wall serves to generate the streamwise wind distribution for the viscous ground boundary layer in which the model buildings are placed.

According to the boundary layer theory, to which Walz (1) has made numerous and significant contributions over the years, the viscous effects are concentrated in these thin layers, while the (larger) remainder of the flow may be considered inviscid and even irrotational. Neglecting the thickness of the wall boundary layers and in order to limit the necessary effort to fully treat the two-dimensional (2D) problem, a linearized set of the incompressible flow equations, domain and boundary conditions is used for simplicity. Thus, the following perturbations to the corresponding trivial one-dimensional flow with constant velocity U_e

$$u(x,y) = U_e + u'(x,y) \quad ; \quad v(x,y) = v'(x,y) \tag{2.1}$$

are introduced for the inviscid and irrotational velocities in the streamwise (x) and the normal (y) co-ordinate. Neglecting higher order terms, the following well known flow differential equations (mass conservation, irrotationality) are recovered

$$u_x + v_y = 0 \quad ; \quad -v_x + u_y = 0 \tag{2.2}$$

in which the primes have been omitted for convenience. The integration domain D then simply becomes a rectangular

$$D = \{x,y \mid 0 \le x \le L, \quad 0 \le y \le H\}. \tag{2.3}$$

As is well known, the above equations are elliptic and one boundary condition may be imposed on each side of the boundary of D. (The conventional, well posed, analysis case). The interesting design problem, here, however, manifests itself by a different set of boundary condition for the perturbations:

- entrance and exit plane: $u(o,y) = 0$
- tunnel floor : $v(x,o) = 0$ (2.4)
  $u(x,o) = u_o(x)$ given
- upper tunnel wall : no condition,

since the (linearised) boundary condition for the impermeable upper tunnel wall

$$dh/dx = v(x,H)/U_e \qquad (2.5)$$

involves the as yet unknown perturbed tunnel shape h(x).
To be specific, let $u_o(x)$ (the wind distribution at the tunnel floor) be given by

$$u_o(x) = U_n \sin n\pi\, x/L \qquad (2.6)$$

where n denotes an integer and U_n a constant velocity amplitude corresponding to the n-th mode. Due to the linearity of the problem one term out of an infinite (and converging) Fourier series will suffice here.
It is worth pointing out that the velocity $u_o(x)$ defined in this way remains finite for arbitrary high wavenumbers $k_n = n\pi\, x/L$. According to the boundary conditions above, our design problem has then the following exact solutions

$$u(x,y) = U_n \cosh k_n y \sin k_n x \qquad (2.7)$$

$$v(x,y) = -U_n \sinh k_n y \cos k_n x. \qquad (2.8)$$

Upon inspection, however, it is immediately realized that these solutions, though exact, do not behave well: for any fixed $y>0$ they grow exponentially fast with increasing wavenumbers. Therefore these solutions are unbounded and obviously they do not depend continuously on the given data, contrary to what can be expected in nature (Violation of Hadamard's continuity condition). The above, exceptional, behaviour renders our design problem "not properly posed" or "ill-posed", assuming existence and uniqueness of the solutions. The possible counter argument, that U_n could decrease exponentially fast with increasing n is not acceptable since polynominal decrease is the best one can hope to get in practice. Finally the fact that this design problem is ill-posed carries another fundamental consequence with it: any conceivable consistent numerical approximation of the equations and boundary conditions inevitably leads to a totally unstable method. For a discussion, see for example Richtmyer and Morton (2). Therefore, ill-posed problems have to be rejected computationally also, unless measures are taken to prevent the (numerical) solutions from unbounded growth (instability).

Of course, there are different ways to prove the ill-posed nature of our design problem; the choice here favours simplicity not calling on more elaborate (and therefore heavier) techniques. The above characteristic features are quite typical of most design problems with partial differential equations, and therefore simply abandoning the problem because of some mathematical difficulties, though fundamental, appears not to be good engineering practice. Since one can hardly do better than to find the exact solutions, some price must be paid to obtain bounded and useful solutions in a certain approximate sense.

3. Regularization with artificial relaxing evolution systems (ARES)

Looking at the solutions, equs. 2.7 and 2.8 it is the high wavenumber components which appear most dangerous. This then suggests the addition of some small but fixed artificial viscosity to the differential equations thereby introducing a dissipative mechanism which is known to efficiently damp the high wavenumber components. The fundamental disadvantage of this (known) approach as a remedy for the solution of ill-posed first order systems appears to be the inevitably increased order of the equations. This requires the imposition of additional boundary conditions which are typically not known. Therefore, it is believed that a different approach to the problem needs to be considered. Introducing for convenience a streamfunction

$$u = \psi_y \quad ; \quad v = -\psi_x \tag{3.1}$$

the following initial-boundary value problem (IBVP) for the solution of the equs. 2.7 and 2.8 can be formulated

$$\begin{aligned} -u_y - \psi_{xx} &= 0 \\ -\psi_y + u &= 0. \end{aligned} \tag{3.2}$$

This system is apparently not an arbitrary one, since it represents the prototype of a class of well known "Streamline Curvature Methods" Katsanis (3), for computing internal flows. We have shown, Wirz (5), that this IBVP is also ill-posed, thereby explaining the often observed numerical instabilities. The ill-posed nature of the above IBVP remains unchanged even if only one condition on each boundary of D is imposed, i.e. the analysis problem is also ill-posed here, due to the specific choice of eq. 3.2. Consider now the following system of artificial evolution equations, where the additionally introduced independent variable could be called a "time",

$$\begin{aligned} \psi_t - \tau c^2(u_y + \psi_{xx}) &= 0 \\ \tau u_t - \psi_y + u &= 0 \end{aligned} \tag{3.3}$$

and where $\tau > 0$, $c \neq 0$ are some constants. Without proof we state here that the above evolution system (IBVP) is well posed for $t > 0$ provided that one and only one boundary condition on each side of D is imposed. The above IBVP has another remarkable property namely that for sufficiently long "times" the corresponding steady state solutions will not only be obtained but are bounded, i.e. they are the exact classical solution of the analysis problem, independent of the choice of the initial data.

Finally, it can be shown that the above artificial relaxing evolution system (ARES), as we have called this class of new evolution equations, converges extremely fast to the desired steady state, provided that the characteristic relaxation time, τ, is chosen optimal. Further details may be found in Wirz (4). The above system, however, is not suitable for the solution of our design problem, since two boundary conditions at the tunnel floor need to be imposed. Therefore, we now generalize the ARES turning the ingoing characteristics of the corresponding non-dissipative (hyperbolic) evolution system

$$\begin{aligned} \psi_t + \alpha c(\psi_y - \varepsilon u) - \tau c^2 u_y - \tau c^2 \psi_{xx} &= 0 \\ \tau u_t + \alpha c\tau(u_y + \varepsilon\psi_{xx}) - \psi_y + u &= 0 \end{aligned} \qquad (3.4)$$

in which $\tau > 0$ denotes the constant relaxation time; c, a constant reference velocity; α,ε constant parameters. Again, without proof we state here that this ARES is well posed and solves extremely efficiently our design problem, provided that the following conditions are met:

$$\tau>0; \quad c>0; \quad \alpha>1; \quad \varepsilon>0; \quad \varepsilon|\alpha|<1; \quad 1-\varepsilon\alpha^2 \neq 0 \qquad (3.5)$$

For $\alpha = 0$ we recover, of course, the analysis case treated before. As will be seen immediately, however, the bounded steady state solutions so obtained are only approximate ones, satisfying a slightly altered steady state operator. This typical behaviour is not unexpected, since we know the exact solutions of our problem do not behave well (unboundness).
It may also be seen that the parameter α and ε cannot be chosen arbitrarily for treating the "design problem"; they may be selected according to a certain measure for the "best" approximate solutions.
Finally, the rate of convergence of this ARES towards the desired steady state may be optimized by choosing an optimal value for the relaxation time, τ.

Thus, it seems that with ARES design problems can be attacked successfully which will, we hope, stimulate further research in this area of considerable engineering importance. The missing proofs together with the general concept of ARES will be published separately.

4. References

(1) Walz, A.: Strömungs- und Temperaturgrenzschichten Braun Verlag, Karlsruhe (1966).

(2) Richtmyer, R.D.; Morton, K.W.: Difference Methods for Initial Value Problems. John Wiley, New York (1967).

(3) Katsanis, T.: Use of arbitrary quasi-orthogonals for calculating flow distributions. NASA TN, D-2546 (1946).

(4) Wirz, H.J.: Relaxation Methods for Time Dependent Conservation Equations in Fluid Dynamics. AGARD LS No 86, Paris.

(5) Wirz, H.J.: Instabilities and ill-posed Solution Concepts for Inviscid Flows. Brussels (1982) to be published.

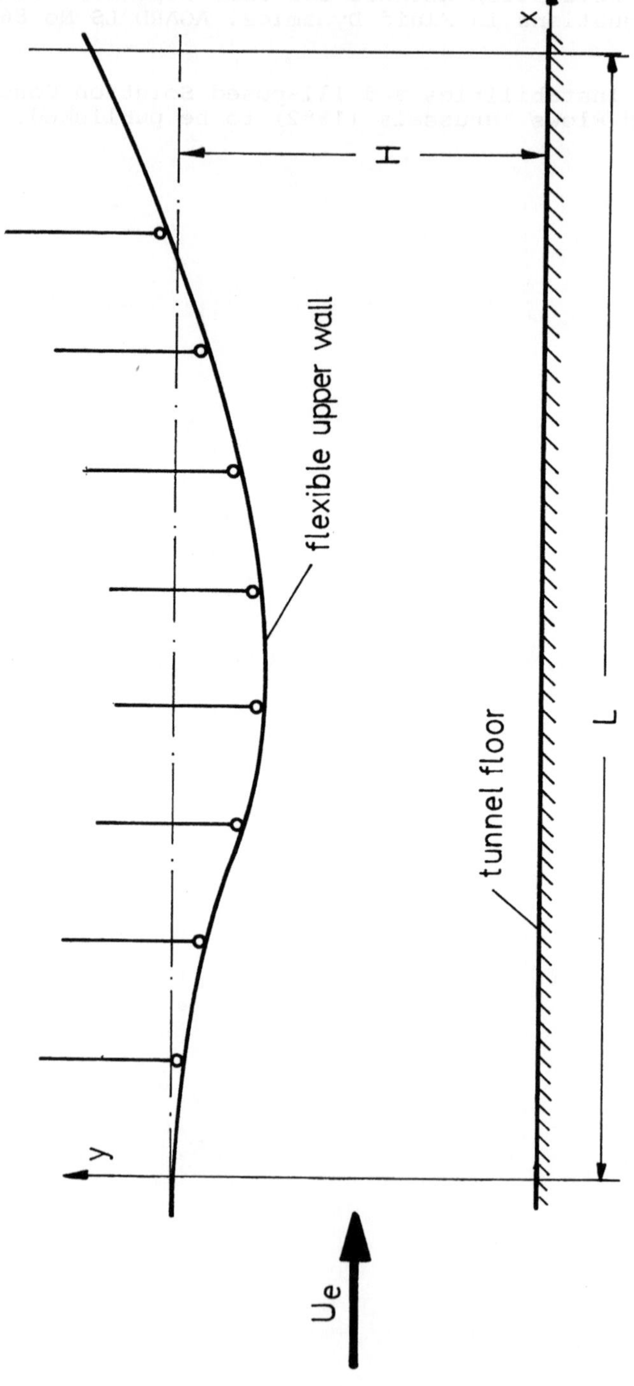

FIG.1 LOW SPEED WIND TUNNEL WITH FLEXIBLE UPPER WALL

What can be Achieved by Similarity Laws?

JÜRGEN ZIEREP

Institut für Strömungslehre und Strömungsmaschinen
Universität Karlsruhe
Kaiserstrasse 12, 7500 Karlsruhe 1

Summary

In the following it has been investigated by means of simple examples, what the expressional power of the dimensional analysis is. The Buckingham theorem will be recalled and it will be shown, how by simple additional physical assumptions in some cases the statements of the dimensional analysis can be made more substantial.

1. Introduction

Similarity considerations have gained substantial importance in theory and practice. Not only in respect to model considerations the analysis allowed many new insights. Nevertheless very often confusion obscures the expressional power of these laws. Further the laws are as often overestimated as underestimated. Always it is rewarding to control the laws by interesing examples. Furtheron there is the question, how, for example, in single cases the statements of the dimensional analysis can be made more rigorous. In general, here are small additional physical considerations needed only, that very often are obvious for the problem treated. Up to now general statements in this direction are completely absent. Now we will recall the statements of the dimensional analysis.

2. The π-theorem of Buckingham /1,2/

2.1 Starting point for the problem being treated are n physical, dimensional quantities of influence:

$$Q_1 , Q_2 , \ldots , Q_n ,$$

among those relations are existent. This contains the physics of the problem treated.

2.2 m base quantities:

$$A_1 , A_2 , \dots , A_m$$

describe the section within that the analysis is done (mechanics, thermodynamics etc.).

2.3 The elements of the dimensional matrix (a_{ik}) are the exponents of the powerproduct representation of the quantities of influence to the base quantities:

	Q_1	Q_2	...	Q_n
A_1	a_{11}	a_{12}	...	a_{1n}
A_2	a_{21}	a_{22}	...	a_{2n}
.	.			
.	.			
.	.			
A_m	a_{m1}	a_{m2}	...	a_{mn}

The rank of this matrix may be $r \leq m$.

2.4 The related linear, homogenious system of equations (m equations, n unknown quantities) produces $(n - r)$ dimensionless solutions (π-numbers).

2.5 The general dependency among the Q_1, Q_2 ... Q_n is transfered to the - numbers.

$$f(\pi_1 , \pi_2 , \dots , \pi_{n-r}) = 0$$

is the solution of the problem.

With the specific application of this π-theorem a number of questions arise. Here we discuss the expressive power of the theorem only. 2.1 and 2.2 cover the physical considerations. By this general treatment there can not be gained more insight, finally, than has been put into the problem from the beginning, because the following points contain mathematical statements only. In special cases however, substantially more exact statements of the

π-theorem are possible. For this, additional (often obvious) physical assumptions about the correlations among the $Q_1, Q_2, \ldots, Q_n$ are given that allow far reaching statements about the form of the function f in 2.5. This will be demonstrated with some examples. A general representation of these, for applications very important intensification, is lacking.

3. Discharge of an incompressible fluid from a vessel

For this simple example we are interested in the mass flow rate $\dot{m}$. We ex-

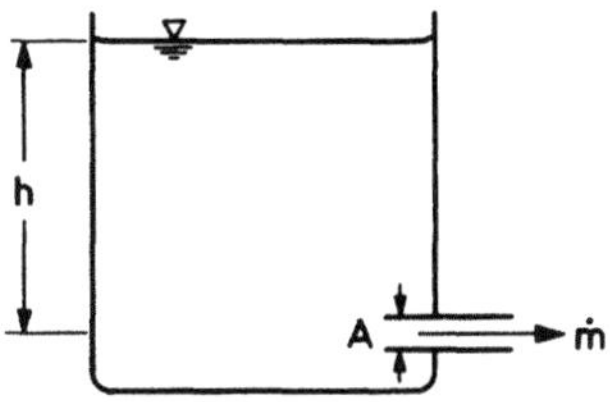

Fig. 1: Discharge from a vessel

stract the quantities of influence for this problem from figure 1.

Q_1	Q_2	Q_3	Q_4	Q_5	Q_6	...
$\dot{m}$	ρ	g	h	A	μ	...

With three basic quantities this set results in:

$$\pi_1 = \frac{\dot{m}}{\rho\, g^{1/2}\, h^{5/2}}\,,\quad \pi_2 = \frac{A}{h^2}\,,\quad \pi_3 = \mathrm{Re} \tag{3.1}$$

and this gives:

$$\dot{m} = \rho \sqrt{g}\, h^{5/2}\, f\left(\overset{\downarrow}{\frac{A}{h^2}}, \mathrm{Re}\right). \tag{3.2}$$

The dependency from the first argument (↓) can be determined immediately. Without the jet contraction, the geometric area (A) is also the fluid filled one, i.e.

$$\dot{m} \sim A.$$

In (3.2) therefore f can be a linear function of the first argument only.

$$\dot{m} = \rho \sqrt{g\, h}\, A\, F(\mathrm{Re}). \tag{3.3}$$

For discharge without losses the formula of Torricelli appears with $F(Re) = \sqrt{2}$.

A similar conclusion appears in connection with the estimation of the pressure drop Δp for fully developed flow in a horizontal pipe (length ℓ, diameter D, roughness k). The invariance of translation in the flow direction yields a linear dependency on ℓ:

$$\frac{\Delta p}{\frac{\rho}{2} w^2} = F\left(\overset{\downarrow}{\frac{\ell}{d}}, Re, \frac{D}{k}\right) = \frac{\ell}{d} f\left(Re, \frac{D}{k}\right).$$

4. Discharge from a vessel of a compressible fluid

We are interested in the time t_{krit}, during that the critical state exists in the throat of the connected Laval nozzle. This quantity limits the time of measurement in a supersonic blow down wind tunnel. With the quantities, shown in figure 2 the following relation can be set up:

$$t_{krit} = f(p_{10}, \rho_{10}, p_a, V, A^*, \gamma). \tag{4.1}$$

The π-theorem gives:

$$\pi_1 = \frac{t_{krit} A^* a_{10}}{V}, \quad \pi_2 = \frac{p_{10}}{p_a}, \quad \pi_3 = \frac{V}{A^{*3}}, \quad \pi_4 = \gamma. \tag{4.2}$$

Here for the fluid at rest the velocity of sound a_{10} is:

$$a_{10}^2 = \gamma \frac{p_{10}}{\rho_{10}}.$$

(4.1) gives with (4.2):

$$t_{krit} = \frac{V}{A^* a_{10}} F\left(\frac{p_{10}}{p_a}, \overset{\downarrow}{\frac{V^2}{A^{*3}}}, \gamma\right). \tag{4.3}$$

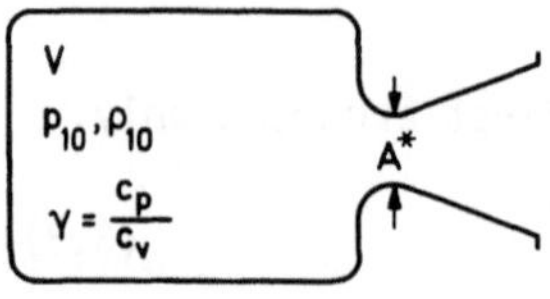

Fig. 2: Discharge from a vessel through a Laval-nozzle.

Here the dependency on the second argument (↓) is of interest. From the balance of mass immediately follows with $\dot{m} \sim A^*$

$$t_{kr} \sim \frac{1}{A^*}.$$

Therefore, in (4.3) F cannot be a function of the second argument and it is:

$$t_{krit} = \frac{V}{A^* \, a_{10}} \, f\left(\frac{p_{10}}{p_a}, \gamma\right). \tag{4.4}$$

Taking into account the balance of mass and the balance of energy, a simple consideration gives /3/:

$$t_{krit} = \frac{V}{A^* \, a_{10}} \, \frac{2}{\gamma - 1} \, \frac{\left(\frac{p_{10}}{p_a}\right)^{\frac{\gamma - 1}{2\gamma}} - \left(\frac{\gamma + 1}{2}\right)^{1/2}}{\left(\frac{2}{\gamma + 1}\right)^{\frac{1}{\gamma - 1}}}. \tag{4.5}$$

Of interest is the discussion about the two limiting cases:

$t_{krit} \to 0$ 1) $\frac{p_{10}}{p_a} \to \left(\frac{\gamma + 1}{2}\right)^{\frac{\gamma}{\gamma - 1}}$, (critical pressure ratio)

2) $\frac{V}{A^* \, a_{10}} \to 0.$

$t_{krit} \to \infty$ 1) $\frac{p_{10}}{p_a} \to \infty$

2) $\frac{V}{A^* \, a_{10}} \to \infty$

5. Expansion of spherical explosion waves /4/

At the time $t = 0$ the amount of energy E is released at the centrum. We are interested in the expansion of the related wave front. For the quantities of influence (fig. 3), the following general dependency is valid:

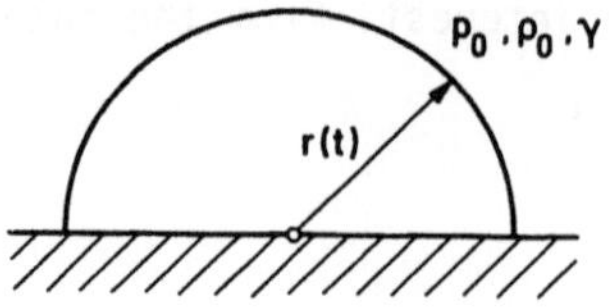

Fig. 3: Expansion of a spherical wave.

$$r = f(t , E , p_0 , \rho_0 , \gamma). \tag{5.1}$$

Connected with those are three π-numbers:

$$\pi_1 = \frac{p_0 \, r^3}{E} , \; \pi_2 = \frac{\rho_0 \, r^5}{E \, t^2} , \; \pi_3 = \gamma. \tag{5.2}$$

With (5.1) it is:

$$r = \left(\frac{E}{\rho_0}\right)^{1/5} t^{2/5} \; F\left(\overset{\downarrow}{\frac{p_0 \, r^3}{E}} , \gamma\right). \tag{5.3}$$

This implicit form is of little use. But the dependency on the first argument ($\downarrow$) can be found easely with an important limiting case. Gasdynamic relations show that for strong shocks ($p/p_0 \gg 1$) no dependency on p_0 exist (but on ρ_0). Therefore the first argument in (5.3) ($\downarrow$) can be dropped completely and we get:

$$r = \left(\frac{E}{\rho_0}\right)^{1/5} t^{2/5} \; f(\gamma). \tag{5.4}$$

With equation (5.4) the simple relation

$$r \sim t^{2/5}$$

is given. The constant $f(\gamma)$ can be deduced from gasdynamic relations. The importance of (5.4) is given by the fact that from photographs of the wavefront, taken at different times, the energy E, released at the centrum can be estimated.

6. Wavelength of the Taylor-vortices /5/

The gap between two concentric cylinders, with the inner one rotating (ω) and the outer one at rest, is filled with a viscous fluid (ρ,μ). When the

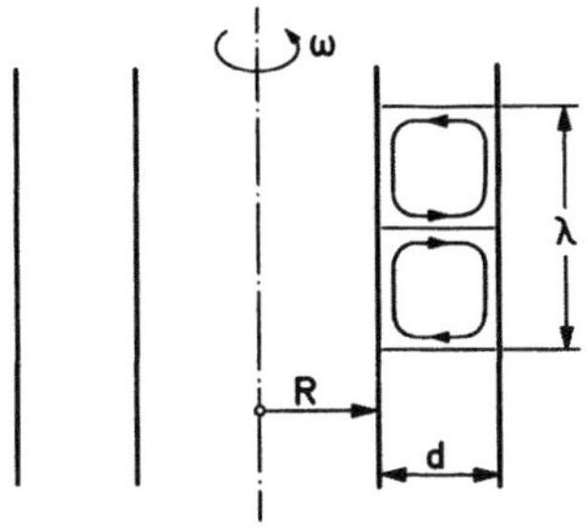

Fig. 4: Taylor-vortices in the gap between two counterrotating cylinders.

rotational speed surpasses a critical value the gap becomes filled with counterrotating Taylor-vortices, having the wave length λ (fig. 4). Of interest is the question, what quantities determine this wavelength λ. To begin with we get:

$$\lambda = f(R, d, \rho, \mu, \omega) \tag{6.1}$$

and therefore:

$$\pi_1 = \frac{\lambda}{d}, \; \pi_2 = \frac{d}{R}, \; \pi_3 = \frac{\rho \omega R d}{\mu}. \tag{6.2}$$

(6.1) and (6.2) combined give:

$$\frac{\lambda}{d} = f\left(\overset{\downarrow}{\frac{\rho \omega R d}{\mu}}, \overset{\downarrow}{\frac{d}{R}}\right). \tag{6.3}$$

This relation is valid generally. That means, this dependency describes as well the laminar flow as the turbulent one. In both cases vortices of this form appear. Specializing on the laminar case, boundary layer considerations show that instead of both arguments in (6.3) only the

$$\text{Taylor-number} = Ta = \frac{\rho \omega R d}{\mu} \sqrt{\frac{d}{R}} \tag{6.4}$$

is relevant. That means:

$$\frac{\lambda}{d} = F(Ta). \tag{6.5}$$

Vortices begin to appear at $Ta_{krit} = 41.3$ with $\lambda/d \approx 2$.

7. Flow of a viscous fluid with homogenious suction or blowing

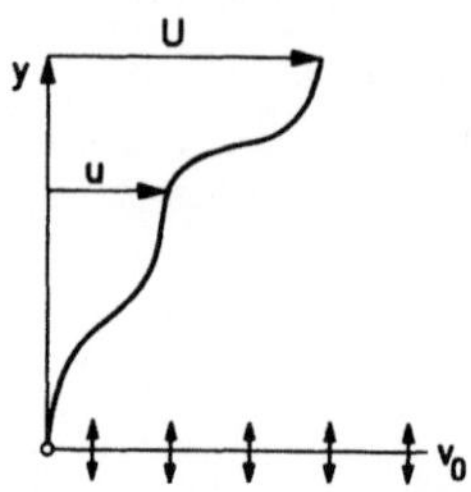

Fig. 5: Flat plate flow with suction and blowing.

We investigate the flow over a flat plate. For the variable velocity u we get with the notations of figure 5:

$$u = f(y\,,U\,,v_0\,,\rho\,,\mu). \qquad (7.1)$$

Here U is the undisturbed outside velocity as assumed for $y \to +\infty$ and $v_0 \gtrless 0$ is the suction or blow velocity.
We get:

$$\pi_1 = \frac{u}{U}\,,\ \pi_2 = \frac{U\,y}{\nu}\,,\ \pi_3 = \frac{v_0}{U}\,. \qquad (7.2)$$

Therefore it is:

$$\frac{u}{U} = F\left(\frac{U\,y}{\nu}\,,\frac{v_0}{U}\right). \qquad (7.3)$$

This dependency can be made more definite by using the Navier-Stokes equations. Another - more lucid explanation - has not been encountered by the author.
The first Navier-Stokes equation leads to the ordinary differential equation:

$$v_0\,\frac{du}{dy} = \nu\,\frac{d^2u}{dy^2}\,.$$

Together with the boundary conditions $u(0) = 0$ and $u(\infty) = U$ this equation gives a solution only for $v_0 < 0$ /6,7/

$$\frac{u}{U} = 1 - e^{\frac{v_0\,y}{\nu}}\,. \qquad (7.5)$$

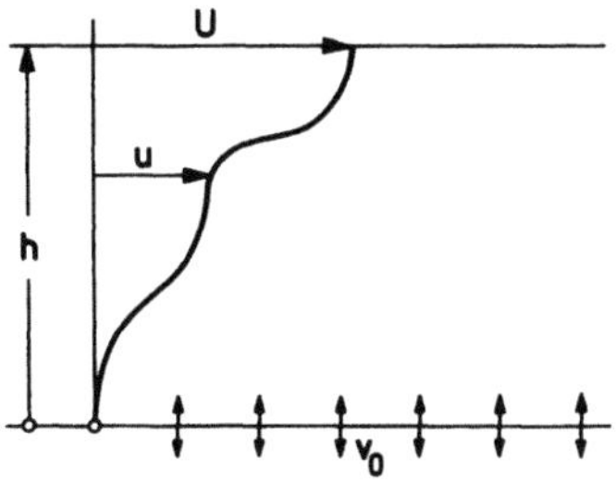

Fig. 6: Flow in a gap with suction and blowing

The second Navier-Stokes equation gives:

$$\frac{dp}{dy} = 0$$

that means, the pressure is constant. The continuity gives $v \equiv v_0 = \text{konst.}$ The comparison of (7.5) with (7.3) and (7.2) shows that obviously only the combination $\pi_2 \cdot \pi_3 = (v_0\, y)/\nu$ appears. Furtheron there remains the question about the solution with blowing. For this case difficulties develop in the infinity. With the boundary conditions $u(0) = 0$, $u(\infty) = U$, for $v_0 > 0$ the solutions of (7.4) don't stay finite. Here a discussion of the flow in a gap of the height h is rewarding (fig. 6). The first Navier-Stokes equation gives with $u(0) = 0$, $u(h) = U$ for $v_0 \gtreqless 0$ the solution

$$\frac{u}{U} = \frac{1 - e^{\frac{v_0\, y}{\nu}}}{1 - e^{\frac{v_0\, h}{\nu}}} . \tag{7.6}$$

In figure 7 the different profiles are plotted for constant h. $v_0 \to 0$ leads to Couette-flow, $v_0 < 0$ gives concave and $v_0 > 0$ gives convex profiles. Going here to the limit $h \to \infty$, it can be confirmed immediately that $v_0 < 0$ only does make sense. This can be seen, too, in the behaviour of the displacement thickness

$$\delta_1 = \int_0^h \left(1 - \frac{u}{U}\right) dy = -\frac{\nu}{v_0} - h\, \frac{e^{\frac{v_0\, h}{\nu}}}{1 - e^{\frac{v_0\, h}{\nu}}} .$$

The non-existence of a (stationary) solution of the Navier-Stokes equations for $v_0 > 0$ is surprising. This may be explained by the fact that the insta-

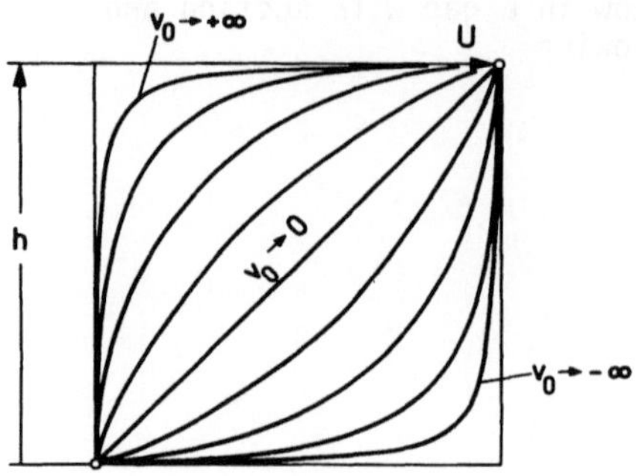

Fig. 7: Velocity profiles in a gap with $v_0 \gtreqless 0$

h	v_0	δ_1
fixed	$+\infty$	h
	0	h/2
	$-\infty$	0
$\to \infty$	< 0	$-\nu/v_0$
	> 0	$+\infty$

tionary Navier-Stokes equations don't give a stationary solution for $v_0 > 0$ in the limit $t \to \infty$. Possibly this is due to the unstable form (convex) of the velocity profiles for $v_0 > 0$ (fig. 7).

References

/1/ Görtler, H.: Dimensionsanalyse. Berlin/Heidelberg/New York 1975.

/2/ Zierep, J.: Ähnlichkeitsgesetze und Modellregeln der Strömungslehre. Karlsruhe 1972.

/3/ Abboud, M.: Private communication 1981.

/4/ Taylor, G.J.: Proc. Roy. Soc. A 201, 159 and 175, 1950

/5/ Taylor, G.J.: Phil. Trans. A 223, 289-293, 1923

/6/ Schlichting, H.: Grenzschichttheorie. Karlsruhe 1965. Fifth ed., 359

/7/ Meredith, F.W. and Griffith, A.A.: In "Modern Developments in Fluid Dynamics". Oxford Univ. Press 2, 534, 1938

MIX
Papier aus verantwortungsvollen Quellen
Paper from responsible sources
FSC® C105338

If you have any concerns about our products,
you can contact us on
ProductSafety@springernature.com

In case Publisher is established outside the EU,
the EU authorized representative is:
Springer Nature Customer Service Center GmbH
Europaplatz 3, 69115 Heidelberg, Germany

Printed by Libri Plureos GmbH
in Hamburg, Germany